Highly-Distributed Systems

Andriy Luntovskyy · Dietbert Gütter

Highly-Distributed Systems

IoT, Robotics, Mobile Apps, Energy Efficiency, Security

Springer

Andriy Luntovskyy
BA Saxony, BA Dresden University of Coop. Education
Dresden, Germany

Dietbert Gütter
BA Saxony, BA Dresden University of Coop. Education
Dresden, Germany

ISBN 978-3-030-92828-5 ISBN 978-3-030-92829-2 (eBook)
https://doi.org/10.1007/978-3-030-92829-2

Responsible Editor: Reinhard Dapper
This Springer imprint is published by the registered company Springer Nature Switzerland AG
The registered company address is: Gewerbestrasse 11, 6330 Cham, Switzerland

It is the supreme art of the teacher to awaken joy in creative expression and knowledge.

(one of the most beautiful quotes of Albert Einstein)

Es ist die wichtigste Kunst des Lehrers, die Freude am Schaffen und am Erkennen zu wecken.

(eine der schönsten Zitate von Albert Einstein)

We, authors, thank our loving wives Lena and Aranka for their constant support.
Andriy Luntovskyy thanks his lovely daughters Anna and Maria.
Then Andriy Luntovskyy dedicates this book to his late beloved father Oleg Luntovskyy

(20 September 1938; † 3 September 2020)*

Foreword of Prof. Dr. habil. Dr. h. c. Alexander Schill, Chairman of Chair for Computer Networks at TU Dresden

This book covers a broad range of leading-edge topics in the area of highly distributed and networked systems. It is suitable as a textbook for advanced lectures in the domains of systems architecture, networking, and distributed platforms. Moreover, it can serve as a basis for research and development seminars and as an inspiration for the interested IT specialist looking for new challenges and developments. The authors have many years of experience as professors and lecturers in computer science and have already published a significant number of textbooks, journal contributions, and conference papers.

The book starts with an introduction into the principles of large-scale distributed systems and then addresses several selected areas of specialization in this context. Recent developments of 5G networks with various advanced optimization aspects are discussed, while also providing an overview of intelligent networking technologies with highly flexible setups. Security-critical aspects also play an important role, namely the emerging blockchain techniques for forgery-proof data management in highly distributed environments.

Another crucial area covered is energy-efficiency and green IT with different optimization techniques like load balancing in large-scale cloud computing environments. Finally, application areas of large distributed systems such as the internet of things domain or advanced robotics are equally addressed.

Last, but not least, a number of student tasks are found in the appendix, providing important control questions and an ideal basis for accompanying exercises at universities and colleges.

Altogether, the content of this book is technically highly interesting while still being rather practically oriented and therefore straightforward to understand. We would like to wish all readers successful studies and lots of inspiration by this useful textbook!

Dresden
28.09.2021

Alexander Schill

Preface

The given book as a monograph and a handbook simultaneousely complements the existing relevant set of books (refer Fig. 1) by A. Schill, D. Gütter, A. Luntovskyy, J. Spillner, T. Hara, T. Springer, I. Braun, which during last 10–15 years by Springer Nature have been published already:

1. Schill, Alexander, Springer, Thomas. Verteilte Systeme: Grundlagen und Basistechnologien: Kompakte Darstellung der Grundlagen und Techniken Verteilter Systeme, Springer-Verlag Berlin Heidelberg, 2012, 2.Ausgabe, ISBN: 9783642257957 (https://www.springer.com/gp/book/9783642257957#otherversion=9783642257964).
2. Schill, Alexander, Springer, Thomas. Verteilte Systeme: Grundlagen und Basistechnologien: Kompakte Darstellung der Grundlagen und Techniken Verteilter Systeme, Springer-Verlag Berlin Heidelberg, 2007, 1.Ausgabe, ISBN: 9783540684718 (https://www.springer.com/gp/book/9783540684718).
3. Luntovskyy, Andriy, Spillner, Josef. Architectural Transformations in Network Services and Distributed Systems: Current technologies, stand- ards and research results in advanced (mobile) networks, Springer Vieweg Wiesbaden, 2017, 344p., ISBN: 9783658148409 (https://www.springer.com/gp/book/9783658148409#otherversion=9783658148423).
4. Luntovskyy, Andriy, Gütter, Dietbert, Melnik, Igor. Planung und Optimierung von Rechnernetzen. Methoden, Modelle, Tools für Entwurf, Diagnose und Management im Lebenszyklus von drahtgebundenen und drahtlosen Rechnernetzen: Planung von Rechnernetzwerke theoretisch anspruchsvoll und praxisnah, Springer/Vieweg + Teubner Wiesbaden, 2012, 415 Seiten, 245 Abb., ISBN: 9783834814586 (https://www.springer.com/gp/book/9783834814586#otherversion=9783834882424).
5. Luntovskyy, Andriy, Gütter, Dietbert. Moderne Rechnernetze: Protokolle, Standards und Apps in kombinierten drahtgebundenen, mobilen und drahtlosen Netzwerken, 500 Seiten, 263 Abb., Springer Nature, Juli 2020, ISBN: 9783658256166 (https://www.springer.com/gp/book/9783658256166).

Fig. 1 Book Grouping

6. Luntovskyy, Andriy, Gütter, Dietbert. Moderne Rechnernetze - Übungsbuch: Aufgaben und Musterlösungen zu Protokollen, Standards und Apps in kombinierten Netzwerken, 150 Seiten, 44 Abb., Springer Nature, Juli 2020, ISBN: 9783658256180 (https://www.springer.com/gp/book/97836582561803658256180).
7. Springer, Thomas, Braun, Iris, Feldmann, Marius, Hara, Tenshi, Wutzler, Markus. Distributed Systems (to be released prospectively in 2022).

It is recommended as a lecture-accompanying book for the teaching modules "Computer Networks", "Radio Networks", "Distributed Systems", "Mechatronics", "Internet of Things and Automation" in ET and IT curricula for students and lecturers at technical universities and study academies in Germany, Austria and Switzerland as well as in other German-speaking areas (especially as a textbook for distance students).

In Germany and the German-speaking countries (Austria, Switzerland), despite the availability of a wide selection of literature on Computer Networks and Distributed Systems (A. Tanenbaum, R. Schreiner, H. Zisler, J. Roth, G. Bengel, C. Baun, G. Schürmann, K.-H. Weiß, M. Weber etc.), topics such as Green IT, Fog Computing, Distributed Applications that guarantee Data Security, the Internet of Things determine that for us as subject experts and actively teaching lecturers, some aspects of these monographs and handbooks are not adequately treated from the perspective of overarching trends and transformations.

The analysis of modern literature on the aforementioned problems also showed that the translated technical and handbooks are often oversaturated with the listing and enumeration of items from existing standards, catalogue passport data and configuration instructions, and that they contain too few concrete implementation examples and use cases.

Some authors present well-known theoretical learning material, e.g. underlying methods for network services, products and standards, but with little practical illustration through scenarios, case studies, modern (mobile) applications, examples, etc. Such evident and tangible use cases are important. In our book, the above-mentioned gaps should be obviously closed, while the addressed problem is resolved and the previous works should be supplemented and expanded.

The given book contains the foreword, preface, introduction and 12 chapters, which are divided into the following parts:

- Part I. Foundations for HDS (Highly-Distributed Systems)
- Part II. HDS For Internet of Things and Robotics
- Part III. Conclusions and Outlook, Learning Tasks and Appendixes.

The book includes a chapter with learning exercises and solution examples aimed to so-called Highly-Distributed Systems. They are referred with the work contents and targeted to the deepening of the material comprehension. These tasks practically illustrate the discussed subjects and can be divided into three essential topics:

1. Basics and Performance in HDS
2. Efficiency and Security in HDS
3. Internet of Things.

The presented book about HDS (Highly-Distributed Systems) possesses a strong modular structure and is used for the set of the following relevant teaching modules such as

1. Application Development for Mobile and Ubiquitous Computing (English)
2. Distributed Systems (English)
3. Internet and Web Applications (English)
4. iOS Programming (English)
5. Mobile Communication and Mobile Computing (English)
6. Service and Cloud Computing,

which all are recommended in ET and IT curricula for students and lecturer at technical universities (Fig. 2).

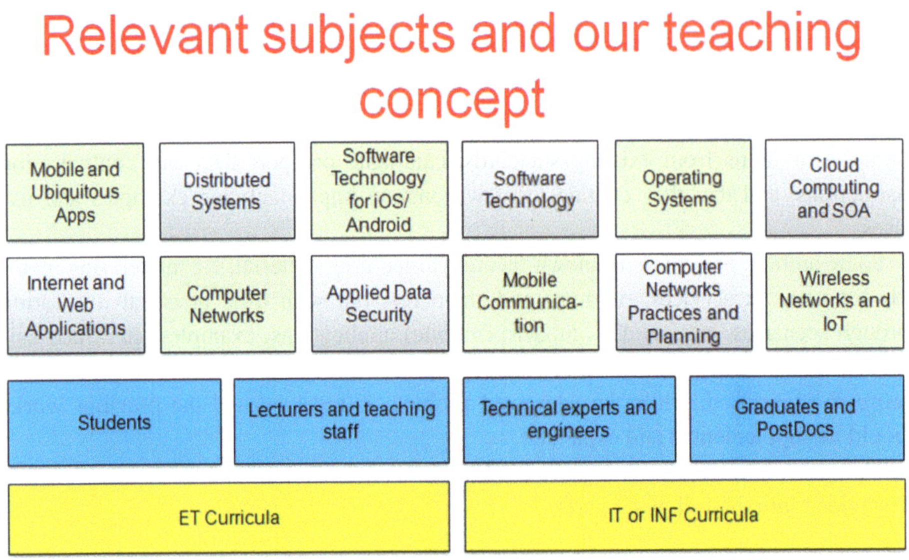

Fig. 2 Relevant subjects and our teaching concept

Each chapter provides the references to further useful literature and web links and, therefore, stimulate to study original sources.

Motivator and Added-Value Effect

Therefore, you are reading a book that aims to cover the field of recent innovations in network services and Highly-Distributed Systems. The book's target audience includes university and technical college students, graduate engineers and teaching staff. If you are someone else, do not worry, the topics covered may still be of interest to you!

This book offers readers individual functionality and added-value effect:

As a monograph … With the given work, we decided to help not only the readers and students, but also ourselves, as the professionals who are actively involved in the networking, telecommunications and systems communities, by understanding the trends, which have developed in the recent decade in distributed systems and networking applications. Important architectural transformations of modern distributed systems are examined and presented in a survey style. Examples of new architectural solutions for HDS software, network apps and mobile services and micro-services are discussed. Among them are the Internet of Services and Internet of Things, Fog and Clouds, compulsoriness based on Blockchain, environment-aware Parallel and Highly-Distributed Computing, to mention a couple of popular concepts.

As a handbook ... Current (software) HDS technologies, provisioned and choreographed decentralized services like SaaS, BaaS, micro-services and other basics, defacto standards and research results for advanced (mobile) networks, connected devices and robotics as well as for higher level network functions and software applications are focused with in this book from a practical aspect. The authors highlight how these technical solutions for our digital time, ubiquitous communication and collaboration infrastructures are widely and highly-distributed and being transformed to reflect society and environmental requirements.

Efficient HDS architectures, principles and systems for mobile and wireless communication, criteria for optimization of networks and highly-distributed systems, as well as central ideas to new system concepts are widely discussed herein.

Use case presentations and studies with in-depth technical descriptions along with a control questions and tests strengthen the nature of this book as handbook to use for courses and projects.

This book also offers special didactic features (e.g. exercises and tasks, solution examples, illustrations etc.):

- The mostly important technological approaches and methods as well as up-to-date models and implementation paradigms for HDS (Highly-Distributed Systems) are examined
- Theory is accompanied by numerous examples in the given chapters, multiple system examples are also available
- Numerous colored illustrations, comparison tables and schemes are listed within the work
- Model tasks and exercises on the problematics of HDS and IoT for consolidating the current material are available in the book with synergy effect as monograph and handbook both.

Dresden
December 2021

Andriy Luntovskyy
Dietbert Guetter

Acknowledgements

We thank our students and multiple researchers and colleagues from TU Dresden, BA Saxony, ZHAW Zurich, LNPU Lviv, NTUU Kyiv "Igor Sikorsky KPI" for your valuable questions, suggestions, case studies, and inspiration in creating and correcting this work.

In the content of this book as a monograph and textbook, the didactically proven examples from lecturers from TU Dresden and BA Dresden have also been adopted. We thank for that Prof. Dr. rer. nat. habil. Dr. h. c. Alexander Schill, Dr. Marius Feldmann, Dr. Iris Braun, Dr. Thomas Springer with TU Dresden, as well as PhD student Bohdan Shubyn, Dr. habil. Taras Maksymyuk, Prof. Dr. habil. Mykhailo Klymash, Dr. habil. Mykola Beshley (all with Lviv National Polytechnic University) and many of our colleagues, like-minded people and fellows.

Authors' acknowledgement goes to the colleagues Prof. Dr. L. Zipfel (BA Dresden), Prof. Dr. F. Fitzek (TU Dresden), Dr. J. Struckmeier (Cloud & Heat), Prof. Dr. T. Horn (IBH Dresden) for inspiration and challenges by fulfilling of this work.

Our acknowledgement belongs to Prof. Dr. D. Gembris (BA Dresden) and Mr. M. Baumgärtner (NoDNA) for offering the pictures of Bioloids.

The authors' acknowledgements we would like to express also to the BA Dresden students and scientific fellows M. Stoll, E. Zumpe, F. Franke, M. Podoprygora, P. Kunick, N. Liebich, T. Zobjack, O. Graetsch, F. Schmidt, W. Salmaier, the colleagues and friends Prof. Dr. habil. Larisa Globa with NTUU "Igor Sikorsky KPI" Kyiv, Dr. habil. Josef Spillner with ZHAW Zurich for technical support, inspiration and challenges by fulfilling of the examples and ideas to the exercises for this work.

A special acknowledgement goes to Prof. Dr. Tenshi Hara with BA Saxony for proofreading and organisational advices, constructive discussions and suggestions by fulfilling of this work.

Introduction

Digitalization (as Movement to Digital Age) is a driving force behind the development of so-called Highly-Distributed Systems (HDS).

Digitalization and its effect on e-commerce in 2030 is represented in Fig. 3. The figure shows which digital means and tools that can be used to draw attention to products, services and goods nowadays, and how trading and marketing problematics can be successfully solved in a modern company [1, 2]. The illustration compares the respective tools in mid-term, in a future fully digitized world. However, one thing is for us clear: digitalization gives an abstract company more opportunities and minimizes its risks in comparison to the rivals.

In addition, the prospect of potential and possible opportunities for future e-shops is generous. The experts mostly tend towards the positive effects of digitalization factors. The connections between digitalization trends, flexible corporate structures and strategic importance for common IT infrastructure development are depicted. The presented overview is intended to give the readers a targeted impetus for the first chapters of the work, to better comprehension of the HDS foundations.

Figure 4 depicts the comparison of typical hierarchical company structures with the agile structures of tomorrow, which will appear due to ongoing digitalization guaranteed [1, 2].

Both representations are to be considered under the aspect of digitalization and from small to medium-sized enterprises (SMEs). Digitalization provides for a company organizational flexibility with cost savings and accompanies us nowadays in modern value chains and manufacturing processes as well as in everyday life or entertainment. Maintaining digital document flows and a seamless process workflow simplifies these processes, optimizes and accelerates them several times. The current slogan is as follows: "Faster on the market", i.e. competitive advantages for companies, more convenience for the users, so-called "the old-the new" organizational form of production and service provision for the population under wide support of distributed and highly-distributed applications.

The above-mentioned changes in the corporate structure due to ongoing digitalization lead to the changes in convenient economy to the digital economy.

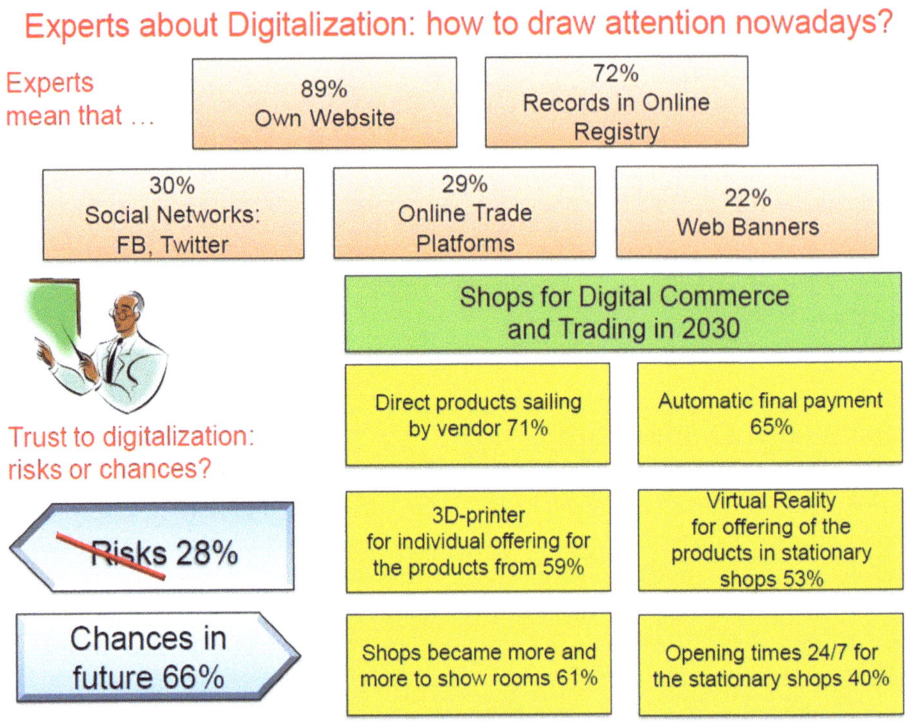

Fig. 3 Digitalization and its effect for e-commerce in 2030 expected

Instead of the convenient analogues, the digital economy is also changing the face of the company, the actual corporate structure. On the first hand, the strictly hierarchical organization is softly replaced by agile forms, which are characterized by a high degree of distribution.

Among other things, roles and forms of employee interaction, company applications, use of IT infrastructure, tools and apps are subject to rapid changes and is being permanently optimized.

Highly-distributed systems (HDS) are namely the modern computer network applications that are distributed over numerous (partially autonomous) computing nodes. These can be characterized under use of advanced decentralized communication models (e.g. P2P, M2M) and be implemented using energy-efficient protocols. In common, HDS guarantees advanced multilateral data security, based on collaborative intrusion detection systems (IDS) and networking (CIDN), as well as blockchaining technology (BC).

Nowadays, multiple forms of HDS exist: fog computing, sensor networking, embedded and IoT, robotics, smart contracting, blockchained and ML-based apps.

Common high-level programming languages (PL), software-technology (SWT) approaches and agile process models with the special focus on (micro-)service choreography are predestined as the implementation basis here.

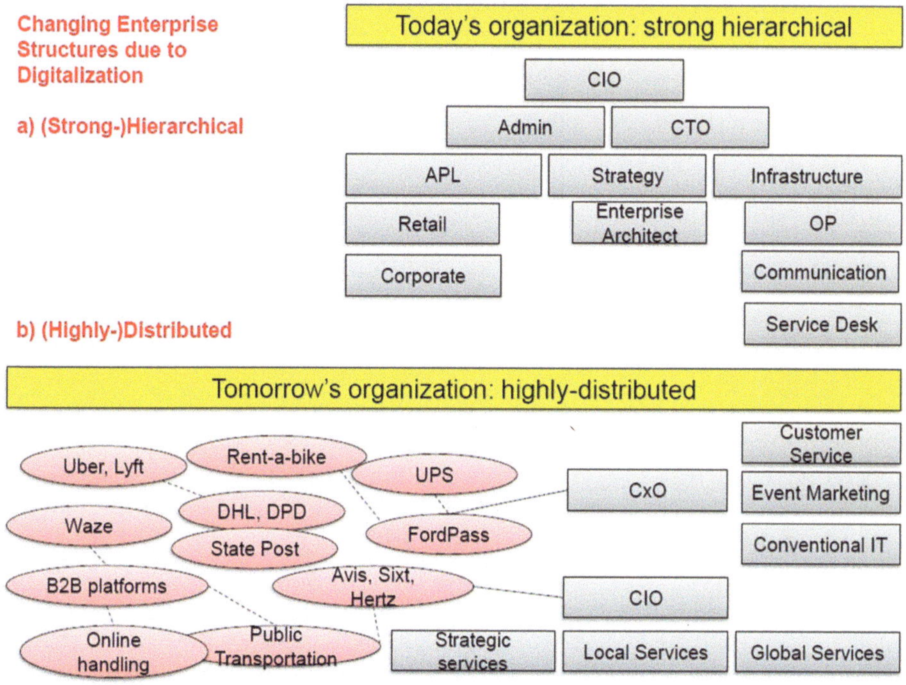

Fig. 4 Changes in the corporate structure due to ongoing digitalization

In this work, the internal organization, paradigms, technologies, architectures and components, process models and software-technological methods for HDS development are analysed. The structure and expansion as well as the inner workings of so-called highly-distributed applications, their data security and workflow compulsoriness as well as suitable data structures are examined.

References

1. Christiane Puetter. 5 Phasen der Digitalisierung (in German), Xing News (Online 28.05.2018): http://www.xing-news.com/.
2. Der ganzheitliche Ansatz zur Optimierung der gesamten Wertschöpfungskette (in German), Siemens (Online 19.06.2018): https://www.siemens.com/global/de/home/unternehmen/themenfelder/zukunft-der-industrie/digital-enterprise.html/.

Contents

Acronyms and Abbreviations

AI	Artificial Intelligence
AMQP	Advanced Message Queuing Protocol
BaaS	Blockchain as a Service
BC	Blockchain and other blockchains
CAPEX	Capital Expenditures, Investments, Procurement Costs
CIDN	Collaborative Intrusion Detection Network
DIDO	Distributed Input Distributed Output Wi-Fi
DR	Data rate
EM	Electro-Magnetic
ERE	Energy Recycling Efficiency
FW	Firewall
GEO	Geostationary Orbit Satellite
HDS	Highly-Distributed Systems
HW	Hardware
IaaS	Infrastructure as a Service
IDS	Intrusion Detection System
IoP	Internet of People
IoS	Internet of Services
IoT	Internet of Things
IPS	Intrusion Prevention System
LEO	Low Earth Orbit Satellite
MEO	Middle Earth Orbit Satellite
MIMO	Multiple Input Multiple Output Antennas
ML	Machine Learning
MQTT	Message Queuing Telemetry Transport Protocol
MS	Micro-Services
NFC	Near Field Communication
NWK, NW	Network
OPC	Open Platform Communications
OPC UA	OPC Unified Architecture

OPEX	Operational Expenditures, Personal, Maintenance and Electricity Costs
OS	Operating System
PaaS	Platform as a Service
PKI	Public Key Infrastructure
PL	Programming Language
PUE	Power Usage Effectiveness
QoE	Quality of Experience
QoS	Quality of Service
RFID	Radio-Frequency Identification
ROS	Robot Operating System
SaaS	Software as a Service
SC	Smart Contracting
SLMA	Server-Less Mobile App or Application
SOA	Service-Oriented Architecture
SW	Software
SWT	Software Technology
VM	Virtual Machine
WPAN	Wireless Personal Area Network
WS	Web Service
WSN	Wireless Sensor Network
XaaS	Everyth(X)ing as a Service

About the Authors

Prof. Dr. habil. Andriy Luntovskyy is professor at the Saxon University of Cooperative Education – State Study Academy Dresden (BA Dresden).

His "alma mater" is the University of Technology Kiev "Igor Sikorsky KPI", Ukraine (diploma with award in 1989).

From 1989 until 2001, he worked at University of Technology Kiev "Igor Sikorsky KPI" Ukraine as PhD student, teaching assistant, lecturer, senior lecturer, as well as private docent. In the same timetable, he worked part-time for several companies and institutions as a software developer and project manager.

From 2001 until 2008, he worked as a PostDoc at the Chair of Computer Networks at Technische Universität Dresden (TU Dresden). Since 2008, he has his position at BA Dresden.

Research Interests and Areas of Teaching:

- Computer Networks and Mobile Communication
- Distributed Systems and Applied Data Security
- Software Technology and Operating Systems
- Fundamentals of Programming
- Fundamentals of Computer Science

Contact:

Andriy.Luntovskyy@ba-sachsen.de

Dr. rer. nat. Dietbert Gütter is lecturer emeritus at Technische Universität Dresden (TU Dresden) and at the Saxon University of Cooperative Education. His "alma mater" is TU Dresden with his doctorate in 1974. He worked at the Chair of Computer Networks at TU Dresden for over 40 years and continues to teach at various Saxon educational institutions.

Research Interests and Areas of Teaching:

- Computer Networks and Operating Systems
- Computer Networks Practices and Planning
- Web Applications
- Software Technology
- Information and Communication Systems

Contact:
Dietbert.Guetter@t-online.de

Part I

FOUNDATIONS FOR HDS

Definition: Highly-Distributed Systems 1

1.1 Motivation: Highly-Distributed Systems

So-called HDS use modern combined fixed, wireless and mobile networks and possess a complex internal construction. They have to be secured (BC—Blockchain, SAML—Security Assertion Markup Language, firewalls, IDS/IPS—Intrusion Detection/Prevention Systems) and provide extended QoS parameters (higher DR and availability, small latency).

The HDS deploy flexible structures, based on SOA (Service-Oriented Architectures) and micro-services, as well as deploy efficient communication models (P2P—Peer-to-Peer, cloud-fog, M2M—Machine-to-Machine), which are able to solve the distribution conflicts in short time and support rapid access to the data analytics.

Such HDS are often developed under use of advanced SWT (Software Technology) process models like DevOps and Scrum and are driven via Blockchain-conform cryptographic structures, which provide compulsoriness of required workflow steps and predictable execution of the deployed modules, services, micro-services and of other components within the internal architecture of the above-mentioned HDS.

Since 2005 the P2P systems (Internet of Things, Fog Computing) in combination with convenient C-S communication model as well as server-less structures (SLMA—Serverless Mobile Apps, robotics) have gained on popularity. Then the Cloud-based solutions became a trend (2011) under predominant use of the load-balanced "thin clients" with functionality delegation to the clouds [1–4]. Under use of fog computing the IoT solutions are constructed. The workload is shifted on the edge [5–12] to the energy autarky and resource economizing small nodes. Finally, what does it mean "Highly-Distributed Systems"?

A. Luntovskyy and D. Gütter, *Highly-Distributed Systems*,
https://doi.org/10.1007/978-3-030-92829-2_1

The term "Highly-Distributed Systems" (HDS) must be deployed for the new mobile, frequently "quasi-offline" or SLMA, which extend the convenient distributed systems. They understand the use of efficient and fast networks under clear cooperation goals, as well as no centralisation in memory access or synchronisation in the clocking. Additionally, they possess more redundancy and possibility for replications due to use of flexible P2P structures, use of cloud and fog services. A very important role the energy autarky plays. Highly-Distributed Systems have more strata and layers in their architecture (better modularity and management with efficient conflict resolving) as well as are better secured especially for privacy and anonymity. For the development of such systems, the agile SWT methods and process models must be used [1, 3, 5, 7]. The distinguishing features of HDS are as follows (Fig. 1.1):

1) Advanced communication models (C-S, i.e. Client-Server, with Clouds, Fog, P2P, M2M)
2) Advanced methods for performance management and optimisation as well as for QoE, Quality of Experience (Fig. 1.2), increasing.
3) Advanced SWT (refer Fig. 1.3 and Fig. 1.4, agile approaches like XP, DevOps, Kanban, Scrum, Micro-Services [4]).
4) Advanced Data Analytics regarding to solving of "Big Data" shortcomings [5, 6].

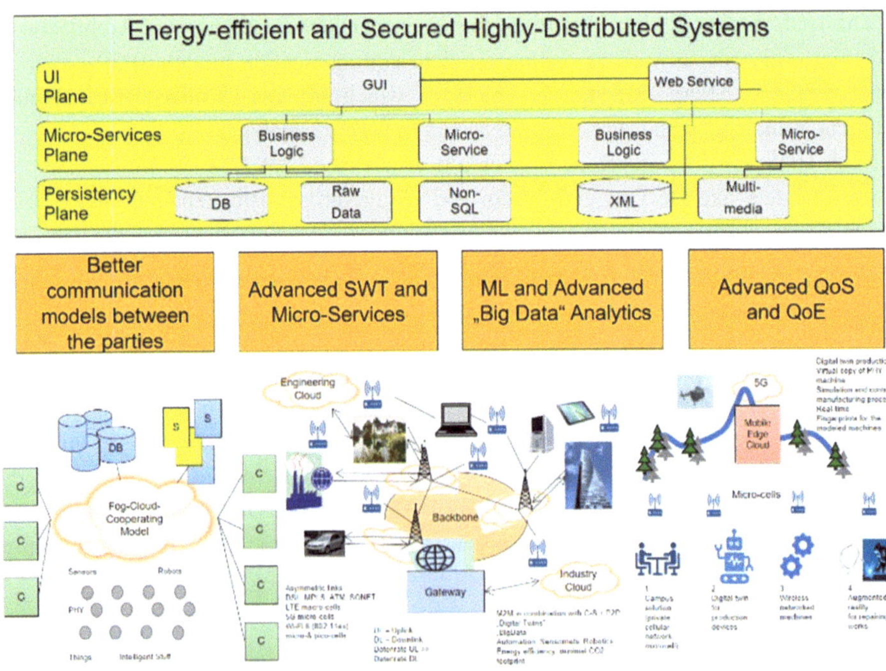

Fig. 1.1 To the motivation on HDS

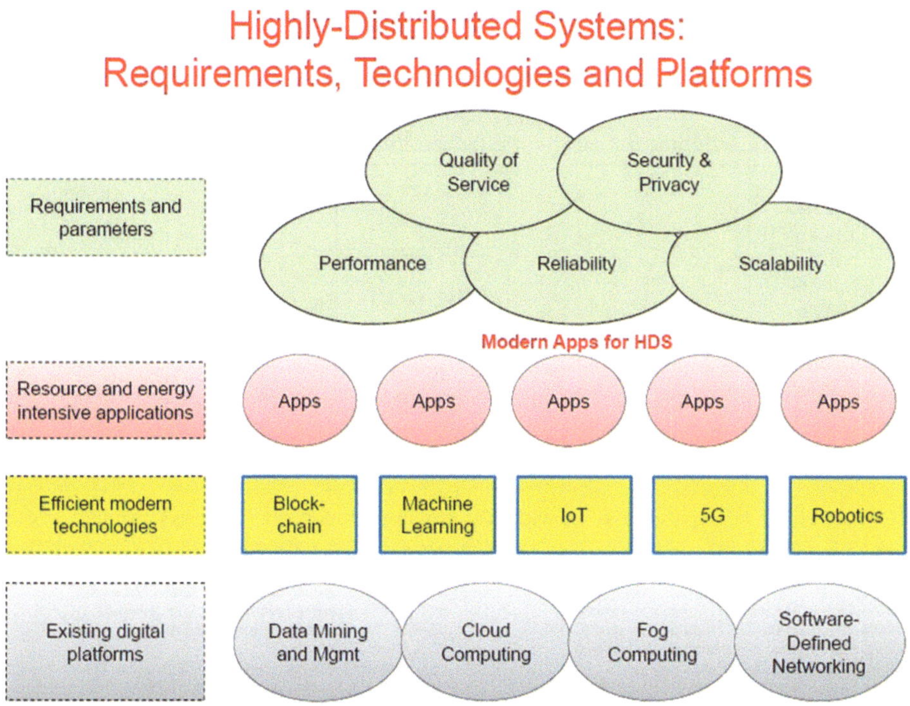

Fig. 1.2 Technologies, Platforms and QoS Requirements for HDS apps

HDS systems can be efficiently used and provide powerful modern apps only if they achieve needed QoS requirements and parameters (refer Fig. 1.2), such as Performance; Reliability; Scalability as well as support necessary security and privacy. HDS cooperate with basic modern technologies and/or offer the apps for such technologies on existing digital platforms, e.g. such subject like Blockchain (Decentralised Transactions, Data Mining, Smart Contracting); Machine Learning and Neural Networks; IoT, 5G and Beyond networks and Robotics. For rapid development of efficient HDS, the novel SWT (Software Technology) approaches and process models are widely used, like Scrum and DevOps (refer Fig. 1.3).

Scrum (1990's) is a virtual analogy to rugby sports and describes the successful teams, which are working on the development of a project or SW product together and involved into close human communication. As an iterative SWT process model, Scrum does not describe classic project phases, but instead of attaches the values of so-called continuously usable results (artefacts) right from the beginning of the project.

DevOps (2009) uses a closer correlation between software development and operation teams. The similar successful agile techniques are known as Kanban (with origins by Toyota) and widely spread Scrum (refer Fig. 1.3).

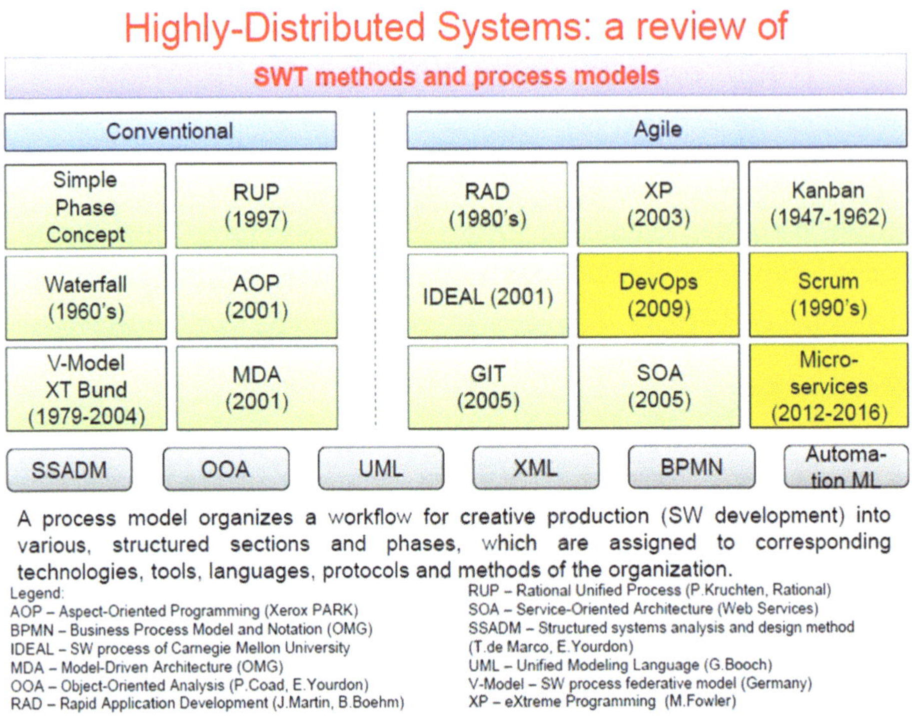

Fig. 1.3 Advanced SWT methods, languages and process models for HDS (especially Micro-Services, Scrum and DevOps)

The appropriate construction basis for HDS are often SOA and Micro-Services, which combine development process agility with flexible communication models [1–7, 9]. Systematically, HDS are involved to the creation of AI (Artificial Intelligence). The neural networks and ML (Machine Learning) techniques are widely integrated into internal structures of HDS. The main purpose of ML for HDS is a special deployment of some workflow steps and learning algorithms, which are performed without human intervention. The time factor and amount of data, which are required for training, belong to the most important indicators of performance and QoS for apps and HDS. Traditionally, ML techniques are divided as follows:

- Supervised learning (SL);
- Unsupervised (UL);
- Reinforcement Learning (RL).

Frequently, to overcoming of misconceptions ML techniques are used by Big Data problematic. The intelligent algorithms can provide clusterisation and data pre-processing too. ML systems, based on cloud-concentrated knowledge (knowledge databases), create

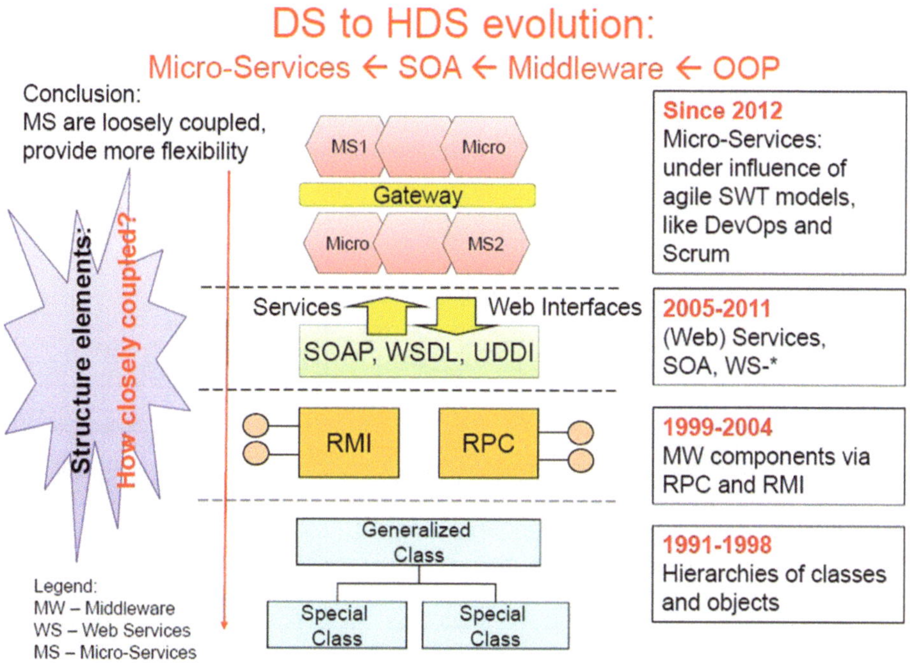

Fig. 1.4 The Evolution Way: OO-MW-WS-Micro-Services

statistics and regressions on obtained voluminous experimental data in background mode. An artificial system is 'learned' from samples and examples and can summarize them after the completion of the study and evaluation (training) phase. The ML system recognizes templates and trends in research data. Thus, the ML systems within HDS can also evaluate data on representativeness and compliance.

1.2 HDS Based on Micro-Services

HDS, which frequently based on Micro-Services, possesses the following successful architecture (refer Fig. 1.1). On the first hand, under use of conventional concepts (n-tier, classic cloud-centric with thin clients), a quasi-monolithic application appears, which is composed from several inflexible modules or macro-services that are not fully independent [1–3, 9]. They are not free-coupled to the application to which they belong and depend one each other in high measure. Their deployment is carried-out only with complexity that means less scalability and complicated reconfiguration. So-called dependencies conflicts occur often [1–3, 9]. On the other hand, the application, which is based on Micro-Services consider the necessary technological trade-offs and offer more flexibility, loosely coupling of the modules and components, as well as is easily configurable (refer Fig. 1.5).

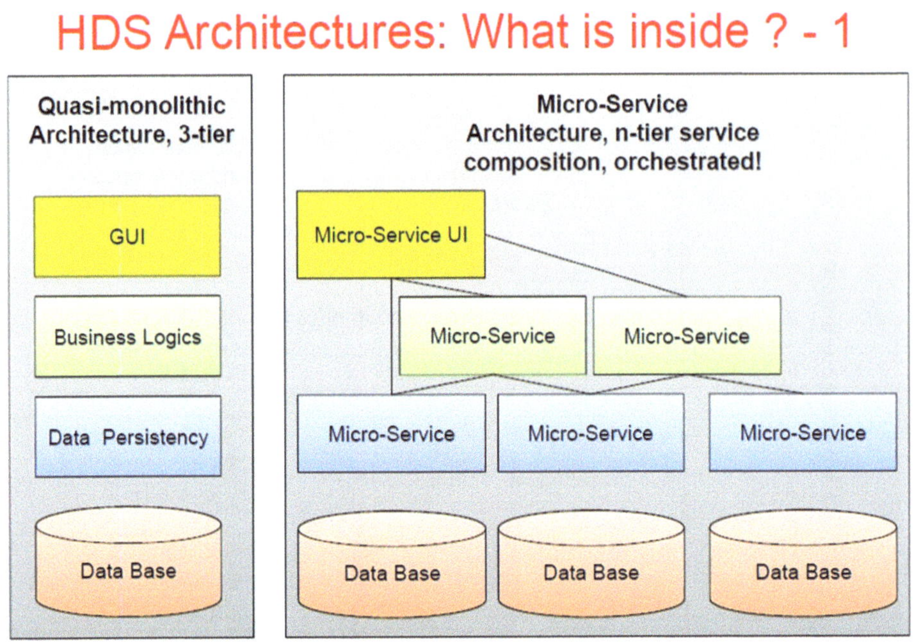

Fig. 1.5 Micro-Services: what is inside?

A comparison of Micro-Services to so-called quasi-monolithic architectures with internal look is given. A demarcation of the Micro-Services to conventional quasi-monolithic architectures with predominantly hardly-coupled modules and SOA and Web Services is evident from the representation.

Under use of agile SWT process models, Micro-Services, available tools and variety of application protocols the modern apps appear which corresponds to HDS paradigms and possess advanced features like QoE (Quality of Experience as a subjectively filled quality of robust services), performance, small latencies and high flexibility as well as are autarky and energy-efficient. Such apps are, in addition, better scalable and more secured via a better management. One of possible options regarding to security, privacy, authentication and compulsoriness of such apps for HDS is use of Blockchain infrastructures [8–10, 13–19]. An option is deployment of Ethereum or a comparable Blockchain structure [13, 14], which provides authentication and compulsoriness of execution (or not execution) of some modules, (macro-)services and Micro-Services within an established SW system. Some examples for such deployment of cryptographic-supported workflow are represented in this chapter in further.

A meaningful example for the use of the architecture and components for Micro-Services is depicted below. Such multi-mode HDS provide n x n MS functionality.

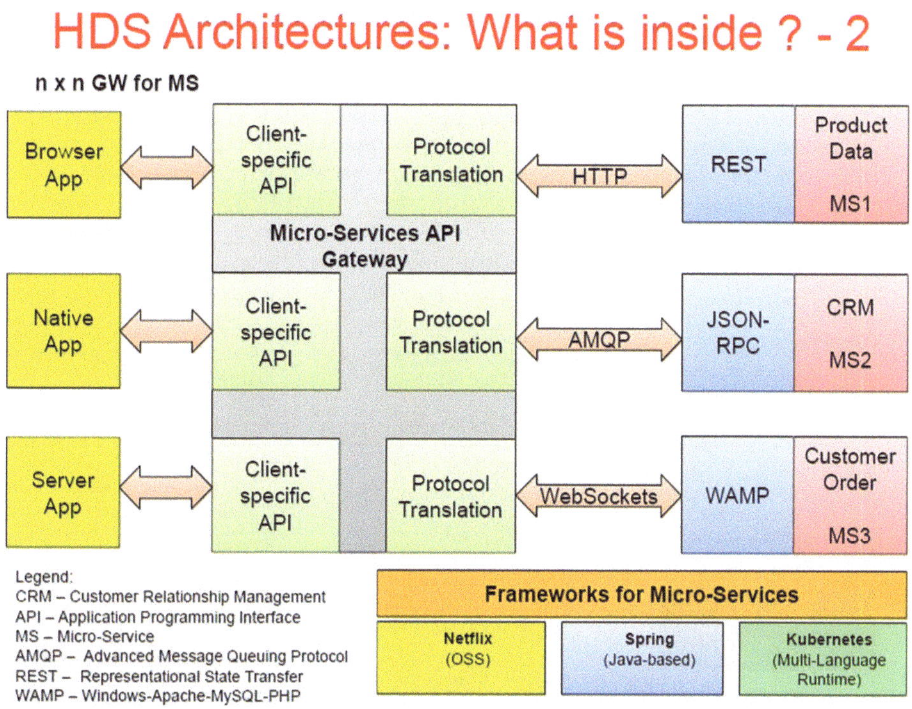

Fig. 1.6 Micro-Services in a gateway

Figure 1.6 depicts use of Micro-services for agile HDS apps under use of variety of efficient protocols like REST and AMQP (Advanced Message Queuing Protocol). The MS1, MS2 and MS3 can use HTTP and WSDL and are interoperable to Web Services. Otherwise, they can be deployed for flexible construction of the IoT app via a gateway (GW) under use AMQP for asynchronous messaging. The depicted GW enables to communicate via widespread HTTP, XML-RPC (JSON-coded) and WebSockets even in real-time. WebSockets (2010–2011) is a L5-7 application protocol, which operates over TCP-connections and is oriented to message exchange between Web-browsers and Web-servers in real-time. As IDE for Micro-Services development Kubernetes, Spring and Netflix frameworks are available (refer Fig. 1.7).

They can provide online tutorials, multi-language support and, even, open source support (OSS). The essential differences between the convenient architectures and Micro-Services are represented in Figs. 1.5, 1.6 and 1.7 on the example of a supermarket with established (macro-)services for customers, products and carts. This example shows the advantages of the discussed approach.

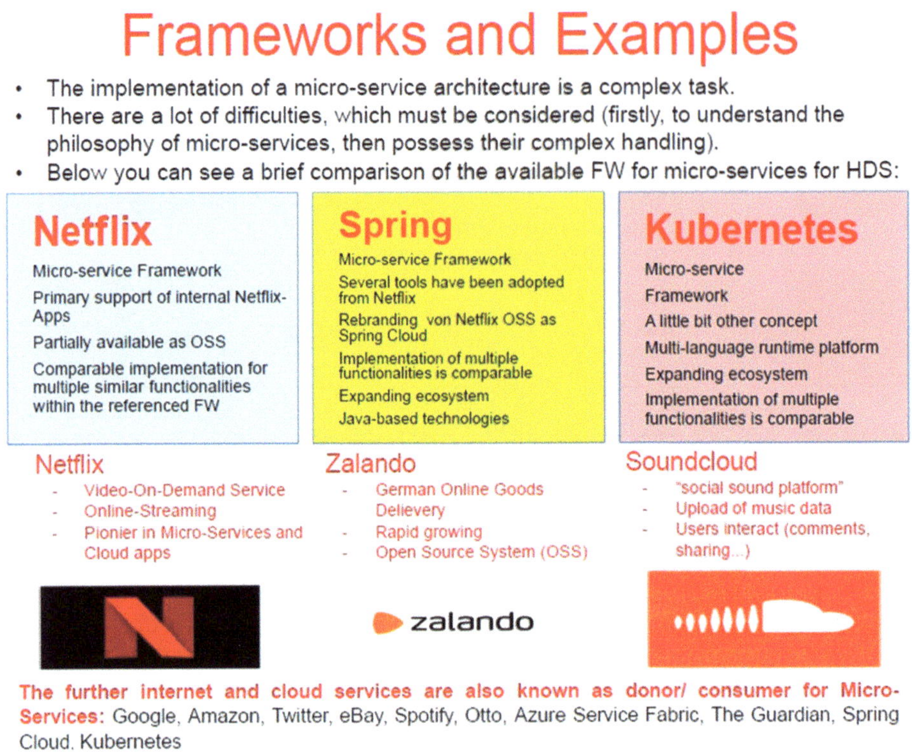

Fig. 1.7 Micro-Services frameworks

1.3 Blockchain as a Decentralised Transaction System

Blockchain is a cryptographically secured and highly-distributed computer network application supporting a decentralised payment system and decentralised financial online transactions in the peer-to-peer (P2P) concept. However, the economic success of this crypto-technology will be evident in the next 10 up to 20 years. The deployment of Blockchain technology speaks mainly for a decentralised financial system. The advantages of such a solution are obvious:

- Sustainability, general transparency and commitment
- Accelerated economic workflows and digitisation processes (so-called IT in the digital age)
- Blockchain cryptotechnology is also well suited to supporting current crypto-currencies (such as Bitcoin, Ethereum, Ripple, Litecoin, ZCash, Monero, Stellar etc.).

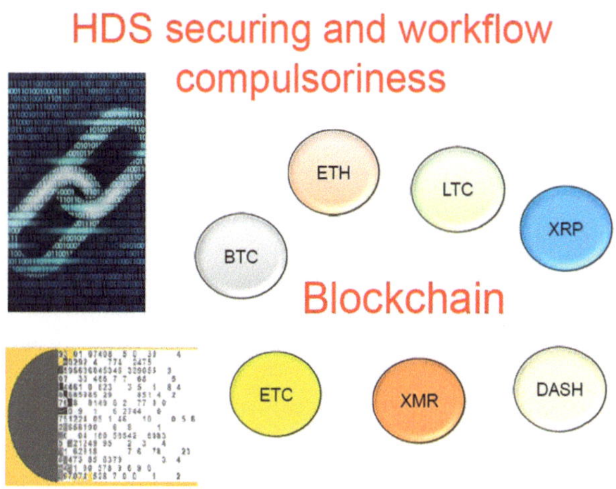

Fig. 1.8 HDS Securing and Compulsoriness

Blockchain architecture contains multiple decentralised, cryptographically secured and unified blocks, their chains and transactions are grouped under a general, global public ledger (account), the structure of which is as follows.

The Blockchain (Fig. 1.8), as a networked Public Ledger, consists of participating nodes that represent an efficient P2P communication model. Typical features of the Blockchain are as follows [13–19]:

- Redundancy and synchronisation;
- Cryptographic hash procedures for integrity assurance and attack safety;
- Decentralised management and control of the Blockchain;
- Network subscribers are also referred to as Nodes (Full-Nodes, Miners, Validators) and run redundantly with mutual synchronisation;

In addition, large block volumes can cause the "Big Data" problem.

Figure 1.9 depicts the structure for an exemplary block chain. The defining block chain (green colour) consists of the longest sequence of secured blocks from the origin (genesis, light blue colour) to the current block; Alternative chains (orange colour) became orphan as soon as they are shorter than another chain. Within the Blockchain architecture between the following basic components can be distinguished: the simple Nodes, the Full-Nodes, and Miner/Validator [8, 13–19]:

- **Nodes:** Each Blockchain participant (computer, smartphones, tablets, or even clusters) is qualified as Node, if he has installed the corresponding software, which runs based on the Bitcoin protocol or the program code of Bitcoin.
- **Full-Nodes:** A Node with full local copy of the Blockchain; Checking for so-called consensus rules.

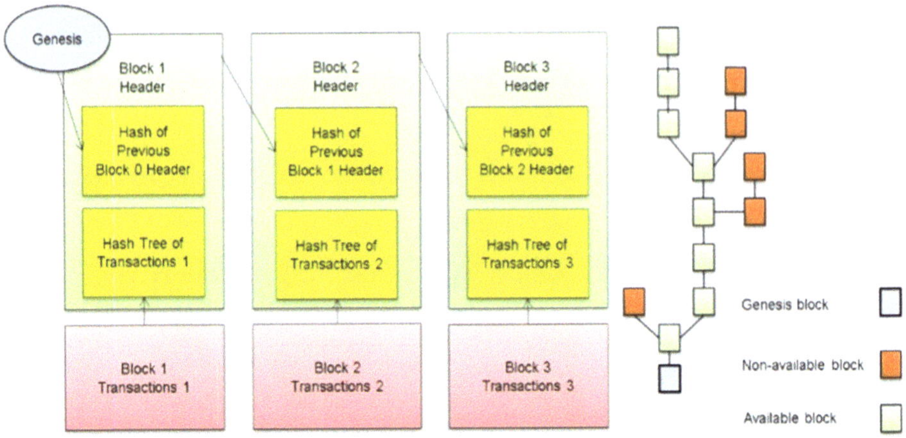

Fig. 1.9 BC basic structure

- **Miner/Validator:** The individual participants or mining pool (high resource requirements regarding hardware and energy consumption); Finalising of blocks (Miner—block generation, Validator—proving); Externally they act each like a large participant, but in fact, many small blocks are generated for payment in fractions of the crypto-currency units.

The basic principle of BC is explained in Fig. 1.10. The technology uses the PKI (Public Key Infrastructure) for multilateral security and compulsoriness based on digital

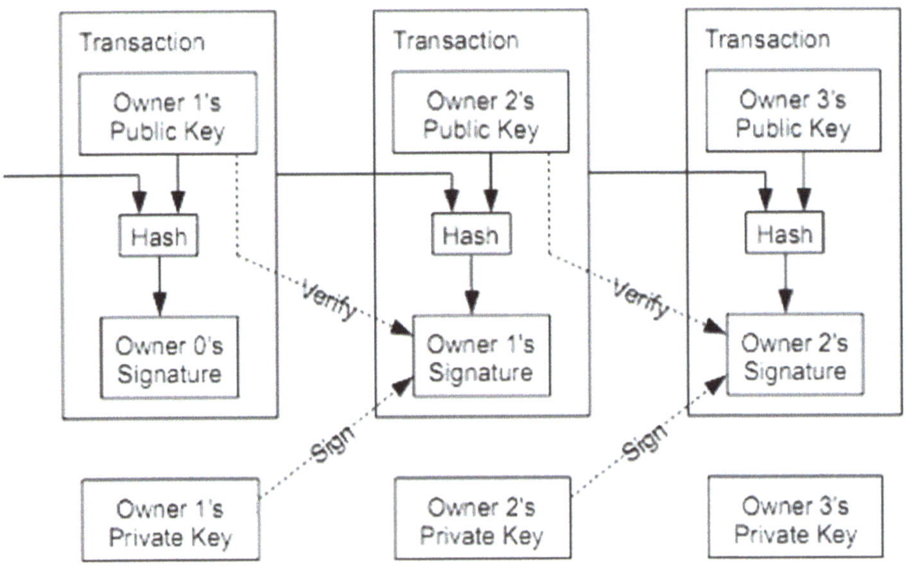

Fig. 1.10 Blockchain Principle

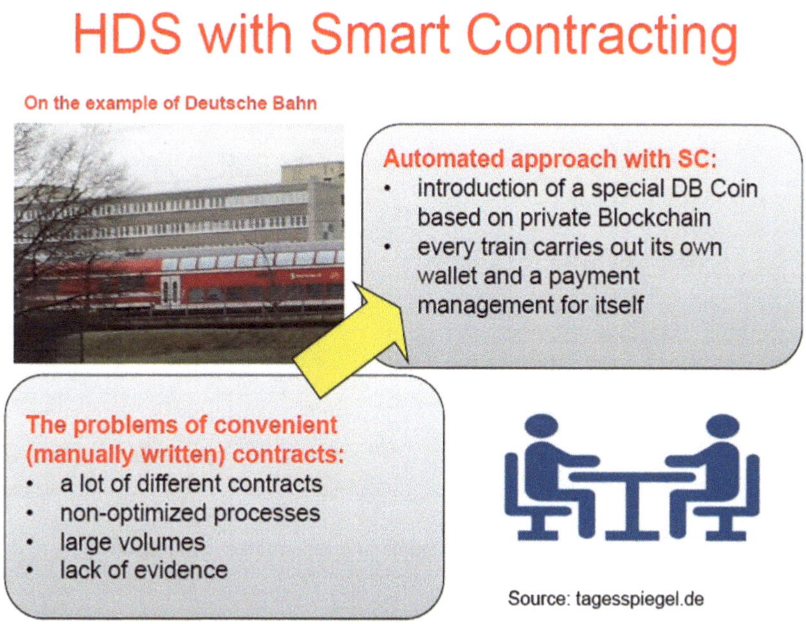

Fig. 1.11 HDS with Smart Contracting

signature and RSA crypto-method. There are many modern BC applications, inter alia: Smart Contracting (SC), refer Fig. 1.11. One of the mostly important Blockchain applications after the mining of the crypto-currencies SC is [15, 16]. Historically, SC does not require exceptionally Blockchain, but certain consensus algorithms (protocols), which are cryptographically conditioned via hashes, private and public keys and signatures.

A possible solution for automated, not manually closed contracting is given in Fig. 1.11. The example is here given, which SC for Deutsche Bahn consider and illustrates distributed transactions between waggons and trains.

Generally, the BC as well as SC technology is widely used in a typical HDS (refer Fig. 1.12), based on Micro-Services.

However, the following problems occur during the Blockchain operation:

- Enormous energy consumption due to mining of crypto-currencies (processing of the hash blocks via its algorithmic complexity);
- Exponential memory growth (including capacity migration between USB media, smartphones, PC, storage media such as SAN/NAS, as well as cloud storages);
- Cryptographic data security is guaranteed, but privacy issues may arise. Out of the way is as follows: no processing the complete Blockchain with all the transactions, but only use of excerpts of the Blockchain without a prehistory;
- Mining of Cryptocurrency and Resource consumption by Blockchain.

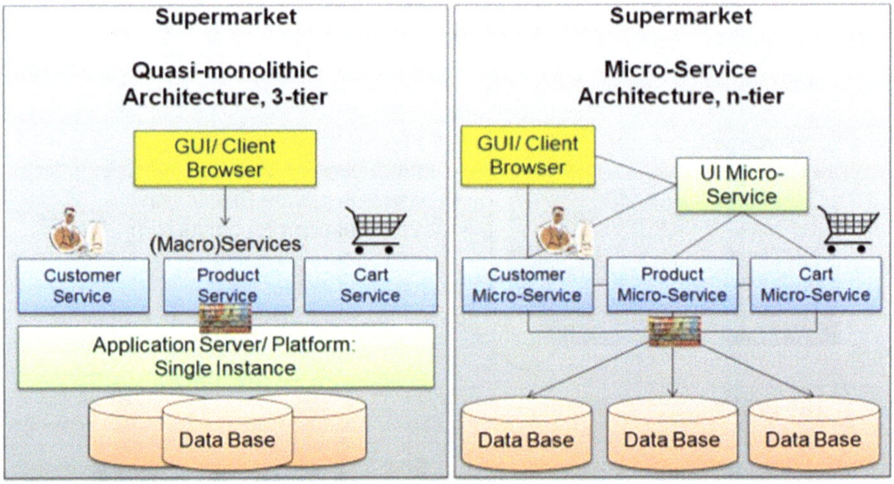

Fig. 1.12 HDS with Micro-Services as an example of a supermarket application

At least five factors are necessary and must be considered when calculating the profitability of the mining [8, 13–19]:

- Investment (basis device costs): hardware investment for a Mining Rig;
- Electricity costs: in EUR per kWh;
- Energy consumption: electrical power of Mining Rig in KW;
- Hash-rate: how much hash values can be computed each second?;
- Mining difficulty: factor has always a new actual value!
- Pool fees: how much in percent belongs to the joined pool;
- Unit reward per Block: Bitcoin amount for each new computed block (or for other units); Unit price: exchange course for the crypto-currency BTC (or other units like XMR, ETH, ZEC, and LTC).

There are different multipurpose and specialised devices for mining available, so-called Mining Rigs. The old good PCs or plain smartphones can be used too but under considering of the energy consumption problems. The following types of devices can be deployed to generate the hash values for mining process:

- CPU mining: powerful processor is required
- GPU mining: powerful graphics card is required
- ASIC mining (Application-Specific Integrated Circuit, s. Table 1.1).

The practical experience has shown that in a lot of cases the mining of the crypto-currencies like BTC, ETH, ZEC, XMR etc. leads unfortunately to “no reward” cases due to a large energy consumption as well as essential CAPEX + OPEX.

Table 1.1 Features of a Mining Device. *Source* amazon.de

Features of AntMiner S15	High-performance	Low-energy mode
Hash rate	28 TH/s	18TH/s
El. power	1596 W	900 W
Power efficiency	57 J/TH	50 J/TH
NW connection	Ethernet	Ethernet
Weight and dimensions	7 kg, 240 mm × 178 mm × 296 mm	

1.4 Integration of Blockhain within the HDS

The integration of Blockhain within the HDS applications is depicted in Figs. 1.12 and 1.13. The following mostly used topics have to be mentioned:

- ML and AI;
- Digital Economy [2, 4, 9, 12];
- Crypto-currencies [13–19];
- 5G Network Slicing [4, 5, 7, 9].

Such HDS contain flexible Micro-Services (refer Figs. 1.12 and 1.13) in interoperability with Web Services as construction components and possess, as a rule, multiple planes (SDN), several strata (digital plat forms) and slices (5G network slicing) in interoperability with Web Services as construction components and possess, as a rule, multiple planes (SDN), several strata (digital platforms) and slices (5G network slicing).

Every of the listed topics are concerning the HDS with built-in crypto technologies and, in particular Blockchain. This important layer or plane provide the confidentiality, authentication and compulsoriness as most important security aims for multiple components of the complex internal structures and workflow steps (refer Fig. 1.13).

However, as the main disadvantage of the offered method, the performance reduction by real-time services as well as energy consumption can be mentioned as a critical position.

The following multiple crypto-platforms can be used for Blockchain integration: Ethereum Classic, MS Bletchley, Codius for Ripple, Hyperledger Sawtooth [8–12].

An important requirement to such platforms is the source security and trust especially by risks of ransomware and other crypto-Trojans like Petya, WannaCry or GandCrab. In opposite to standard solutions, based on PKI and combined symmetric-asymmetric encryption (RSA) and digital signatures Blockchain provide its own security decentralised infrastructure, which distinguish from centralised PKI or bilateral Web-of-Trust incorporated in convenient distributed apps. As one of the most important solutions, Framework MS Bletchley serves.

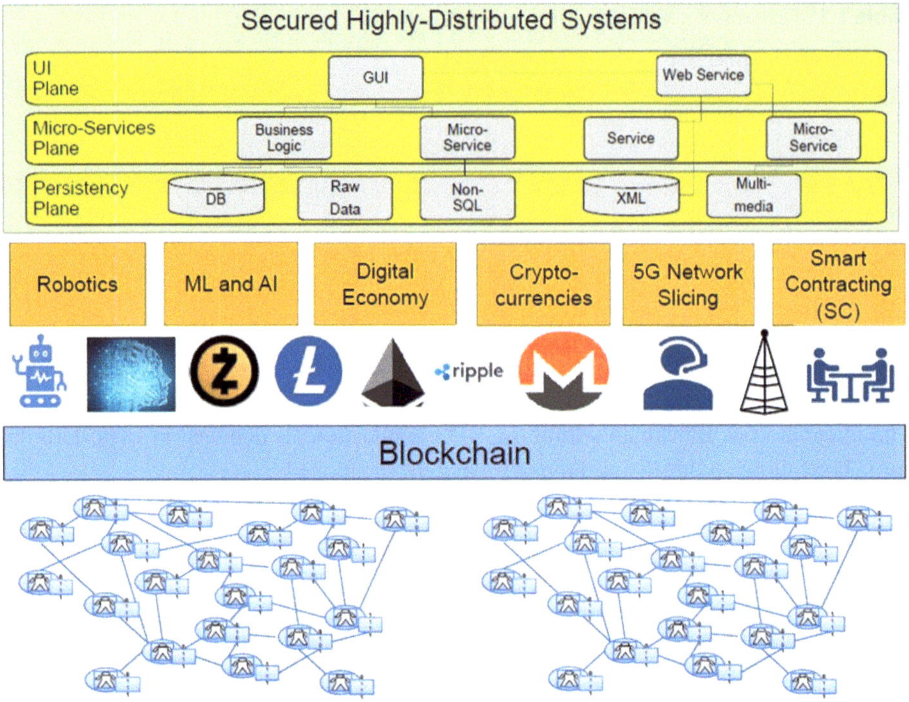

Fig. 1.13 Blockchaining in HDS

1.4.1 Example: Framework MS Bletchley

MS Bletchley was launched as a specific BaaS (Blockchain as a Service) and as a part of the Azure cloud platform (Fig. 1.14). The main goal of the framework introduction [17] was the acceleration of the development of the practical Blockchain applications aimed to financial institutions, manufacturing, retail, healthcare, public sector, media on a common platform. The above-mentioned platform contains three following layers:

- Base platform tier;
- Middleware (MW) tier;
- Industrial solutions layer.

The framework MS Bletchley introduces a new concept in particular: Blockchain middleware (MW) with so-called cryptlets. To SaaS and BaaS belong so-called ML and BI functionality (ML and Business Intelligence). The Blockchain MW components support the core features of services in the clouds, such as identity management, analytics, or machine learning. Based on the Azure cloud, these components can work with different Blockchain-based technologies aimed to creation of the practical applications.

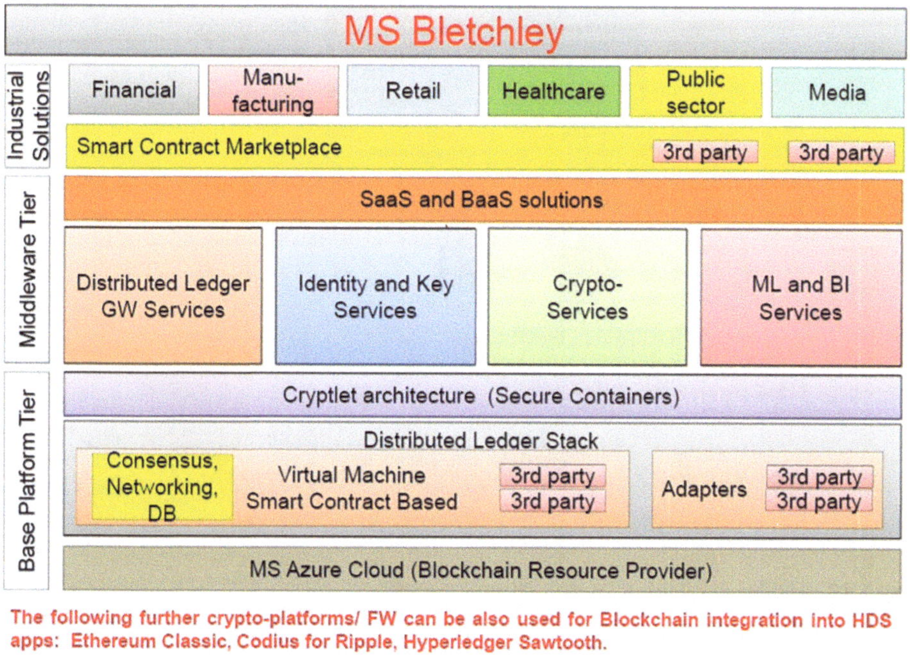

Fig. 1.14 Blockchain Platform MS Bletchley

The cryptlets are the building blocks for the Blockchain technology. They are designed to ensure secure communication between the Azure cloud, the MW ecosystem and each specific customer's technology. The interoperability of the applications with Azure cloud and Azure stack is secured, as well as with the third parties, the clouds like AWS, Google and further private clouds.

1.4.2 Compulsoriness via BC

The Blockchain must guarantee the secured transitions between the interlinked Web Services, Micro-Services and further modules within the slicing structures and provide their compulsoriness. In the best way, the solution is possible based on Blockchain. In opposite to standard solutions, oriented to PKI and combined symmetric-asymmetric encryption (RSA) and digital signatures Blockchain provide its own security decentralised infrastructure, which distinguish from centralised PKI or bilateral Web-of-Trust, incorporated in convenient distributed apps [8, 13].

The main goal is to transfer service data between 5G Network Slicing using a Blockchain that will introduce obligatorily and irreversibility by organizing linear chained structure. In [13] this idea realised by using Key Distribution Centre (KDC), which is cooperating with the Authentication Server Function (AUSF) and for a given

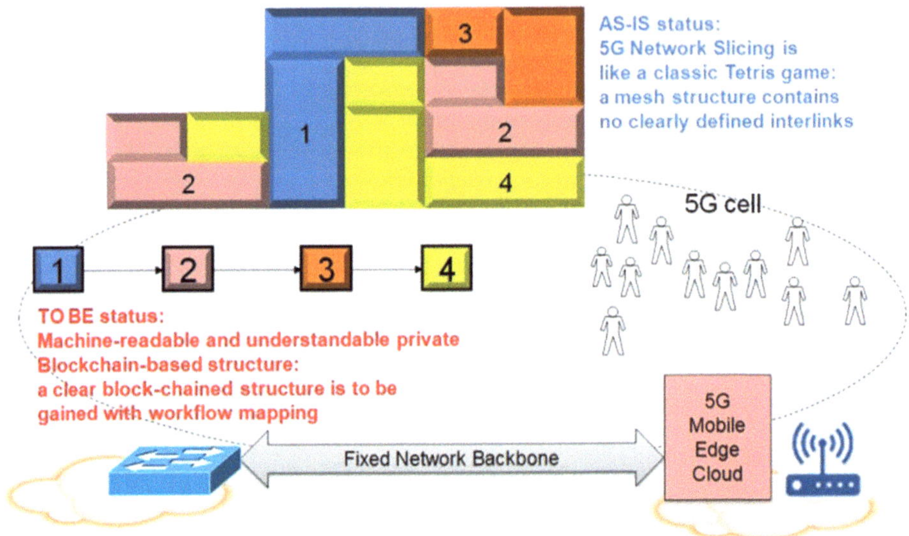

Fig. 1.15 Multivariate Tetris vs. a blockchained clearness

network slice that serves n devices (or n distinctive use cases), KDC generates a key pair (d, e) for the El-Gamal cryptosystem: d and e are private and public keys. In our opinion, this method will not be effective enough. Using Blockchain we can win in terms of security and productivity even unless more complicated and resource consuming character of hash computing.

Let us to compare the both status and the advantages of the offered solution (refer Fig. 1.15):

- AS-IS status: 5G Network Slicing is like a classical Tetris game. There is a mesh structure with no clearly defined interlinks. The workflow is not mapped.
- TO-BE status: Machine-readable and understandable private Blockchain based structure must be constructed instead of them. A linear chained structure must be gained. The compulsoriness of the steps and algorithm blocks is guaranteed as well as the further security aims: confidentiality and authentication.

1.4.3 Blockchained ML and AI

When talking about artificial intelligence (AI), we should also mention Machine Learning (ML) as an important basis for HDS (Fig. 1.16). To begin with, we would like to represent what the difference between the convenient deterministic algorithms

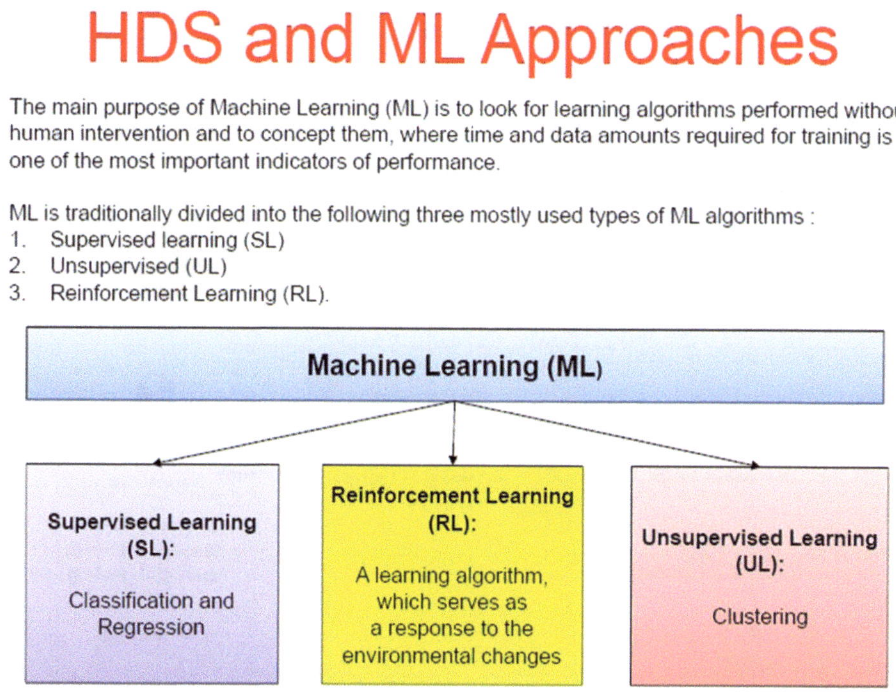

Fig. 1.16 HDS and ML approaches

(cp. classical flowcharts, like SSADM or PAP by DIN 66001) and Machine Learning. Figure 1.17 shows that in classic algorithms we have Input Data and an algorithm that allows us to get results. In ML, we have Input and Output Data, which help us to get a neural network a learning algorithm and it will help in the future to make the neural network more powerful via training.

Nowadays, artificial intelligence is widely used in all spheres of life without stopping in its development. In some cases, such as 5G networks, we need to provide artificial intelligence at different levels of architecture with a clear sequence and compulsoriness of the algorithm steps (Fig. 1.18).

In order to achieve this, we will use Blockchain, which provide the necessary hierarchy and sequence of actions. It will also allow to users of such a system to pay for clearly allocated resources without the need for overpayments. It increases the overall QoE in up-to-date HDS. The entities 1–7 are the agile Micro-Services (i.e. implemented classic algorithms) or separate small neural networks with own intern structure. They are integrated one each other with necessary compulsoriness via Blockchain (refer Fig. 1.18):

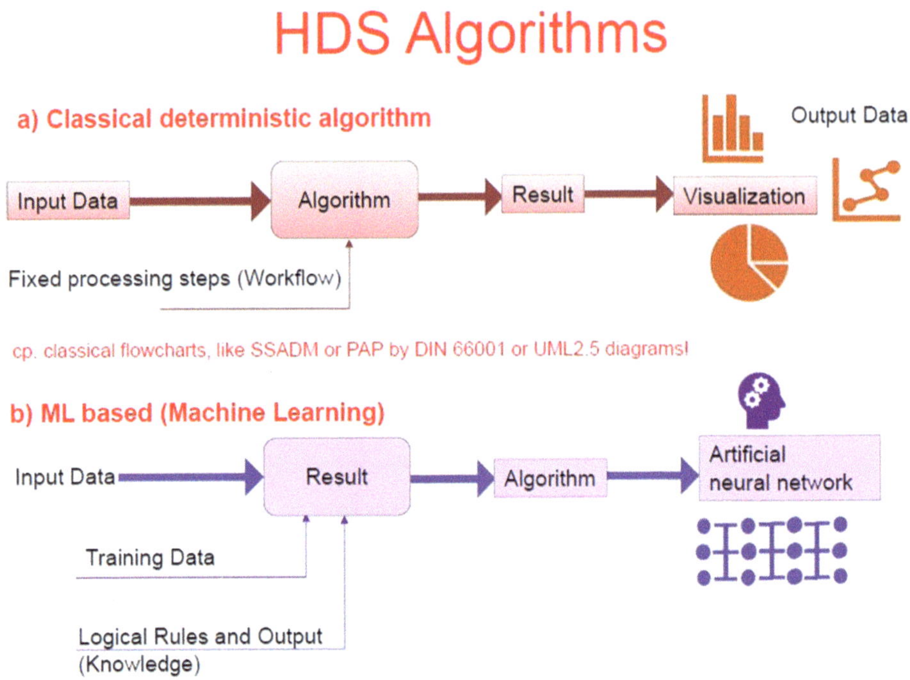

Fig. 1.17 Comparison of ML to deterministic algorithms

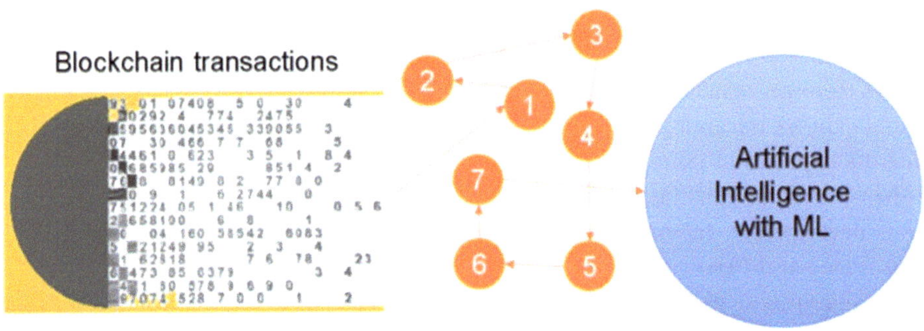

Fig. 1.18 ML with Blokchain transactions: a clear sequence and compulsoriness of the algorithm steps

1.5 Paradigms to HDS Development

The Construction Paradigms for so-called Highly-Distributed Systems (HDS) are examined below.

- P1: The first paradigm is the use of advanced architecture models (M3-M5) for efficient communication, providing advanced QoE and energy autarky within HDS applications.
- P2: The next one is the deployment of modern SWT process models (like Scrum) and use of flexible Micro-Services, leading to the efficient HDS, which provide better scalability, reliability and reconfiguration.
- P3: The further HDS paradigm regards to security, privacy, authentication and compulsoriness of HDS workflow steps, modules and services under use of Blockchain technology. However, as the main disadvantage of the Blockchained paradigm, the performance reduction by real-time services as well as energy consumption become to a critical position.
- P4: In addition to the classic algorithms, ML must be also integrated within HDS applications imperatively as a Paradigm. This work can be positioned as a Work-In-Process. There is a lot of innovative techniques and unclear options, which are to be accurately handled: Blockchaining, Micro-Services themselves, mastering of development tools and frameworks.

1.6 Conclusions and Outlook

The given work represents a short overview on techniques for creation of modern Highly-Distributed Systems under use of Micro-Services and crypto-technology Blockchain and investigates the opportunities for such system development under use of proven agile process models. The case studies and some deployment scenarios are provided. The discussed technologies promise a new technological breakthrough for widespread modern Highly-Distributed Systems, which detach convenient distributed applications.

HDS with Blockchain provide the confidentiality, authentication and compulsoriness as most important security aims for multiple components of the complex internal structures and workflow steps on the Blockchain basis. The HDS in telecommunication, industry, entertainment and education include ML/AI, Digital Economy, support multiple crypto-currencies and 5G Network Slicing, Smart Contracting and more else apps, are oriented to Web Services and Micro-Services, which are placed on different planes, strata and slices.

The deployment scenarios on 5G and Beyond networks contains nowadays so-called digital twins and provides slicing architecture. Blockchain with cryptography provide compulsoriness of the steps and algorithm blocks and guarantee the further security aims: confidentiality and authentication.

In mid-term, the standards for ML and AI will accompany the industries, digital economy and everyday life over the world and for each institution. The basic principles of use of ML/AI with Blockchain structures in behavioural models for 5G networks as deployment scenarios were represented.

The perspectives of so-called block-chained HDS solutions are discussed. As the main disadvantage of the offered method, the performance reduction by real-time services as well as energy consumption can be mentioned as a critical position.

Definition for HDS was given, as well as the demarcation to conventional distributed systems offered. The distinguishing features for HDS are clearly formulated.

References

1. Sam Newman. Building Micro-Services, Publishing by O'Reilly Media, USA, 2015, ISBN: 978-1-491-95035-7, 473p.
2. Schill, Alexander, Springer, Thomas. Verteilte Systeme: Grundlagen und Basistechnologien: Kompakte Darstellung der Grundlagen und Techniken Verteilter Systeme, Springer-Verlag Berlin Heidelberg, 2012, 2.Ausgabe, ISBN: 9783642257957 (https://www.springer.com/gp/book/9783642257957#otherversion=9783642257964).
3. Martin Fowler. Advanced Software Architectures (Online): https://martinfowler.com/.
4. Luntovskyy, Andriy, Spillner, Josef. Architectural Transformations in Network Services and Distributed Systems: Current technologies, standards and research results in advanced (mobile) networks, Springer Vieweg Wiesbaden, 2017, 344p., ISBN: 9783658148409 (https://www.springer.com/gp/book/9783658148409#otherversion=9783658148423).
5. A. Luntovskyy. Advanced Software Technological Approaches for Mobile Apps Development, 14th Int. IEEE TCSET-2018 Conf., Lviv-Slavske, 2018, 6 p. (IEEE Xplore: https://ieeexplore.ieee.org/document/8336168/), DOI: https://doi.org/10.1109/TCSET.2018.8336168.
6. A. Luntovskyy. SLMA and Novel Software Technologies for Industry 4.0, 21-st Int. Conf. ACS-2018, Szczecin-Miedzyzdroje, 2018, in: J. Pejas, I.El Fray, T.Hyla, J.Kacprzyk (Eds.). Advances in Soft and Hard Computing, Springer Int., 12p. (Part of the AISC book series, vol. 889, DOI: https://doi.org/10.1007/978-3-030-03314-9-16, ISBN: 978-3-030-03313-2).
7. A. Luntovskyy, D. Guetter, M. Klymash. Up-to-date Paradigms for Distributed Computing, Int. IEEE Conf. AICT-2017, Lvyv, pp. 113–119 (IEEE Xplore), ISBN: 978-1-5386-0638-4, DOI: https://doi.org/10.1109/AIACT.2017.8020078.
8. A. Luntovskyy, D. Guetter. Cryptographic Technology Blockchain and its Applications, in "Advances in Information and Communication Technologies", Springer (ISBN: 978-3-030-16769-1), LNCS "Processing and Control in Information and Communication Systems (Int. Conf. UkrMiCo-2019)" (eds.: M. Ilchenko, L. Globa et al.), 2019, pp. 14–33 (https://link.springer.com/book/https://doi.org/10.1007/978-3-030-16770-7).
9. A. Luntovskyy, D. Guetter. Moderne Rechnernetze: Protokolle, Standards und Apps in kombinierten drahtgebundenen, mobilen und drahtlosen Netzwerken, 500 Seiten, 263 Abb., Springer Nature, Juli 2020, ISBN: 9783658256166 (https://www.springer.com/gp/book/9783658256166).
10. A. Luntovskyy, D. Guetter. Moderne Rechnernetze - Übungsbuch: Aufgaben und Musterlösungen zu Protokollen, Standards und Apps in kombinierten Netzwerken, 150 Seiten, 44 Abb., Springer Nature, Juli 2020, ISBN: 9783658256180 (https://www.springer.com/gp/book/9783658256180).
11. Mahmood Zaigham (Ed.): Fog Computing: Concepts, Frameworks and Technologies, Springer 2017, London, ISBN 978-3-319-94890-4.
12. Jamil Y. Khan, Mehmet R. Yuce (Eds.). Internet of Things (IoT): Systems and Applications, 2019, New York, Jenny Stanford Publishing, ISBN 9780429399084, 366 p.

13. MIT Blockchain Course (Online): https://executive.mit.edu/course/blockchain-technologies/a056g00000URaa7AAD.html.
14. A. Antonopoulos, G. Wood. Mastering Ethereum: Building Smart Contracts and Dapps, 2019, O'Reilly Media, 345p. ISBN: 978-1491971-949.
15. Smart Contracts (Online): http://www.icertis.com/.
16. Survey on Crypto-platforms 2019 (Online): https://hackernoon.com/top-blockchain-platforms-to-watch-out-in-2019-aa80e336a426/.
17. MS Bletchley (Release & Roadmap), Cryplets Deep Dive (Online): https://azure.microsoft.com/.
18. Codius: Open Source Hosting Platform for Smart Programs (Online): https://codius.org/.
19. Hyperledger Sawtooth (Online): https://www.hyperledger.org/projects/sawtooth/.

5G Networks Deployment and Service Modeling. New Generation Networks. 5G and Beyond

2

In given chapter the backgrounds for the deployment of 5G in EU countries are discussed. The most important requirements for 5G networks are examined like slicing, wide network interoperability (with Wi-Fi 6, IoT devices, LoRa WAN etc.). The operation in 5G micro-cells of small private providers is favoured. With the aim of compulsoriness so-called Blockchained 5G Slices are used. The authors investigated the appropriate Blockchain (BC) architectures, recent BC types, BC applications, platforms as well resource consumption. So-called Smart Contracting for 5G networks and DAO technology can be taken into account for use as 5G Blockchained slices. Intelligent 5G handover and roaming within the 5G Networks can be provided using ML standards. Artificial Intelligence for Handover Optimisation is discussed. To the demarcation of our research area is given. Recurrent neural networks as well as their modelling and training processes are represented. Modelling with a neural network and prediction of the user number in a micro-cell are investigated. Recurrent neural networks, based on GRU cell, can optimize the handover in 5G cellular structures. The method is extended for the hierarchical roaming in backhauls, mega- and giga-cells.

New Generation Networks, such as 5G and Beyond, 6G and Starlink are prospectively examined.

2.1 Deployment of 5G in EU Countries

In July 2019 in Germany after testing, the 5th generation mobile network was launched (Deutsche Telekom, Vodafone, Telefonica Germany) [1]. Suisse has assigned in 2019 the 5G frequency bands to three major providers: Swisscom, Sunrise and Salt. In July 2019, 334 antenna towers with 5G frequencies began to operate in the country [2]. Through the program 'Horizon 2020', the European Commission is investing 700 million e in

A. Luntovskyy and D. Gütter, *Highly-Distributed Systems*,
https://doi.org/10.1007/978-3-030-92829-2_2

research and innovation funding, related to 5G nets. The 5G nets provide IoT (Internet of Things), remote industry and events monitoring, plants automation, facility management, safety in the cities, assist health-related care (ambulance, surgery), touristics and more else. The distinguishing features of 5G nets are as follows [1–5]:

- use of conventional 4G nets;
- DR for the first time being up to 20 GBit/s;
- use of higher frequency areas (24.5–29.5 GHz) and still up to 60 GHz;
- real-time transmission for telepresence by short latency of approx. 1 ms;
- QoE and compatibility of multiple machines and devices (so-called ubiquitous connectivity);
- energy consumption decreasing per transmitted bit (1/1000);
- 90% lower power consumption per mobile service.

Nowadays wireless and mobile communication is occupied with provision in general of IP-services and transmission of multimedia content from one place to another, but tomorrow the new 5G will be able to control a wide range of objects in real time with only insignificant human intervention in the frame of IoT and other attractive apps. The quote of the day: "5G automation network will change our world relatively more than the cars (1920), the telephony (1950), the computers (1980) at the time of their appearance!" (by F. Fitzek) [3, 4].

2.2 State of the Art: Goals and Technologies

The first experiments with 5G nets began already in 2014. The main 5G requirements are as follows [1–7]:

- use of existing 4G-infrastructure with augmentation via flexible WLAN conform communication everywhere under international voting and conventions;
- to medium term obtaining of DR = 10 GBit/s;
- this DR corresponds to up-to-date needs to MM-content download;
- tiny latencies, real time, interoperability, services without human intervention;
- wide use of available frequency bands: mm-Band F = 30 up to 300 GHz (partially and questionable);
- interoperability with further mobile and wireless radio networks.

The advanced antenna technique MIMO (Multiple Input/Multiple Output) has been already deployed in diversity network technologies since 2008 like WiMAX 802.16a/d/e/m, WLAN 802.11n/ac/ad/ax, LTE etc. MIMO-antennas allows nowadays communication in3D-mode with NTx = 16–64 transmitting and NRx = 16–64 receiving antennas (Fig. 2.1). This scenario depicts so-called multi-user MIMO (MU-MIMO) in

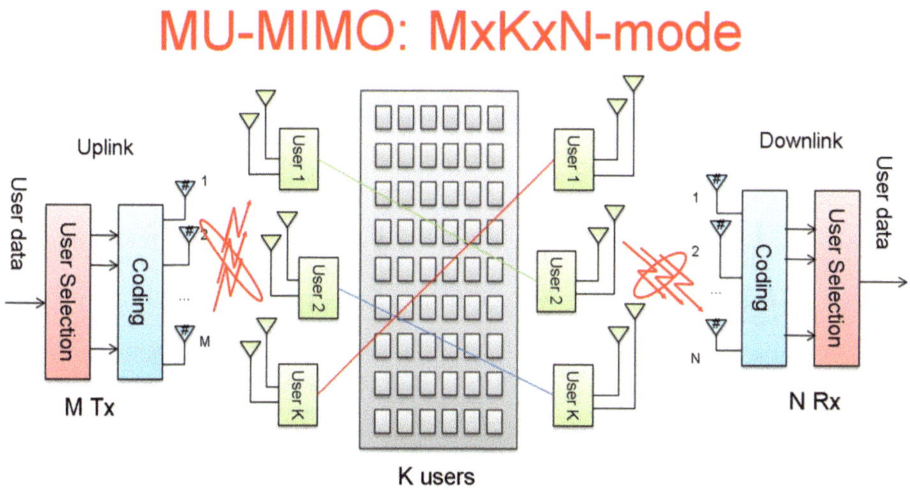

Fig. 2.1 Up-to-date 3D-MIMO antennas for radio networks in MxKxN mode

MxKxM-mode [6–12]. The downlink (DL) with DR = 10 GBit/s and uplink (UL) with 1 GBit/s became available. This DR = 10 GBit/s is about × 100 more in contrast to DR = 100 Mbit/s of LTE.

The prognosis is as follows: in 2025, up to 100 milliards devices will be IPv6-driven working, partially with 5G [1–4].

The architecture and components for 5G Mobile Network are shown in Fig. 2.2 In the given 5G scenario, the multi-modal access and SDN core are represented. Based on the representation the 5G construction points of gratitude are considered as follows. The wide interoperability to "Wi-Fi world" is foreseen [8–12]. Furthermore, the 5G architecture is steady based on predecessors: 4G and 3G with appropriate hierarchical cell structure and integrated satellite links. The technology will augment the existing GERAN, UTRAN, SAE and IMS (refer Fig. 2.2). Some prominent companies make the preparations to further symbiosis of 5G and Wi-Fi 6. Thus, Microsoft intends it soon, to provide the access to the 10 million Wi-Fi hotspots. Through its Internet telephony subsidiary Skype Microsoft offers already the Wi-Fi access to about two million hotspots worldwide. Under the label "Microsoft Wi-Fi" the access rights will be granted to the Office and Skype customers [8, 9].

Use of SDN for software implementation of provider core in practice for 5G networks enables to enterprises and providers to receive vendor-independent functions for management and control of network components and services from any type of unified providing center, which will greatly simplify their operation. The use of SDN as part of the 5G/IMT 2020 is a determined position [5–10]. The system deploys advanced RAT (Radio Access Technology) and RLAN (Radio LAN) as well as provide backwards compatibility to existing systems like 4G/SAE, 3G/UTRAN, 2G/GERAN and SAT-links (e.g.

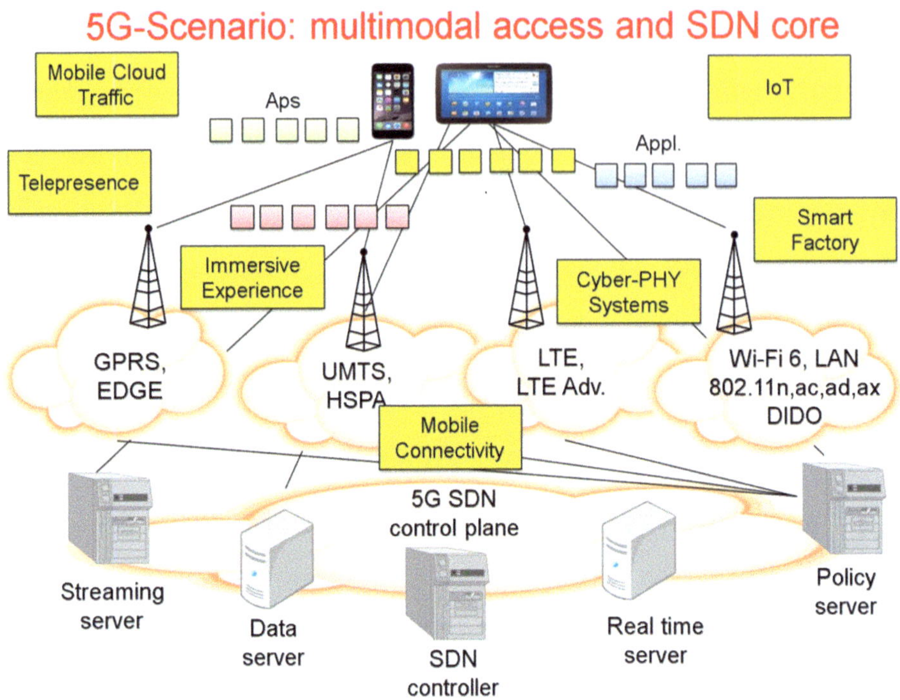

Fig. 2.2 Architecture components of 5G Mobile Network: Legend: SDN—Software-Defined Networking; IoT—Internet of Things

ICO RTT). Advanced adaptive modulation techniques for 5G were developed [1–10]. They are aimed to more dense and robust covering of the geographic areas as well as the increasing of so-called QoE (Quality of Experience) in 5G nets. Under QoE (Quality of Experience), we understand system performance using subjective and objective measures of customer satisfaction. It differs from quality of service (QoS), which assesses the performance of hardware and software services delivered by a vendor under the terms of a contract. The origins of the term went from 5G white papers in last years, which discuss the advantages of IMT2020 from, e.g. LTE with typical QoS requirements [1–3].

In contrast to QoS, QoE not only depends on the technical performance, but also on a wide range of other factors, including content, application, user expectations and goals, and context of use. Understanding QoE, thus demands for a multidisciplinary research approach that goes beyond the network level. As a very promising application for 5G (IMT2020) networks IoT (Internet of Things) is considered which is based on interoperability of different physical types of radio networks as well as virtualisation technology for the core services to interact with each other and with the external environment (6LoWPAN, SDN).

The following scenarios of 5G deployment regarding to IoT/Ubiquitous Computing became realistic [1–3, 6, 7, 13–15]: Smart Home, Smart Manufacturing, Smart Health; Smart Retail, Smart Transportation, Smart City; Remote Surgery, Hazardous Work, and Remote Driving.

2.3 5G Inter-Operability

As we already have constituted it, 5G/IMT 2020 is interoperable to LTE, Wi-Fi 6 and SAT-Radio. Figure 2.3 compares 5G/IMT2020 networks and their predecessors with some Wi-Fi protocols, which provide mutual interoperability [6–11]. One the most important is a new standards IEEE 802.11ax, or so-called Wi-Fi 6.

The networks IEEE 802.11ac and, especially Wi-Fi6 (2019) provide DR = 1–11 GBit/s, better WPA3-security, MU-MIMO, advanced OFDMA. Wi-Fi 6 possess higher spectral and energy efficiency for the used frequency bands due to use of a set of combo-advantages (refer Fig. 2.4). The networks IEEE 802.11ad possess the following properties: in contrast to traditional WLAN is only for a few meters; this results from the high absorption of oxygen at 60 GHz; a large BW to the higher DR = 7 GBit/s; the 60 GHz band is from 57 to 66 GHz, and is divided by a channel spacing of 2160 MHz in four channels having a bandwidth of 1760 MHz. So-called piconets WSN, Bluetooth, 6LoWPAN etc. are also interoperable to 5G [1, 6, 7, 11].

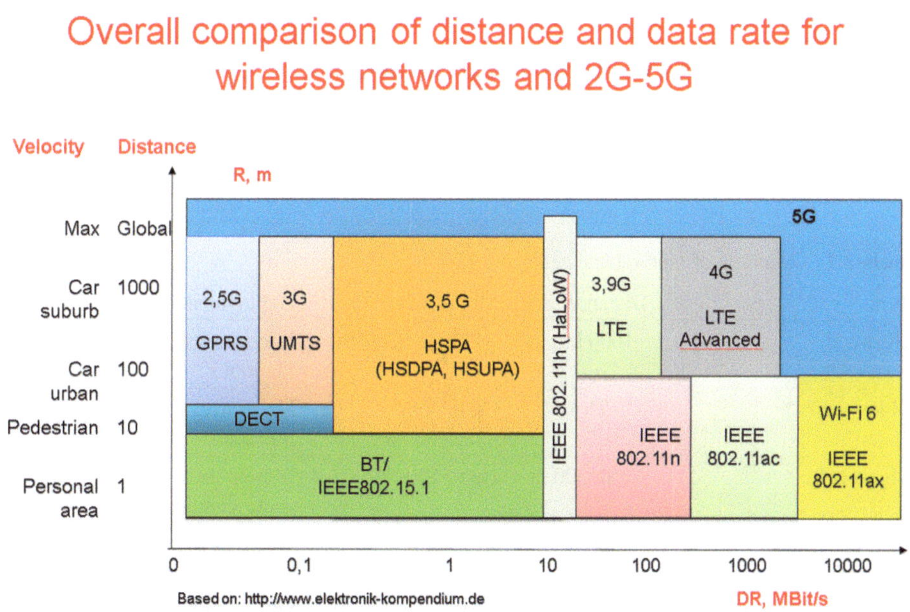

Fig. 2.3 Overall comparison of distances and data rates for wireless and mobile networks

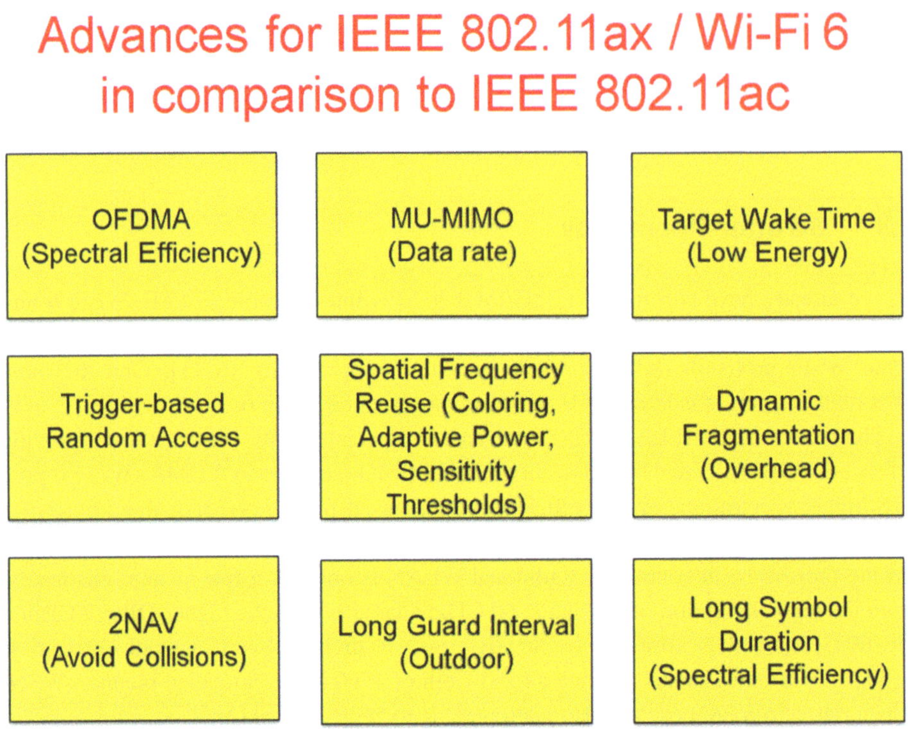

Fig. 2.4 Advances for IEEE 802.11ax/Wi-Fi 6 in comparison to IEEE 802.11ac/Wi-Fi 5

2.4 5G Networks with the Hierarchical Cell Concept

5G networks use the convenient hierarchical cell concept (pico-cells, micro-cells, mega-cells and giga-cells). These networks are completely inter-operable (Table 2.1).

Table 2.1 Overview of hierarchical cells

Type	Distance	Mobility (km/h)	Deployment by ...
Giga-cell	~1000 km	~4700 (~1.3 km/s)	International providers, satellite radio
Macro-cell	~1–5 km	Up to 500	National and regional providers
Micro-cell	~100–300 m	Up to 120	Metropolitan areas, city districts, campus, office area
Pico-cell (Indoor)	~10 m	~10	Hotspots at railway stations, airports, hotels, restaurants, clubs, home area
Pico-Cell (Personal area)	~10 cm	Stationary	PAN, Wearable, Smart Stuff

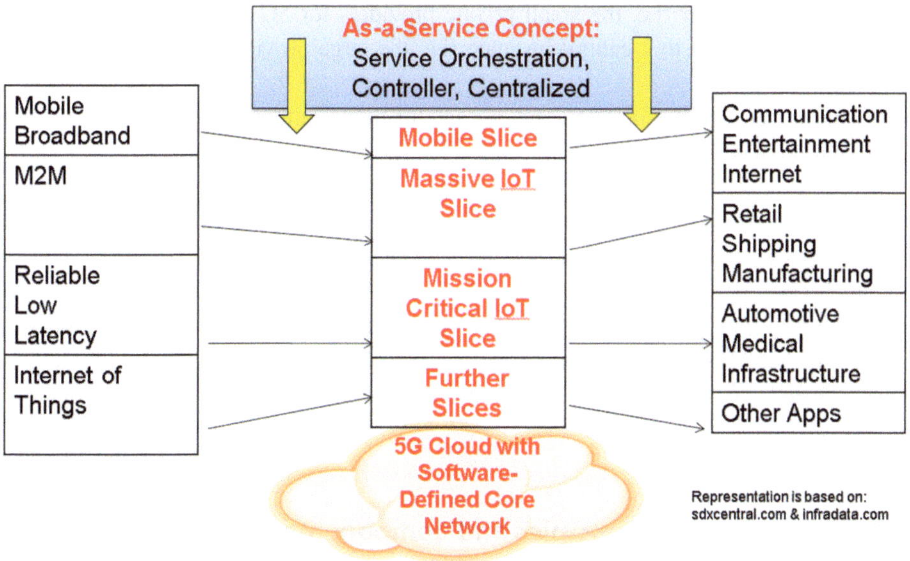

Fig. 2.5 5G Network Slicing: example

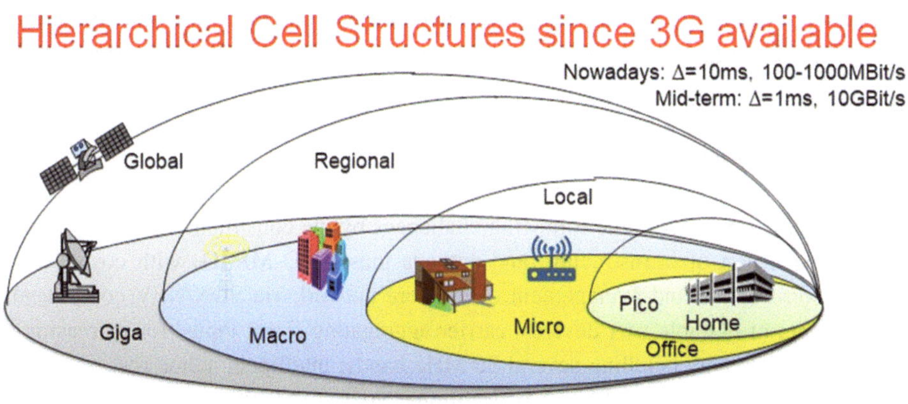

Fig. 2.6 5G networks with the hierarchical cell concept

The above-mentioned known 5G Slicing concept (Fig. 2.5) can be also realised on the following tiers (compare with Fig. 2.6):

- Macro provider tier: a central instance, it means a large provider with large cells and backhauls, the central instance controls the cell hierarchy (referenced on giga- and macro-cells)

- Micro provider tier: i.e. the small private providers for 5G networks in micro-cells, which are ordered to a central instance (i.e. the large providers with large cells and backhauls)
- Partition tier: slices, modules and virtual functions within a private network with decentralised P2P and M2M organisation [5–7].

Usually, 5G systems are designed in such a way that virtual networks can be sliced on so-called As-a-service basis. These services can be scaled up and down quickly and easily. Each network slice can be customized to provide the necessary components for the architecture it requires. For example, 10% of 5G resources can be reserved exclusively for IoT devices. Additionally, each slice is isolated and comprises the device, access, transport, and core network, thereby increasing reliability and network security (by infra-data.com).

2.5 Construction Principles and Components of 5G

The construction of 5G networks requires the deployment of the following important components and technologies like SDN (Software-Defined Networking), NFV (Network Functions Virtualisation) via OpenStack, SDR (Software-Defined Radio), optimal coding and network topologies [1–12]. Therefore, there are the following most important questions (Table 2.2) regarding 5G to the network engineering and construction principles:

2.6 Challenges for 5G Networks

The technology introduces the following challenges by the construction and deployment [1–12]: expansion of infrastructure and antenna masts (3D-MIMO) with costs economising; construction and management of the core network via SDN/NFV components; development of 5G-able user devices; carrier aggregating 'Data Pipes' for increasing of spectral efficiency by bundling several 20 MHz bands; intelligent adhoc reuse and reallocation of channel bandwidths in conventional areas (3.1/4.99 GHz) and higher frequencies (24.5/29.5 GHz) and above. 5G offer flexible service combination from big providers, which use hierarchical cell concept (giga-, macro-, micro-cells), with multiple small providers, which are oriented mostly to micro-, pico- and femto-cells.

In Fig. 2.7 so-called "5G atom" is shown. This representation depicts three strata (atomic orbits) of innovative 5G networks under considering of necessary trade-offs between:

- 1—scientific novelties
- 2—available technologies
- 3—proven concepts [3, 4].

Table 2.2 FAQ Regarding 5G Networks

Question	Solution
What is the goal of the so-called Network Functions Virtualisation (NFV) in the frame of 5G?	The main goal of NFV is to reduce the number of proprietary hardware and consolidate it for running of network services. Routers, firewalls, load balancers and other dedicated hardware have been responsible for the network functions. These components would be hosted on virtual machines (VM) in in the frame of 5G!
What are the two convenient ways to solve the frequency allocation problematics by 5G?	Way 1: Reuse of the old assigned lower frequencies (3.1/4.99 GHz) Way 2: Use of the new higher areas of frequencies migrating to 24.5/29.5 GHz
Which other intellectual method do you know for optimisation of reuse and reallocation of rare frequency resource?	Way 1: Ad-hoc exchange between a big diversity of available micro-cells of a 5G provider in real time considering temporary nonmobility of users. These users can be provided via satisfying QoS under alternative use of an efficient 5G-interoperable corporate Wi-Fi 5.6 within the campus (so-called network migration from micro-cells to the pico-cells, refer Fig. 2.5) Way 2: Intelligent adhoc reallocation of channel bandwidths on the underloaded micro-cells (so-called frequency resource migration)
What is the relationship between frequency, range and bandwidth in radio networks?	Frequency can be increased aimed to bandwidth extension, but the distance (range) is reduced quadratic. On the other hand, the bandwidth can be increased, but path losses are increased automatically too!

The efficient coverage is provided via 3D-MIMO antennas with advanced coding, what, however, requires more masts and CAPEX/OPEX. An important rival for 5G in micro-cell are so-called Wi-Fi 6 (IEEE 802.11ax) networks which offer with 1–11 Gigabit/s bandwidth and better range due to use of convenient dual band 2.4 & 5 GHz under limited mobility (Fig. 2.8). The standard benefits due to combining of OFDMA, MU-MIMO and interoperability to 5G. Wi-Fi 6 possess higher spectral and energy efficiency for the used frequency bands.

2.7 5G Network Slicing

5G-SDN is based on open interfaces and proven industry standards for hardware like Open Stack, which was used for LTE, WLAN, NFC, BT, ZigBee, 6LoWPAN too.

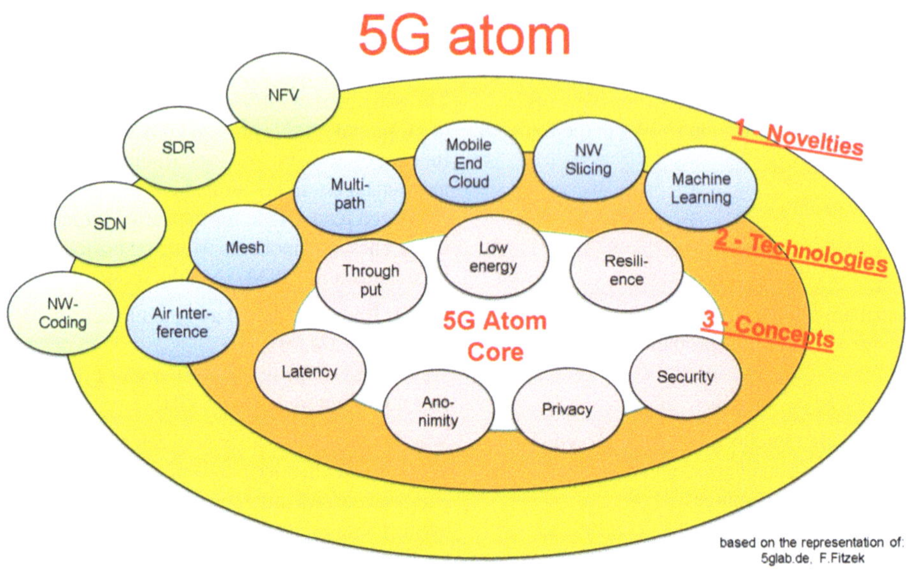

Fig. 2.7 5G atom: challenges of 5G nets [3, 4]

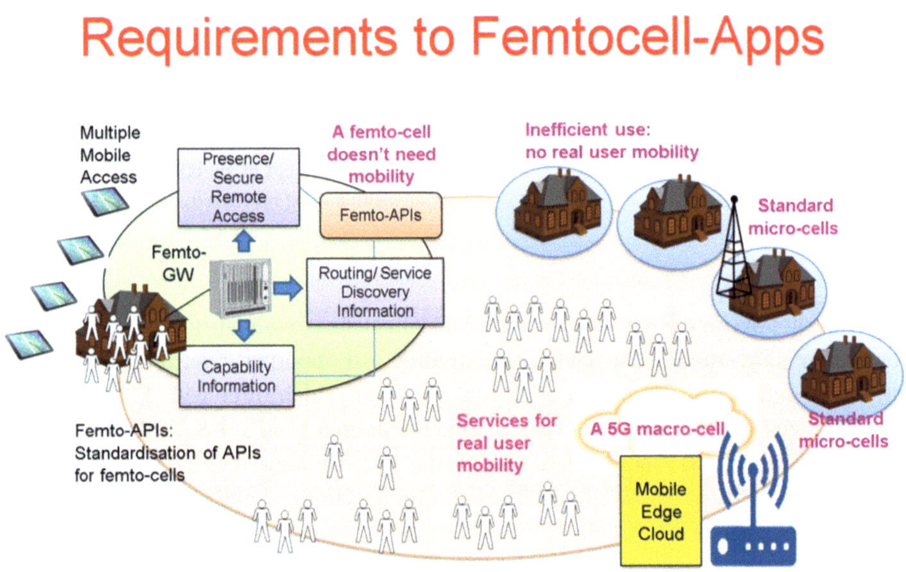

Fig. 2.8 The requirements to Femto-cell Apps

The next one is so-called "Network Slicing" (refer Fig. 2.5). The access and core networks control the creation, orchestration, deployment and operation of 5G and are divided to the so-called "slices" (or partitions). The "Network Slicing" concept means

[1–5]: Shorter latency for mobile network access (under 1 ms) for real-time scenarios; Quickly and efficiently creation and providing of differentiated access for different industry requirements (like L3 DiffServ); Better QoE management due to flexible deployment and modification of the necessary 'slices'(partitions); Seamless and fast mobile broadband connections. The principle of Network Slicing is depicted in Fig. 2.5.

Network Slicing is guaranteed via the used SDN (Software-Defined Networking) and wide NFV (Network Functions Virtualisation). The individual profiled 5G apps can be flexibly deployed under guaranteed QoE. For example, the rescue services can start these apps during an avalanche shift or other critical event [1–5]. Different apps are supported via the available slices, which are involved, taken into account the situation and QoS requirements.

2.8 Operation in 5G Micro-Cells

5G networks are widely operated in micro-cells with limited supply radius about 150 m. This means a big advantage for use in urban areas with higher user density. The 'hot-spots' should be supported. Such 5G microcells can be placed on lanterns, light gas columns or house walls and roofs for sparing of the CAPEX. As deployment examples in micro-cells the festivals, stadiums, event centres and concert halls, theatres, culture sites and institutions can be mentioned.

On this place, we would like to notice, that the use of the "small-cell concept" was also considered in 2005–2010 by some midrange providers for WiMAX/WiBro based services in e.g. South Korea, Slovakia, and Australia. Unfortunately, the technology was too expensive (CAPEX, OPEX) and less efficient than ongoing LTE. Thus, the above-mentioned technology was based on purely hardware solutions in opposite to the new 5G nets and micro-cells, which are implemented under use of SDN, NFV as well as further intellectual software.

Considering the new development of 5G apps and software solutions this might be unbeatable argument for the concept. The current application fields in this regard are as follows [1–7]:

- Robots and "5G digital twins" are supported by 5G;
- 5G-Agriculture with selfpiloted combined harvesters;
- Self-driving cars (refer Google's Waymo, Tesla Autopilot);
- Non-stop driving over intersections in the city;
- Car Navigation and Truck Platooning under shortened distances, control systems for simultaneous acceleration and braking of trucks;
- "Smart Grid", i.e. combination of energy-efficient IT with so-called "green electricity" under reduced risk of phase losses.

The minimal phase shift in the AC is reduced due to a short latency. Compare it: 1 ms means an insignificant phase shift $\Phi = 17°$; 10 ms means a huge phase shift $\Phi = 180°$. Let us to discuss some examples of the deployment of 5G-based "digital twins" in the frame of digitalisation challenges.

2.9 Case Study 1: Engineering and Productive Industry via 5G

The depicted scenario (Fig. 2.9) discusses the deployment of 5G in engineering and productive industry (Swisscom AG).

The campus solutions, "digital twins for production" devices, networked machines and augmented reality for repairing and hazardous works are supported. For such kind of the solutions, 5G macro-cells are used. A large set of 5G masts are placed. The constructors have to consider carefully the factors like mountain relief, enabled transmission power and possible propagation effects (dispersion and diffraction). The overlapped waves can lead to local signal amplifying and impairments for the health of humans and living beings.

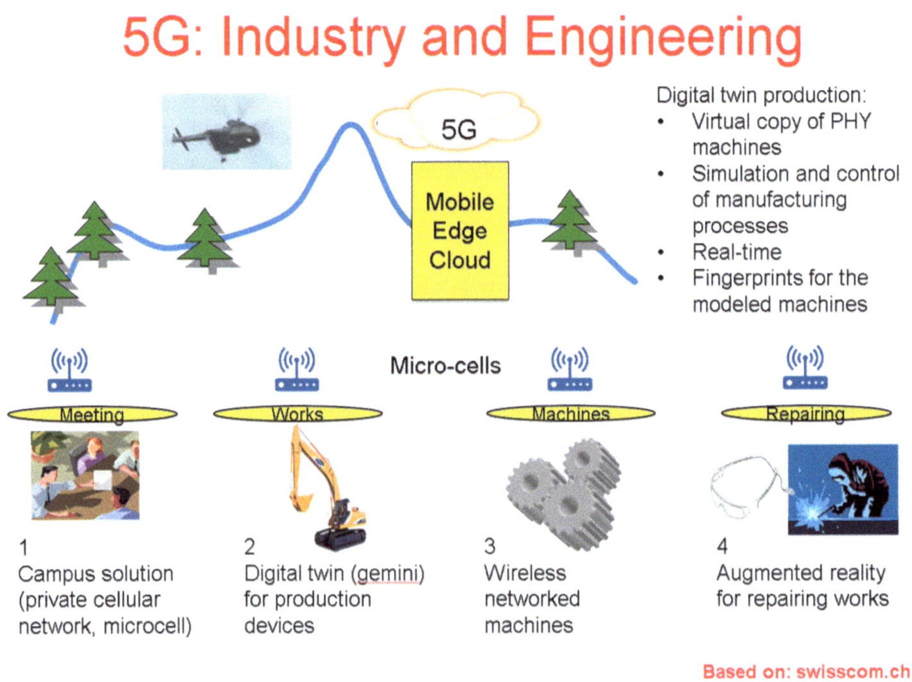

Fig. 2.9 Swisscom AG (Suisse): Engineering and productive industry via 5G

2.10 Case Study 2: Media, Events and Entertainment via 5G

The depicted scenario (Fig. 2.10) discusses the deployment of 5G in media, events and entertainment industry (Swisscom AG). The used slicing architecture provides mobile video recording, remote control of used cameras and content production. For such kind of the solutions, 5G micro-cells are used. Mobile video streaming and real-time creation of "digital twin events" can be offered.

2.11 Further Rivals and Alternatives: DIDO/PCELL, MS WI-FI

DIDO (Distributed Input Distributed Output) was developed by Rearden and Artemis companies with the aim to provide flexible multi user wireless LAN everywhere under international voting and conventions for the used frequencies amongst other mobile technologies towards 5G networks.

The technology is based, inter alia on MU-MIMO and Software-Defined Radio (SDR), which uses interference to create individual virtual cell sites for each wireless user (Fig. 2.11). The technology is in addition very energy efficient and can run the transmitters by approx. 1mW (equivalent power to the class 3 of Bluetooth).

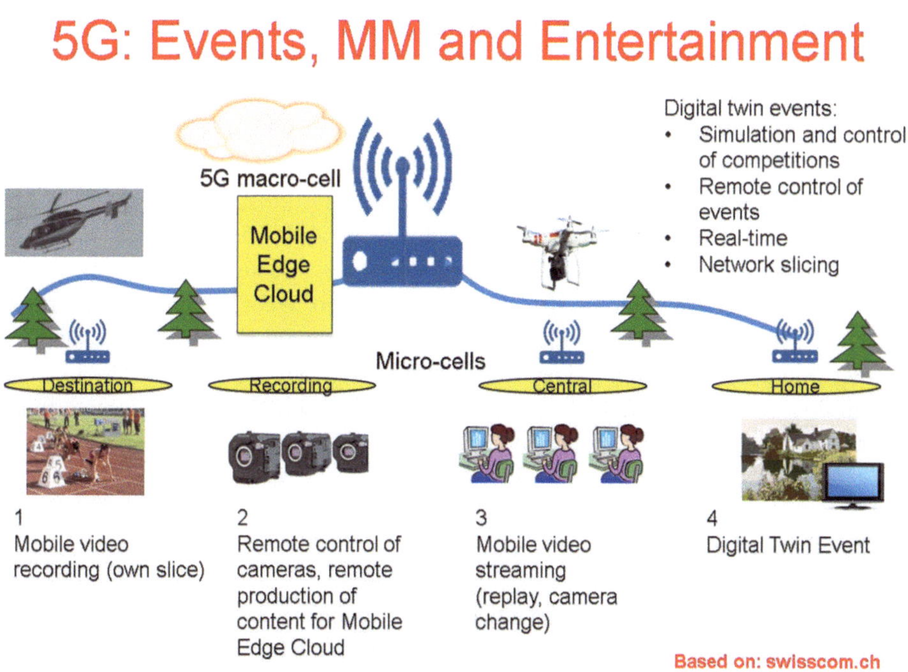

Fig. 2.10 Swisscom AG (Suisse): Media, events and entertainment via 5G

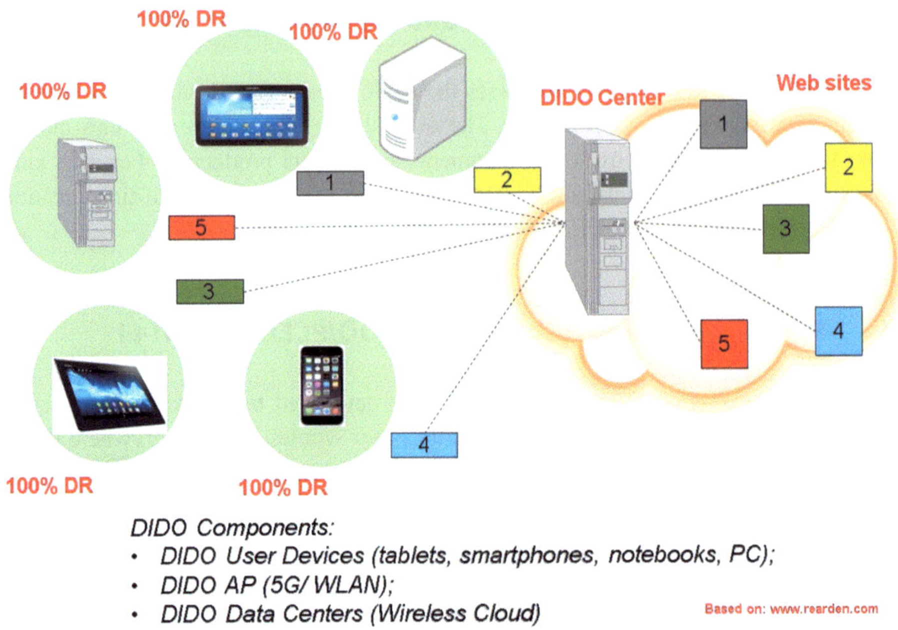

Fig. 2.11 DIDO technology [9, 10]

Furthermore, the Rearden DIDO rebranding followed under the concept of "Artemis pCell". pCell and DIDO [9, 10] are oriented to flexible resource allocation for Wi-Fi, which can use the wide non-traditional frequencies spectra between 3 MHz and 30 GHz independently with full bandwidth and DR. It means, such networks can augment mobile radio cells and operate under the distances between 100 m and 1000 km.

The technology (Fig. 2.11) will create a niche for the existing GERAN, UTRAN, SAE and IMS (2G-4G) with a flexible worldwide WLAN. The mentioned WLAN is operated under use of special databases, called DIDO Data Centers [9, 10]. Furthermore, the prominent companies make also the preparations to 5G. The company Microsoft intends it soon, to provide the access to the 10 million Wi-Fi hotspots. Through its Internet telephony subsidiary Skype Microsoft offers already the Wi-Fi access to about two million hotspots worldwide. Under the label "Microsoft WLAN" the access rights will be granted to the Office and Skype customers [8–11].

2.12 Recent Problems

First of them, some governments and citizens are not quite sure that rapid deployment of 5G is useful. Therefore, the activists in Suisse talk about the impairment of human health [2]. Several Swiss cantons, inter alia Geneva and Fribourg, delayed the construction

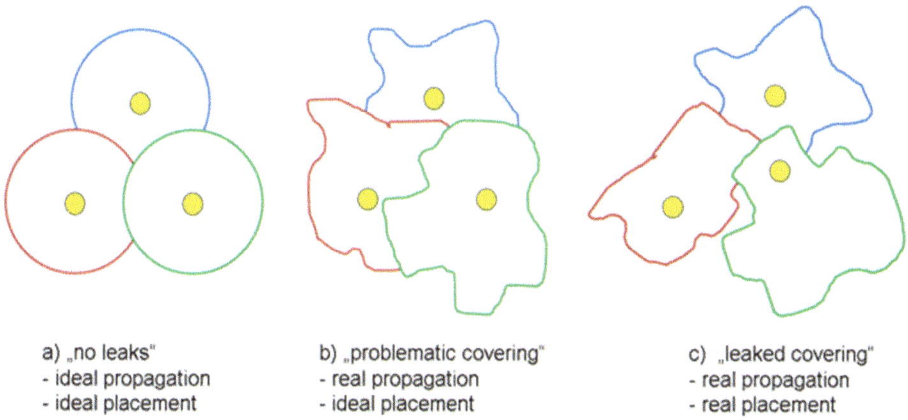

Fig. 2.12 Sub-optimal propagation: real cell configurations—(**a**), (**b**), (**c**)

of antennas for 5G as a precaution. The specialists on Eastern Europe speak about the necessity of finalising of the 3G and 4G construction first in their countries. Multiple expert studies must be conducted for the impact of new 5G technology without the problems for human health under supervising of the World Health Organisation. The Swiss government appointed in 2019 an expert group to study the risks associated with the implementation of 5G. One of the further most known problems for cellular radio networks is suboptimal placement and aberrations of covering in real world conditions (Fig. 2.12a–c).

The depicted problems lead to the complications in handover and roaming in 5G networks, to overloading and service failure, as well as to impairments for real time applications. The range improvement can be realised with classical methods via additional terrestrial relay stations (refer Fig. 2.13). Typically, the radio networks use the elaborated standard planning methods sets and CAD tools [1–4, 11]. For example, based on the clearance model: considering of curvature of the Earth and 1th Fresnel diffraction/scattering zone (i.e. one of a series of coaxial ellipsoidal regions of space between the transmitter Tx and receiver Rx); by transmission frequency $f = 3$ GHz; for distance $d = 15$ km and e.g. tilt $\beta = 30$. The minimal required antenna height equals $Hmin = 47.5$ m.

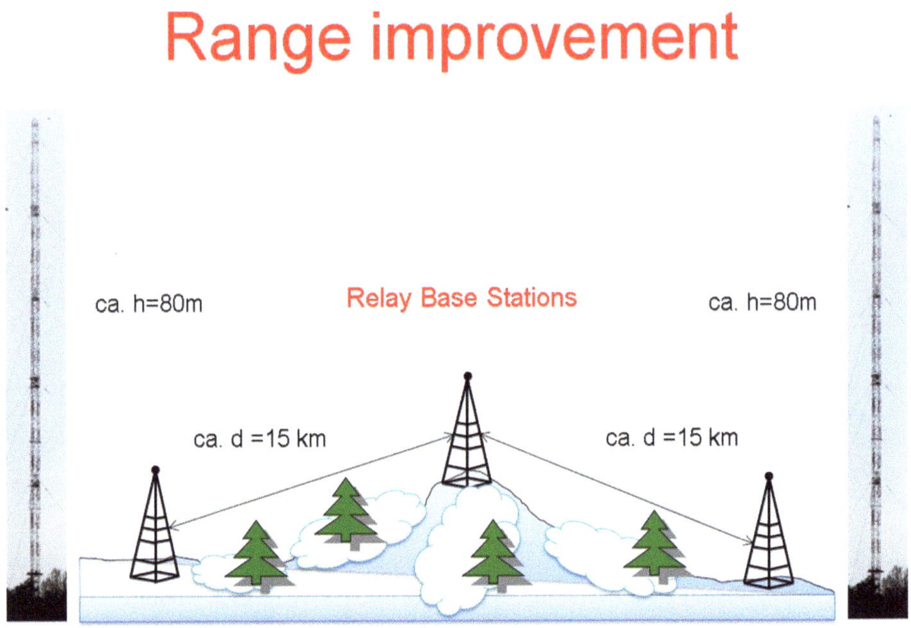

Fig. 2.13 Sub-optimal propagation: range improvement

2.13 To the Demarcation of the Research Area Reagarding 5G

A useful combination of the conventional methods of radio networks planning [1, 4, 12–14] with new intelligent approaches, based on neuronal nets; ML (Machine Learning); AI (Artificial Intelligence), is offered and discussed below.

Herewith the short facts about our research area (refer Fig. 2.14). ML "learns" under use of the acquired examples or training data. We speak about Deep Learning (DL) when the learning process is can be made under support of neural networks. The neural networks we us together with ML but separate the modelling process into two sub-processes:

- First, a model is "trained" on the basis of available training data.
- Second, the model performs later the ML tasks and specific 5G algorithms.

For the modelling of 5G Slicing we use some metrics. The usage purpose for so-called metrics are as follows:

- Understandable parameters criteria for comparing processes and products within a discussed model can be provided.
- Quality and model properties can be quantitatively described.

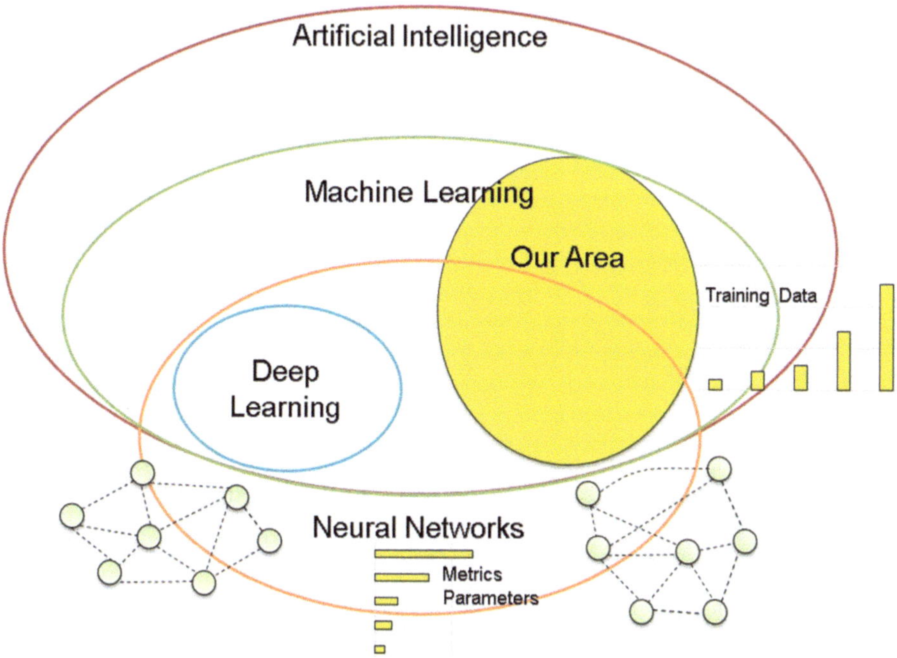

Fig. 2.14 Demarcation of the research area

The metrics are the model variables and numerical values used to measure the quality of the used models for ML and neural networks. A metric can be identified under use of so-called Confusion Matrix (Fig. 2.15) for the relevance. This matrix records how many errors and hits the discussed model has, when the model is executed. The following metrics can be e.g. identified in the context: handover, roaming, DR, delay Δ, jitter.

A method of secured and intelligent handover process between micro-cells of small 5G micro-providers as well as roaming to the integrative macro-cells of the big providers in 5G hierarchical cellular structures under use of ML, neural networks as well as Smart Contracting via Blockchain technology is offered.

The method guarantees adaptive control of QoS for the 5G users independently from the type of provider cell as well as safety of billing and compulsoriness of workflow within 5G slicing structures (Fig. 2.16). Multi-Layered Handover and Roaming based on 5G Cell Hierarchy is implemented within three strata:

- 5G Gateway Stratum
- Megacell Backhaul Stratum
- Microcell Basic Stratum.

The method of secured and intelligent Handover and Roaming is shown below (refer Fig. 2.16). The simulations and performance analysis based on a recurrent neural network are discussed below.

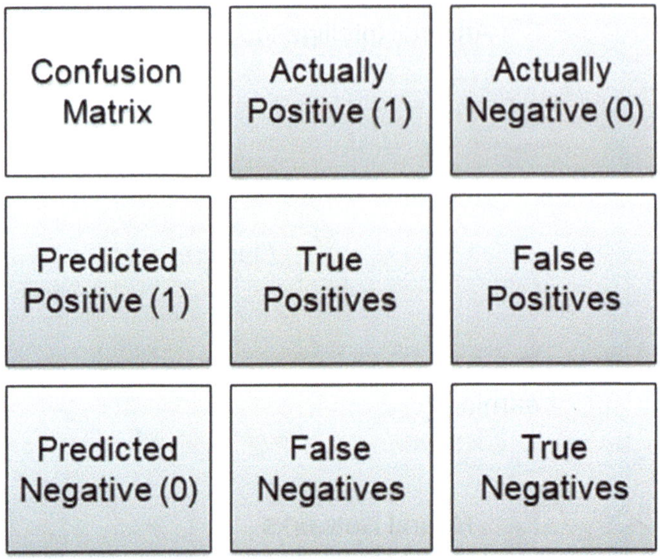

Fig. 2.15 Confusion Matrix for relevance of the metrics

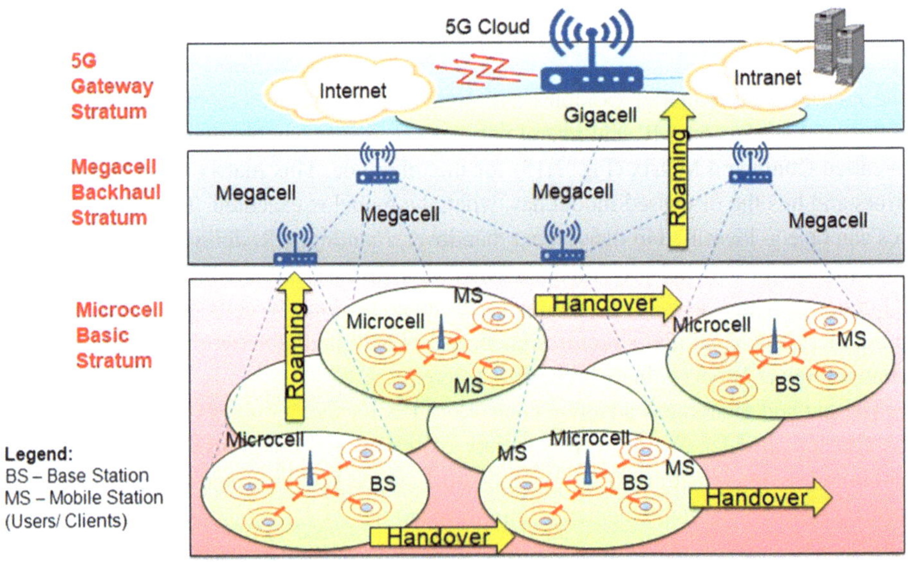

Fig. 2.16 Roaming and Handover Strata in 5G

2.14 Wide Deployment of Standards for ML and AI

The last development on the area of AI (Artificial Intelligence) is mainly based on ML (Machine Learning) processes, which deal with so-called adaptive algorithms. Such algorithms can be systematically improved over the time through the learned experiences (training data). ML can provide the following main approaches: (1) (partially) supervised, (2) unsupervised and (3) reinforcement learning [13].

In midterm, the standards for ML and AI will accompany the industries, digital economy and everyday life over the world and for each institution. Surely, they will find it deployment in 5G networking, and surely, in 5G Slicing. Therefore, let us to give a more detailed overview below. The main purpose of ML is to look for learning algorithms, performed without human intervention, where the time and amount of data required for training is one of the most important indicators of performance.

The main ML approaches are traditionally divided into Supervised Learning (SL), Unsupervised Learning (UL), and Reinforcement Learning (RL). Some authors distinguish this categorisation, for instance, by adding of partially supervised learning as a combination of supervised and unsupervised approaches (so-called evolutionary learning, refer Fig. 2.17):

- In supervised learning, the algorithm collects a series of training data with corresponding labels. The algorithm learns independently a context between training data and labeling i.e. finds patterns to match input and label and then can generalise this to any input data.

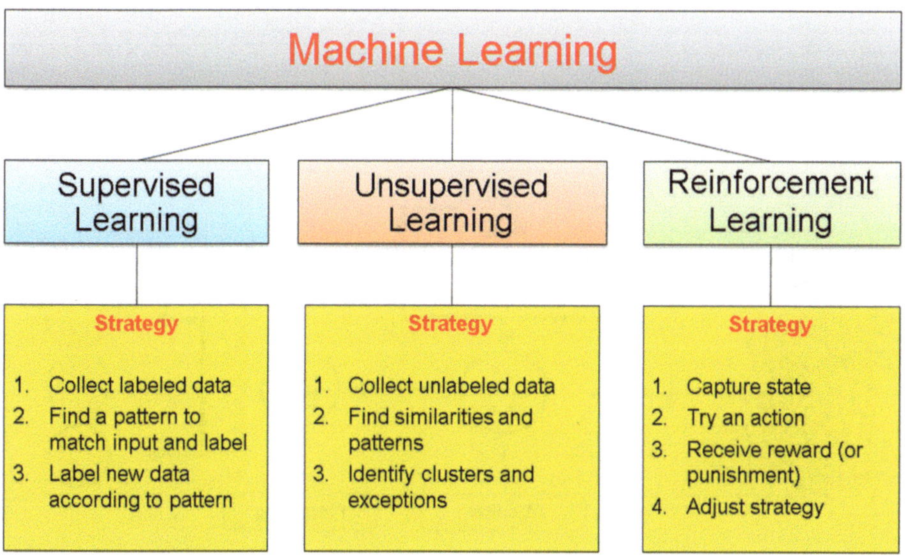

Fig. 2.17 ML Paradigms in Comparison

- The strategy is different with unsupervised learning: the algorithm only collects unlabeled input data. The algorithm identifies similarities and patterns in the input data in order to classify the data or to identify the clusters and exceptions (founded outliers).
- Reinforcement learning is a strategy of ML or usual paradigm, inspired by psychology: the agent learns through reward and punishment. He learns to master a given task autonomously by interacting with his environment, trying to maximize the possible reward.

In mid term, the standards for ML and AI will accompany the industries, digital economy and everyday life over the world and for each institution. Surely, they will find it deployment in 5G networking. Therefore, let us to give a more detailed overview below. The main purpose of machine learning (ML) is to look for learning algorithms, performed without human intervention, where the time and amount of data required for training is one of the most important indicators of performance. The main ML approaches are traditionally divided into Supervised Learning (SL), Unsupervised Learning (UL), and Reinforcement Learning (RL).

Supervised Learning (SL) is an ML method that takes training data (organised into an input vector (x) and a desired output value (y)) to develop a predictive model for the traffic classes. It is easy for the humans, e.g. to understand, where in some types of data we have video, music or pictures, but for a robot it becomes a more difficult task. So, we translate these data into a robot understandable language, and indicate, were we have each kind of data. This will be our training data, that the algorithm memorises and can already analyse the data flow. Knowing what type of data, the user applies, we will give the algorithm the ability to classify them correctly. It enables to provide the bandwidth and other QoS, which are required for its type of data traffic class (refer Fig. 2.18).

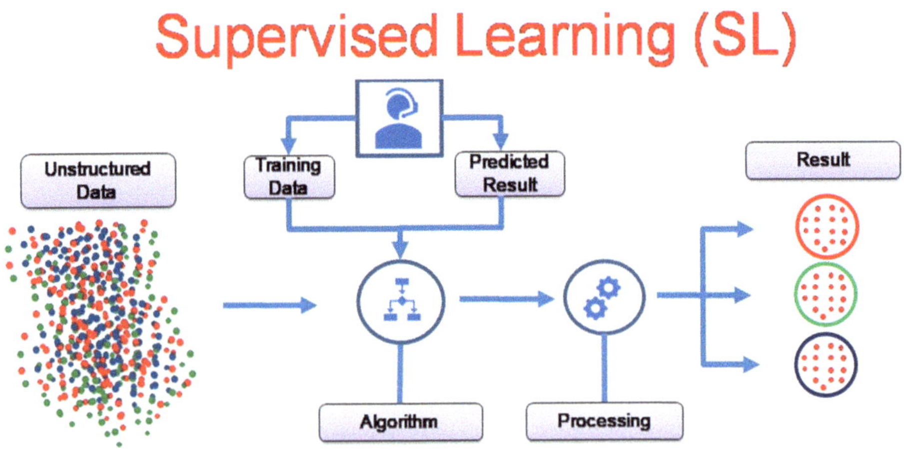

Fig. 2.18 Supervised Learning for 5G traffic classes

By UL (Unsupervised Learning), we allow the robot to learn on its own manner without giving the correct answers to the problem we want to solve. The aim is as follows: without knowing what kind of data we have, the data could be divided via the algorithm into several classes (Fig. 2.19).

Reinforcement Learning (RL) algorithm try to understand how to correctly configure parameters to achieve a specific goal. Many problems and sequential tasks cannot provide a clear answer to a problem, and when it performs poorly, RL has already proven effective in many real-world applications such as robotics, standalone helicopters and drones, fixed network routing, automated industry and hazardous works etc. Therefore, a big option will be use of so-called 5GRLA (Reinforcement Learning Algorithms) for prediction of the services distribution and QoE considering the used traffic classes and behaviour of each of the 5G users (Fig. 2.20). The above discussed 5GRLAs will find augmented deployment in modern 5G service modelling.

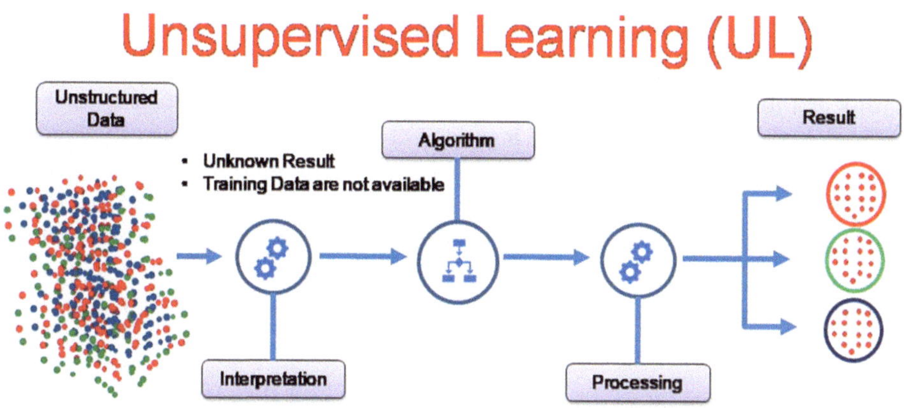

Fig. 2.19 Unsupervised Learning for 5G traffic classes

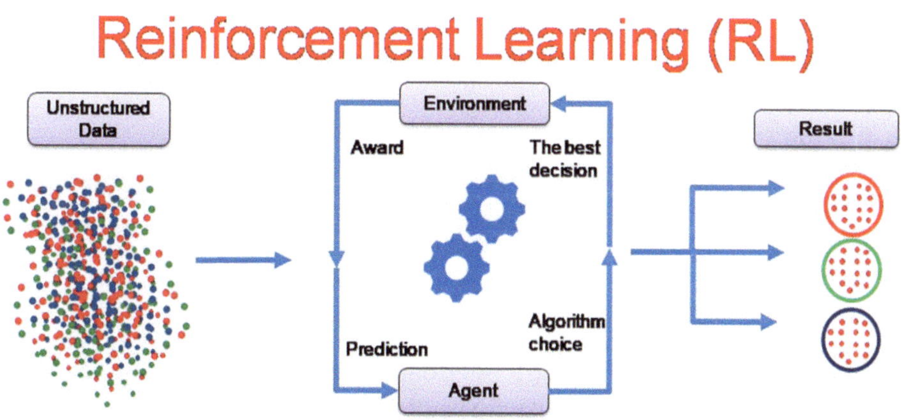

Fig. 2.20 Reinforcement Learning Algorithms aimed to 5G

2.15 Handover Optimisation in 5G

In 5G networks, we can optimise Handover by using intelligent methods. When we are talking about handover, it also needs to take it into account offsets of MLB and MRO, because they are the basis for the decision to carry out the handover. In some case we need to use offsets of MLB, when we see that some cell is overloaded, but in other case, when we don't have overloaded cell is better to use MRO offsets. Therefore, we want to introduce a new coefficient k, which will show us which offset is more relevant in each situation. This decision will be taken by the neural network, based on knowledge of the mobility of subscribers and on the load capacity of the cells, later we will show how neural network can predict mobility of users. So, we introduce a new coefficient k. k = [0…1]. When we have overloaded cells, neural network will react and the k will be in the area of [0.1, 0.3], this means that we will consider the MLB offsets to be near 80%, and we will pay attention to the offsets of the MRO, that is, the influence of its offset will be considered only by 10%. When we have normal loaded cells, neural network accepts the average value when the offset value from MRO and MLB is considered equally and the coefficient k is 0.5 and the handover is triggered at point k (as shown in Fig. 2.21a). In cases if we have a high intensity of subscribers in the cell, the coefficient k is 0.8–1, and the handover is triggered at point k. As shown in Fig. 2.21b, we have overloaded cell 2, so we move the point of triggered handover from k-1 to k (in the side of overloaded cells), it means that it is harder to make handover in overloaded cell 2 but easier make to handover into cell 1. Finally, when we have free cells, based on the knowledge of the neural network about subscriber mobility, we can predict in which cell

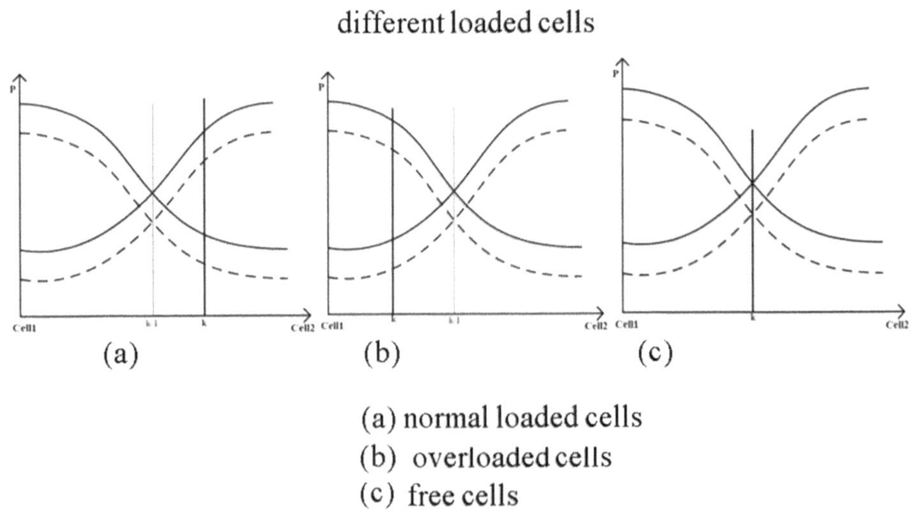

Fig. 2.21 Simulations with different loaded cells

it is necessary to carry out handover subscribers, and as soon as they receive the required signal level, we initiate a handover to provide the proper QoS. In this case, the coefficient k will be equal to 0.1–0.3 for subscribers who move into the cell 2, what helps to make handover easier (Fig. 2.21c).

2.16 Further Optimisation Methods

Evidently, there are two different ways to improve this approach [13–15]. On one hand, the prediction behavioral models for each user, based on only his own mobility were elaborated. On other hand, we can also use the cloud centric solution, i.e. the stored behavioral data from offered knowledge bases, which provide real time monitoring and learning. In the first case, we analyze the mobility of the users over a fixed time, examining their behavior model and their daily routine. We want to know, which macro-, micro- and femto-cells are mostly visited via the users and where the typical users wish themselves the most mobile traffic. In the second case, we want to introduce the stored behavioral data from offered knowledge bases, which provide realtime monitoring and learning. In this case, we will collect the data of each user and send them to a common knowledge database. This will be an analogy as a common knowledge base for Tesla's autopilot. When our network can automatically predict the mass movement of subscribers into specific micro-cells to one of macro-cell [13–15].

2.17 New Generation Networks for HDS

For HDS construction new network technologies and generations play a steady growing role.

The deployed in mid term 5G network technologies are being extended as follows:

- 6G cellular networks will be expanded on the basis of 5G networks by 2030 and beyond
- SpaceX's Starlink project will probably be developed until the final stage by 2027.

2.17.1 Space and Telecommunication Company SpaceX

Space Exploration Technologies Corporation was founded in 2002. The aerospace and telecommunication company with headquarter in Hawthorne (USA) is conducted by Elon Musk (CEO and CTO).

A famous industrialist Elon Musk (born 1971 in Pretoria) is simultaneously co-founder for the space company SpaceX, payment service PayPal, as well as Tesla, known as electric car manufacturing company. SpaceX has been conducting manned space

Fig. 2.22 The heroes and visionaries of the XX and XXI centuries both

Falcon 9 (2010)
spacex.com

Elon Musk (* 1971)

Nikola Tesla (1856–1943)

flights to the ISS (International Space Station) for NASA since 2020. Since September 2021, private individuals will also be transported into space (refer Fig. 2.22).

Starlink is a satellite project for the network, operated by the US space company SpaceX, which is to provide a worldwide Internet access in mid term. The Starlink's slogan in mid term is as follows: 11.927 satellites by year 2027 and then another 30.000 satellites for satellite based broadband Internet!

With the above-mentioned Starlink's project for global satellite Internet access, SpaceX Company is—measured by the number of satellites—the world's largest satellite manufacturer and provider [15].

2.17.2 Worldwide Internet Supply with LEO SAT: Starlink Project

In January 2015, the American companies Fidelity Investments and Google invested a total of USD 1 billion in SpaceX. They held 8.3% of the company. It was believed, that Google was interested in SpaceX's new plan to build a network of satellites to serve the Internet. The execution will cost 10 billion USD and take around five years (Fig. 2.22).

In November 2016, the company submitted initial plans for such the concept to the US regulator, the Federal Communications Commission. SpaceX plans firstly to station 11,927 satellites in orbits between 340 and 1325 km (LEO). In May 2019, a Falcon 9 racket with 60 prototypes, which do not yet have the full-intended functionality, was launched in orbits of up to 550 km orbit height.

It has been in beta testing since 2020. The core business of Starlink includes the provision of Internet access with particularly short packet turnaround times and the provision in areas in which no or insufficient Internet connection was previously available.

With 1433 Starlink's LEO satellites in low earth orbit, SpaceX is by far the largest satellite operator worldwide (as of April 29, 2021).

Overall, there are limited approvals for the launch of a maximum of 11,927 satellites until 2027, as well as applications from SpaceX for a further 30,000 satellites. Taken together, this fact corresponds to approx. five times the total of all satellites launched between 1957 (refer Sputnik-1, USSR) and nowadays.

As end devices are used the SpaceX own terminals (manufactured in USA). The frequency band: 10.95 GHz (Uplink) and 29.1 GHz (Downlink), i.e. in Ka-Band and Ku-Band. The advanced antenna are used, it means the automated Phased-Array-Antennas with diameter D = 59 cm approx. The transmission power PTx is equal to 4 dBm ~ 2.5 mW. With antenna gain GTx = 34 dBi, the maximum radiated power for the user terminals: 6.3 W. The terminal procurement costs are relatively high nowadays: $ 500. The starlink service rental is approx. $ 100/month.

The following QoS parameters are provided:

- Downlink DR = 35–430 Mbit/s
- Uplink DR = 10–50 Mbit/s
- Latency: 30–90 ms.

As a usual criticism to the project, the following arguments are used:

- Possible massive creation and accumulation of space garbage?
- Disturbance of the night sky and astronomy?

Do you agree with these opinions too? The Starlink Germany GmbH was founded in in Frankfurt am Main in November 2020.

2.17.3 Sixth Generation for Mobile Radio Networks

6G is the planned successor to 5G mobile radio networks, but will be significantly faster: 430–1000 GBit/s. 6G networks will be even more heterogeneous than their predecessors, i.e. 4G and 5G, and will be largely supported by SDN/NFV. As far as apps for 6G

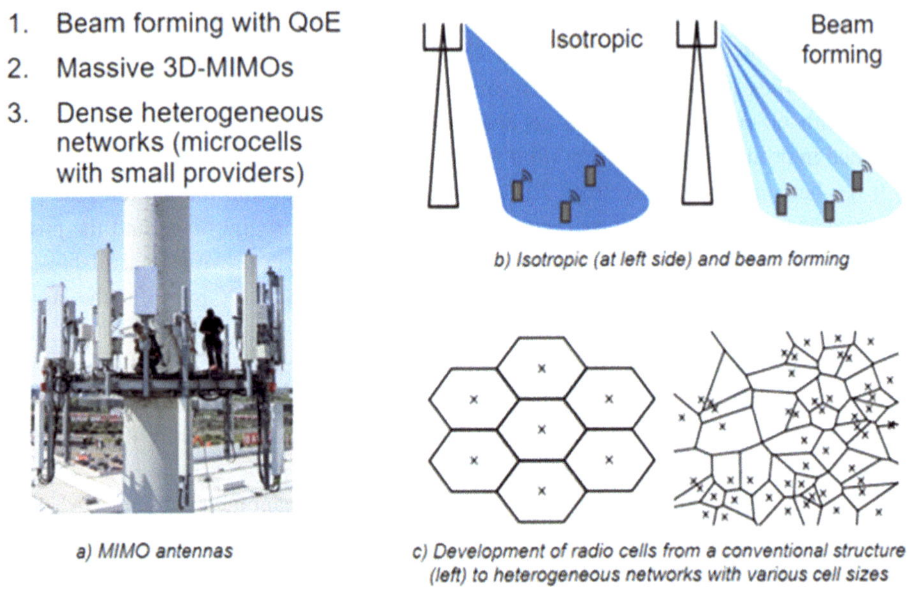

Fig. 2.23 Cellular structures for 6G/NET-2030 (by D. Wasutinski and V. Wasutinski)

networks are concerned, this still includes virtual and augmented reality (VR/AR), pervasive intelligence and, surely, the IoT. The small and medium-sized mobile network operators for 6G use flexible decentralized business models, with local frequency licensing and allocation, infrastructure release and intelligent automated management, supported by mobile Edge Computing, AI and Blockchain technology. It is assumed that 6G cellular networks use also sub-THz and THz frequency bands (100 GHz to 3 THz) and have an even lower latency than 5G/IMT-2020 networks (under 1 ms).

One of the technologies that can be implemented in the 6G mobile communications is the use of digital antenna arrays at base stations in combination with massive MIMO technology.

At the same time, variants of base stations with antenna systems are considered, which form around 250 beams in the working sector (beamforming for targeted QoS). Among the most important features for 6G networks, experts also name the management systems with the use of Artificial Intelligence (AI) and Machine Learning (ML).

As with their predecessors, 6G networks are organized as cellular networks in which the service area is divided into small geographical areas (called cells). Hierarchical and heterogeneous cell concept combines terrestrial and SAT-based routes (Fig. 2.23).

As a next development stage, the interoperability with Starlink's LEO satellites [15] can be provided (SpaceX by E. Musk).

2.17.4 Technological and Organizational Advances for 6G

The telecommunications giants such as Nokia, Ericsson, Huawei, Samsung, LG, Apple and state institutions and organizations in several countries (EU, USA, China, South Korea, Japan) have already shown their interests in development of 6G networks. In 2018, China announced the development of a 6G cellular standard. In November 2020, China launched the first satellite for testing 6G technologies in the THz range. Totally, it was planned 12 satellites in the THz frequency band. In 2020, the researchers from Nanyang Technological University of Singapore and Osaka University of Japan announced that they had developed a chip for THz waves that could be used in 6G networks.

A research group at Santa Barbara University of California made significant progress and reported on the high electron mobility n-polar gallium nitride transistor (HEMT). In October 2020 the Alliance for Telecommunications Industry Solutions (ATIS) launched a "Next G Alliance". This alliance consists of "big players" such as AT&T, Ericsson, Telus, Verizon, T-Mobile, Microsoft, Samsung for the purpose of technology leadership at 6G in the next decade (Fig. 2.24).

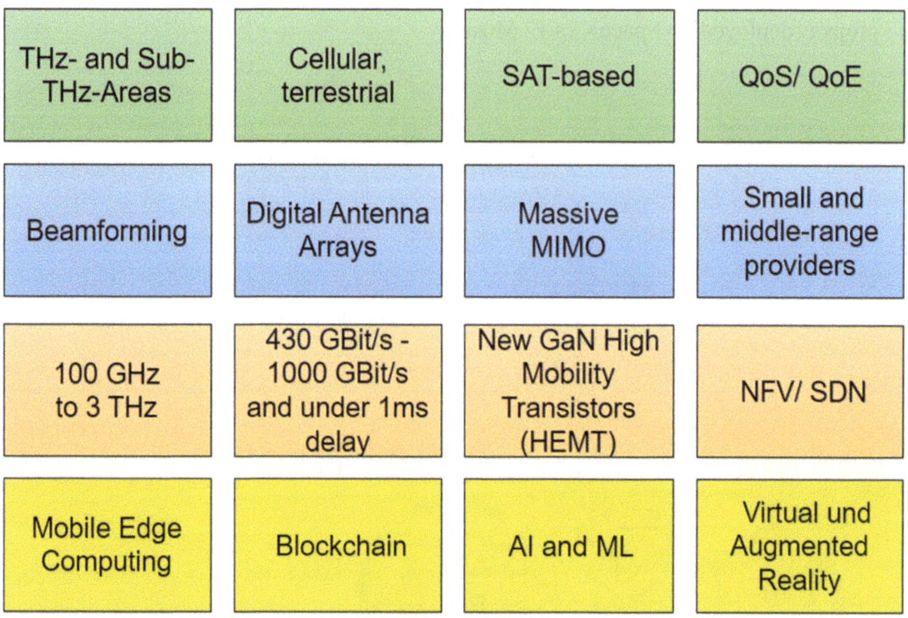

Fig. 2.24 The most important terms for 6G/NET-2030

2.18 Conclusions and Outlook

1) The given chapter represents an overview on challenges for 5G-6G mobile radio networks and multiple deployment scenarios.
2) The recent 5G slicing scenarios are nowadays frequently implemented with security and compulsoriness via Blockchain and with intelligence via Machine Learning (refer Fig. 2.25).
3) The to be used in short term 5G mobile radio promises "a new technological breakthrough" in communication, industry, entertainment and education. Among the scenarios on 5G networks deployment, so-called "digital twins" play a standing increasing role.
4) The following main subjects are examined:
   - 5G Deployment Scenarios with Cell Hierarchy and Inter-Operability are examined.
   - 5G Slicing via Blockchain in Cell Hierarchy is discussed.
   - 5G Intelligent Handover and Roaming in Cell Hierarchy is provided under use of ML paradigms and recurrent neural networks.
5) The case studies on optimising of slicing and handover within the hierarchical 5G networks based on ML and neural networks are offered.
6) 5G and 6G (in mid-term) will provide new attractive and a lot promising applications. As terrestrial services for HDS both have to guarantee interoperability with Starlink project (deployed by SpaceX of E. Musk).

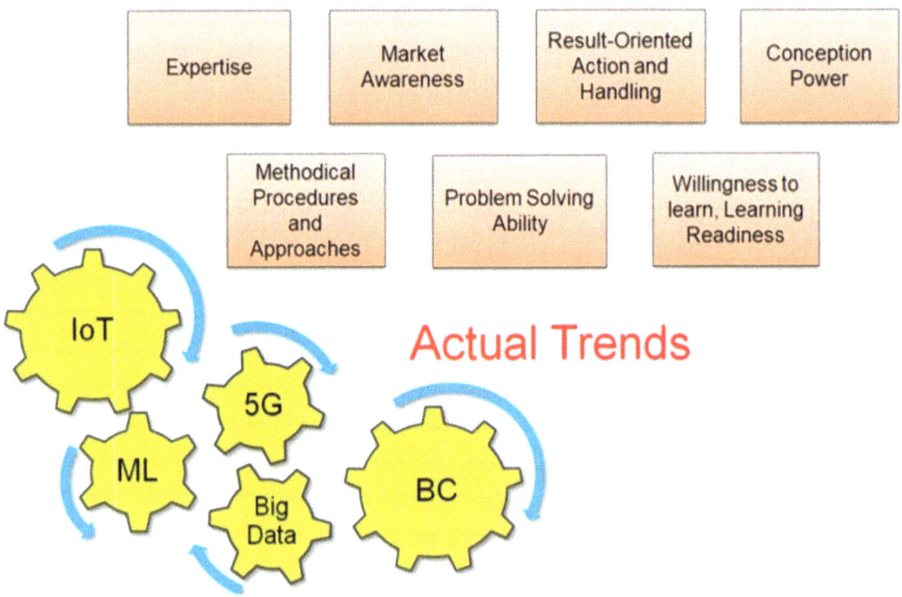

Fig. 2.25 New fields of competence for 5G Slicing Scenarios for 5G micro-providers

References

1. 5G Whitepaper of Vodafone (Online): https://www.vodafone.de/media/downloads/pdf/5G_Whitepaper.
2. 5G Suisscom (Online): https://www.swisscom.ch/de/about/unternehmen/portraet/netz/5g.html.
3. IEEE 5G World Forum in Dresden (Online): https://ieee-wf-5g.org/.
4. G.Fettweis, N. Franchi. Flächendeckende Versorgung mit 5G-Mobilfunk in Deutschland (Online): https://5glab.de/wp-content/uploads/2018-10-22_Diskussionspapier_5GLab_final.pdf.
5. Luntovskyy, Andriy, Gütter, Dietbert. Moderne Rechnernetze - Übungsbuch: Aufgaben und Musterlösungen zu Protokollen, Standards und Apps in kombinierten Netzwerken, 150 Seiten, 44 Abb., Springer Nature, Juli 2020, ISBN: 9783658256180 (https://www.springer.com/gp/book/9783658256180).
6. What is Microsoft Wi-Fi, and Will it Matter To You? (Online): https://www.howtogeek.com/222876/what-is-microsoft-wi-fi-and-will-it-matter-to-you/.
7. DIDO Technology (Online): http://www.rearden.com.
8. pCell Wireless White Paper (Online): https://www.artemis.com/.
9. Luntovskyy, M. Klymash. A Perspective Resource Allocation Method for Future WLAN, Int. Research Journal Telecommunication Sciences, Kiev, 2017.
10. L.Globa, V. Prokopets, N. Gvozdetska. Prognostic-Reactive NFV Resource Allocation Method for Implementation in Virtuali sed Mobile Network EPC of Ukraine, BlackSeaCom-2018, Batumi, Georgia, 8 p.
11. T. Maksymyuk, L. Han, S. Larionov, B. Shubyn, A. Luntovskyy, M. Klymash. Intelligent Spectrum Management in 5G Mobile Networks based on Recurrent Neural Networks, IEEE CADSM 2019, IEEE Xplore: https://ieeexplore.ieee.org/document/8779301.
12. P. Porambage, Y. Michey, A. Kalliolay, M. Liyanagez, M. Ylianttila. Secure Keying Scheme for Network Slicing in 5G Architecture, 6p. (Online): https://www.researchgate.net/publication/335716113/.
13. Yueyue Dai, Du Xu, Sabita Maharjan, Zhuang Chen, Qian He, and Yan Zhang. Blockchain and Deep Reinforcement Learning Empowered Intelligent 5G Beyond, by Intelligent Network Assisted by Cognitive Computing and Machine Learning, 2019, 10p.
14. Tesla: Electro Cars, Solar Systems and Clean Energy (Online): https://www.tesla.com/.
15. Starlink (Online): https://www.starlink.com/.

Blockchain and Its Applications

3

An important trend since approximately the year 2000 is the use of modern crypto-technology "Blockchain" for acceleration, transparency and decentralisation of financial transactions and as a promising digital payment instrument: e.g. crypto-currencies Bitcoin, Monero, and Ethereum.

This chapter focuses on the problem of using Blockchain technology and their networked cryptographic applications and apps, i.a. HDS.

The influencing factors and sources of Blockchain crypto-technology were discussed, the comparison of centralised bank systems vs. decentralised systems was carried out, the mining process for cryptographic currencies, the concept of a public ledger, the validation principles PoW and PoS are represented, as well as profitability of cryptographic currencies was analysed. Furthermore, important applications of Blockchain crypto technology were shown (such as Smart Contracting, Bletchley) were considered as well as the accompanying risks, their advantages and disadvantages were discussed. In addition, the malicious applications were discussed such as the ransomware (extortion Trojans).

Finally, the potential and future perspectives of Blockchain crypto technology for real business applications were assessed.

3.1 Motivation: Payment Instruments in Past and Future

Blockchain is a cryptographically distributed computer network application supporting a decentralised payment system and decentralised financial online transactions in the peer-to-peer (P2P) concept. However, the economic success of this crypto technology will be evident in the next 10 up to 20 years. Figure 3.1a, b depicts the historical development of the payment instruments from archaic shells and early coins to e-cash and

A. Luntovskyy and D. Gütter, *Highly-Distributed Systems*,
https://doi.org/10.1007/978-3-030-92829-2_3

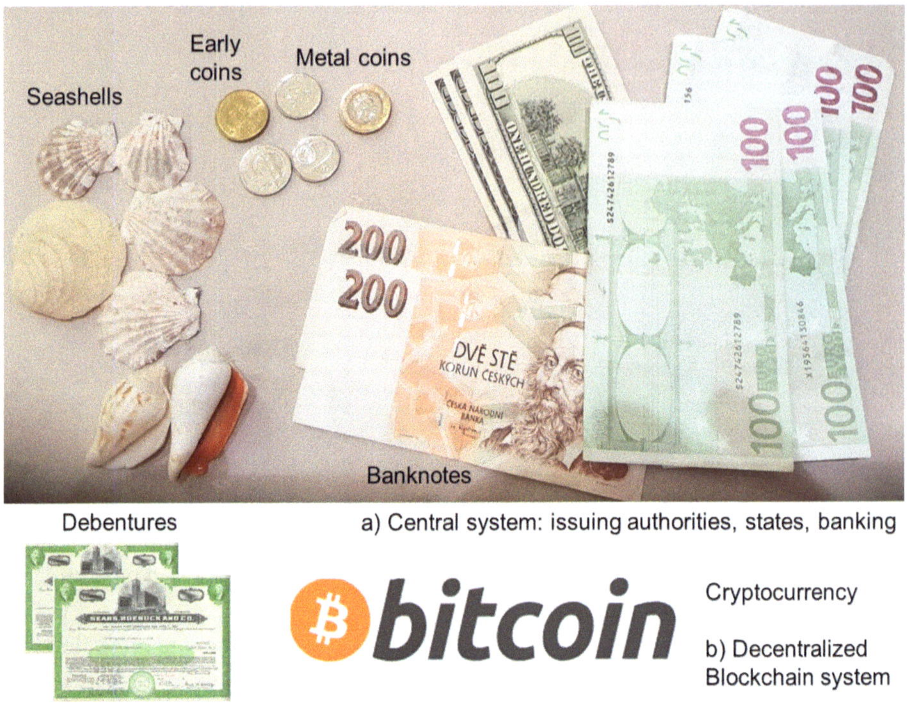

Fig. 3.1 Payment instruments in past and future

crypto-currencies, which can accelerate financial transactions and significantly reduce the cash mass. The important milestones of Blockchain technology are as follows [1–16]:

- 1991—The basic principles from S. Haber and W. Scott: cryptografically secured chaining of individual blocks
- 2000—Theory for cryptographic blockchaining of Stefan Konst as well as some implementation solutions
- 2008—"White Paper" Bitcoin: "A Peer-to-Peer Electronic Cash System" for the conception of a distributed database system BTC created by so-called "Satoshi Nakamoto" (a pseudo of the known developing group?).
- 2009—Launch of the first publicly distributed worldwide Blockchain.
- 2010—Multiple further Blockchain technologies (private blockchains) and crypto-currencies
- 2012—Mining of crypto-currencies, smart contracting, deployment by highly-distributed systems
- 2021—Legalisation for BTC in some countries, like e.g. in El Salvador: "Law about BTC and crypto-currencies mining"

Therefore, an important question can be asked on the edge: who did actually invent and create Bitcoin?

- Assumption 1:
  Bitcoin was combined by the names of the prominent companies Samsung, Toshiba, Nakamichi and Motorola (=SaToshi NakaMoto).
- Asumption 2:
  as well as there are numerous speculations about the BTC/ BC developer name, i.e. Elon Musk was mentioned, the founder, CEO and CTO of the companies like PayPal, SpaceX and Tesla, the leading world industrialist and inventor.

A graphical comparison of the decentralised chaining of the secured blocks with a centralised banking system can be seen in Fig. 3.2a, b. The deployment of Blockchain technology speaks mainly for a decentralised financial system. The advantages of such a solution are obvious:

- Sustainability, general transparency and commitment
- Accelerated economic workflows and digitisation processes (so-called IT in the digital age)
- Blockchain crypto-technology is also well suited to supporting current crypto currencies (such as Bitcoin, Monero, Ethereum).

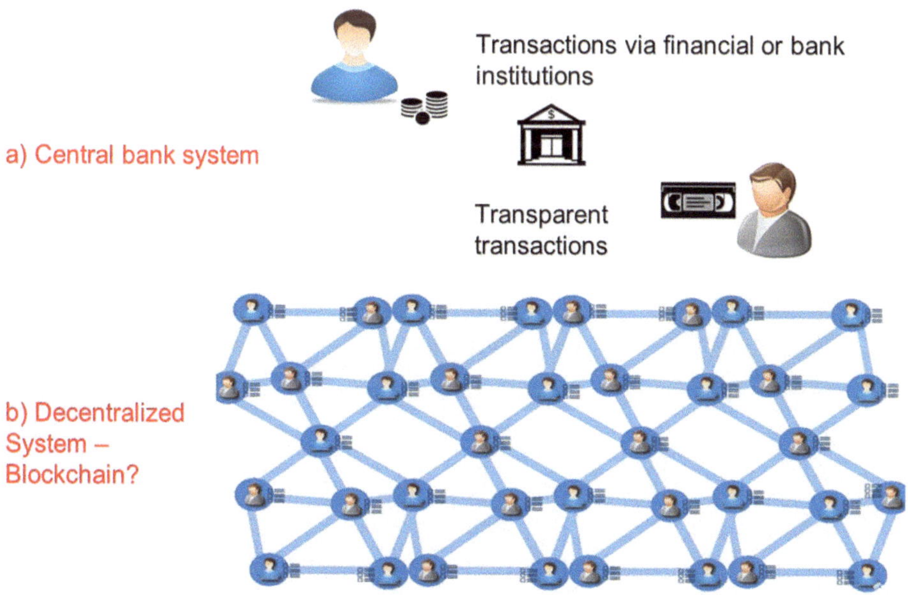

Fig. 3.2 Decentralised chaining of blocks instead of a central bank system?

3.2 Blockchain Architecture

Decentralised, cryptographically secured and unified blocks, their chains and transactions are grouped under a general, global public ledger (account), the structure of which is as follows (Fig. 3.3).

The Blockchain, as a networked Public Ledger, consists of participating nodes that represent an efficient P2P communication model. Typical features of the Blockchain are as follows:

- Redundancy and synchronisation
- Cryptographic hash procedures for integrity assurance and attack safety
- Decentralised management and control of the Blockchain
- Network subscribers are also referred to as Nodes (Full-Nodes, Miners, Validators) and run redundantly with mutual synchronisation
- In addition, large block volumina can cause the "Big Data" problem, refer Figs. 3.2b and 3.3.

Block chaining of Headers and Blocks within a Hash Tree is depicted in Fig. 3.4 schematically. Here a structure for an exemplary block chain was given. The defining block chain (yellow colour) consists of the longest sequence of secured blocks from the origin

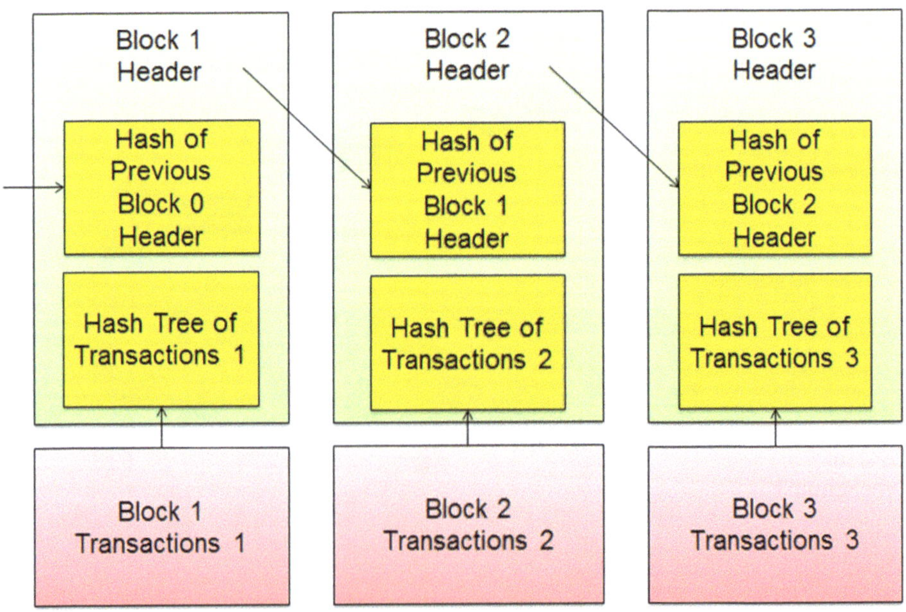

Fig. 3.3 Blockchained Decentralized System

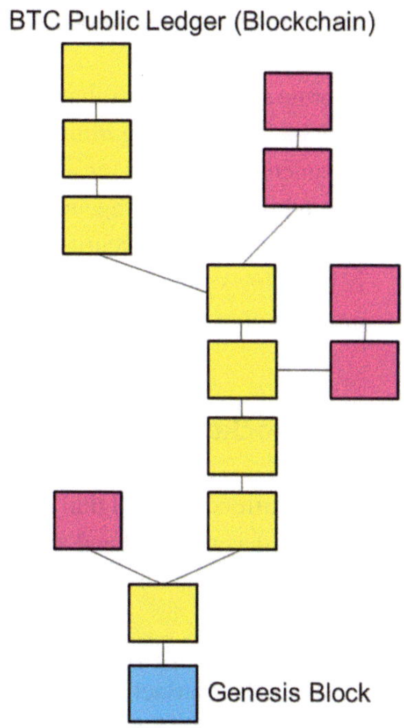

Fig. 3.4 BC as a Distributed Public Ledger

to the current block (blue). Alternative chains (pink colour) became orphan as soon as they are shorter than another chain. Therefore, within the Blockchain architecture between the following basic components can be distinguished: the simple Nodes, the Full-Nodes, and Miner/Validator:

1. Nodes:
   - Each Blockchain participant (computer, smartphones, tablets, or even clusters) is qualified as Node, if he has installed the corresponding software, which runs based on the Bitcoin protocol or the program code of Bitcoin.
2. Full-Nodes:
   - A Node with full local copy of the Blockchain
   - Checking for so-called "consensus rules"
3. Miner/Validator:
   - The individual participants or mining pool (high resource requirements regarding hardware and energy consumption)
   - Finalising of blocks (Miner block generation, Validator proving)
   - Externally they act each like a large participant, but in fact many small blocks are generated for payment in fractions of the crypto currency units.

However, the following problems occur during the Blockchain operation:

- Enormous energy consumption due to mining of crypto currencies (processing of the hash blocks via its algorithmic complexity).
- Exponential memory growth (including capacity migration between USB media, smartphones, PC, storage media such as SAN / NAS, as well as cloud storages).
- Cryptographic data security is guaranteed, but privacy issues may arise. One way out is as follows: no processing the complete Blockchain with all the transactions, but only use of excerpts of the Blockchain without a prehistory.

3.3 Blockchain and Crypto-Currencies

A crypto currency is a digital payment instrument under use of cryptographic methods to realise a decentralised and secure payment system. The following questions can occur for describing of the discussed functionality:

- How is new money created?—Create a new block
- How can the transactions be stored?—Creator of a block selects certain transactions
- How does the respective credit balance come about?—A credit is the sum of all procressed transactions of a user.

Example 3.1. The examples of the crypto-currencies: Bitcoin, Ethereum, Ripple, Bitcoin Cash, Litecoin. One of them, the crypto-currency Monero (XMR), has the following properties:

- Proof-of-Work with CryptoNight
- Memory intensive, but relatively low computational effort
- Supported WebAssembly and Coinhive (Coinhive.com) with JavaScript Mining for Monero Blockchain
- The userscan run the miner directly in their browsers and 'mine' the XMR for anad free experience.

Table 3.1 contains an overview of current crypto-currencies.

Digital payment instruments under use of cryptographic principles, which built a fast-growing market nowadays, are often created via deployment of the hash algorithms titled SHA256 (FIPS NIST 2008). Such algorithms (refer Table 3.2) were introduced since 2002. The first representant was titled SHA-1. The successor was SHA-2 (including the further modifications like SHA-224, SHA-256, SHA-384, SHA-512, SHA-512/224 and SHA-512/256). The hash size is variable between 224, 256, 384 and 512 bits by the parametrised round number 64 or 80.

Table 3.1 Overview of crypto-currencies and market growth (based on: sources [3, 4, 13], https://coinmarketcap.com/)

#	Currency	Shorthand symbol	Launch	Mining available
1	Bitcoin	BTC	2009	Yes, SHA-256
2	Ethereum	ETH	2015	Yes, Ethash
3	Ripple	XRP	2013	No
4	Litecoin	LTC	2011	Yes, Scrypt
5	Ethereum Classic	ETC	2016	Yes, Ethash
6	Monero	XMR	2014	Yes, CryptoNight
7	Dash (formerlyDarkcoin)	DASH	2014	Yes, X11

Table 3.2 Overview of NIST hash algorithms [17]

Hash algorithm	Message size (bits)	Block size (bits)	Word size (bits)	Message digest size (bits)
SHA-1	<264	512	32	160
SHA-224	<264	512	32	224
SHA-256 (Blockchain)	<264	512	32	256
SHA-384	<2128	1024	64	384
SHA-512	<2128	1024	64	512

The crypto-application Bitcoin founded in 2009 and based on the Blockchain is today considered as one of the most important crypto currencies (shorthand symbol BTC). The value growth is very impressive as well as energy consumption per anno by BTC is immense:

http://www.heise.de/forum/,
http://www.statista.de/,
https://digiconomist.net/bitcoin-energy-consumption).

However, there is also criticism ([7, 9, 15, 16], http://www.heise.de/forum/):

- Blockchain is itself the biggest problem for Bitcoin (due to energy and memory consumption).
- Bitcoin was primarily intended as a decentralised payment system. But due to the exponential growth (storage requirements) of the Blockchain, anyone can still hold the blocks on its local HDD or other storage device. On the smartphones or other

mobile devices, the retrieving is also unthinkable anyway. This may affect i.e. the obligation and traceability of transactions.

- Due to critical requirements on ever growing memory for the Blockchain the users hold the blocks on centralised servers and cloud services and then access it with their online clients. This leads to a certain centralisation and is again depending on the central Blockchain servers offered by bank analog financial institutions and stock markets. The original advantage of decentralisation and thus security against manipulation and server failures is no longer available!

3.4 On Profitability of Bitcoin: is questionable?

As it was shown the crypto currency Bitcoin is one of the most popular crypto- currencies with a stable growth. The real profitability of BTC production is difficult to consider or predict due to enormous algorithmic complexity and different rising energy prices.

Example 3.2. Equation (3.1) describes the profit of crypto-currency generation (mining). Equation (3.2) describes the complexity of the mining process via Blockchain [1].

$$\text{profit} = \text{revenue} - \left(\text{cost}_{\text{electricity}} + \text{cost}_{\text{difficulty}}\right) \tag{3.1}$$

$$\text{cost}_{\text{difficulty}} = \frac{\text{maximum} - \text{difficulty}}{\text{current} - \text{difficulty}} \times \frac{2^{32}}{\text{hashrate}} \tag{3.2}$$

It should be noted that the variable exchange rates of these crypto currencies are not related to the realities of the world industry and trade, so investments in these domains is highly risky (refer Fig. 3.5).

Table 3.3 contains the most important questions regarding modern crypto-currencies:

3.5 Operation and Validation of Blockchain

Let us consider the BC in detail (refer Fig. 3.6). The structure of the BC is P2P. This is a growing ledger with strongly decentralised character.

The common Blockchain operation is given in schematic workflow in Fig. 3.7. The system is growing by finalizing of the chained transction blocks. The formation of a transaction block is depicted in Fig. 3.8.

Blockchain is a continuously expandable list of data records (hashes and transactions) in individual chained blocks. It serves in many applications as a decentralized ledger and is a secured alternative to the convenient PKI method for multilateral communication models.

BC is a continuously expandable list of data records (hashes and transactions) in individual chained blocks.

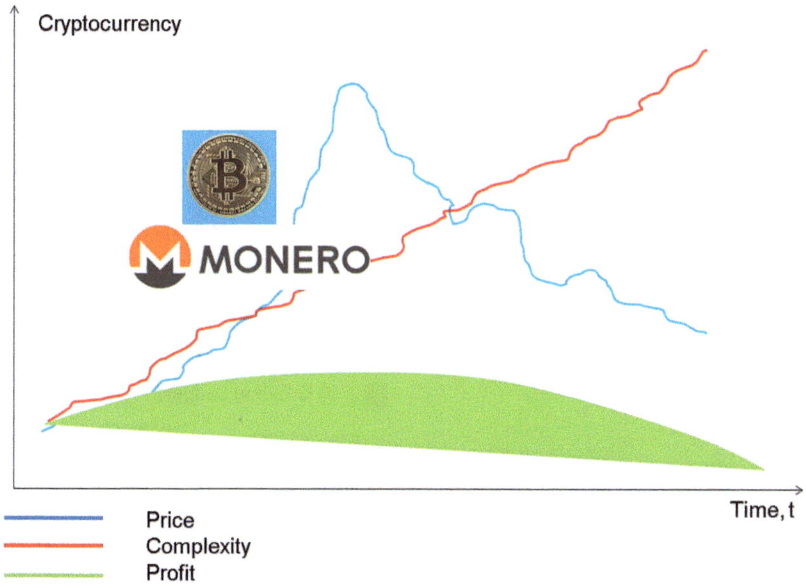

Fig. 3.5 Profitability Estimation for BTC and other crypto-currencies [1, 15]

Table 3.3 FAQ regarding modern crypto-currencies

Question	Answer
How does it operate?	See below
How is new currency generated?	You have to generate a new block for the Blockchain system
How can I save transactions?	The blocker chooses certain transactions
How is the loan balance provided?	Credit is the sum of all user transactions

BC can be decentralized used for multilateral compulsoriness instead of convenient PKI, in opposite to PKI, which operates pairwise, centralized via a CA. Modern HDS are mostly very complex, and consist of multiple communicating parts. The potential benefits of this technology:

- Software solutions decentralisation
- Compulsoriness of the workflow steps for critical applications
- Multilateral communication models like e.g. M2M.

PKI and BC are based both on RSA digital signatures with hash value generation and encryption under use of private key. Decryption with public key provide the signature (block) verification.

Nowadays, multiple BC applications can be used:

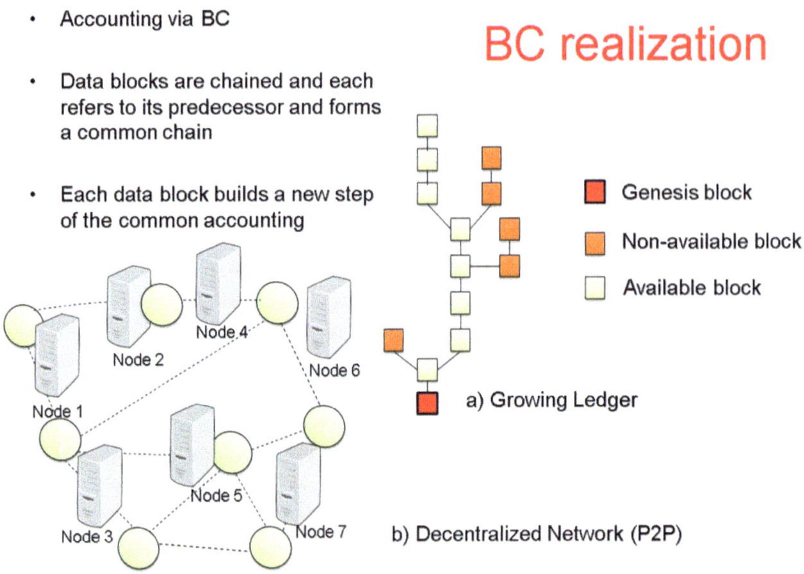

Fig. 3.6 BC common structure and use

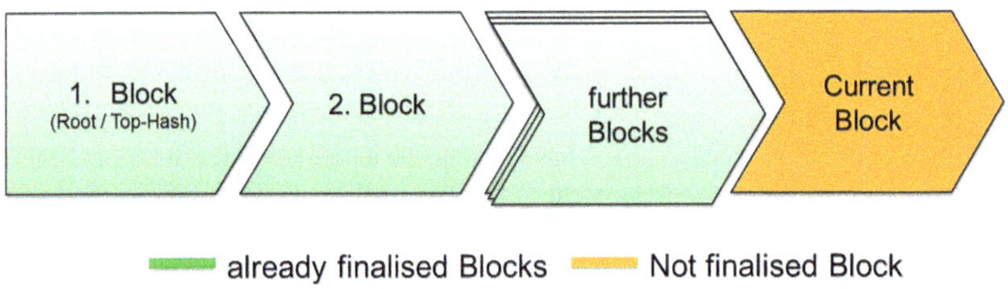

Fig. 3.7 Blockchain Operation: schematic Workflow

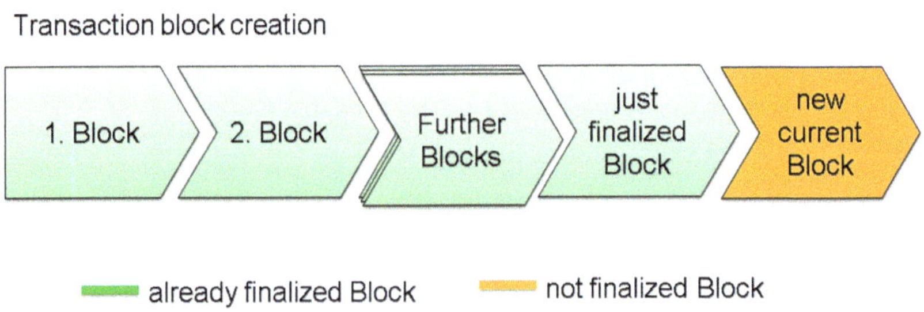

Fig. 3.8 Blockchain Operation: block construction

- Crypto-currencies like Bitcoin, Monero, Ripple.
- Smart Contracting
- Supply Chains
- 5G and Beyond applications
- Applications for HDS (Highly-Distributed Systems)
- Ransomware (as a negative example of BC use): the dangerous crypto-Trojan can encrypt the data on the HDD and then require "a ransom" in crypto-currencies (normally in BTC) for release!

The assignment of the above-mentioned protection goals to the security mechanisms known for IoT solutions is given below (Table 3.4):

Consensus protocols for BC in simple terms are concerning to so-called "rules to reach an agreement". There are many different ways, how Blockchains can organise communication between any two IoT devices while they are can be located in a different part of the world without any third party (centre), which can control their communication [1–17]. This solution certainly has many advantages, but with such a M2M-communication, the BC can be vulnerable in cases where attackers on intermediate nodes have not falsified information. For these cases, we need consensus protocols, which reduce the chances of failure of distributed information and forgery of information at intermediate nodes. The consensus protocol largely serves two purposes: to maintain a single version of the truth and to prevent sick actors from interfering with data already stored in blocks.

Which BC consensus algorithms can be used? There are several so-called consensus algorithms for creating blocks, the most important of which are listed in Table 3.5 below.

During the typical Blockchain operation, the transaction block is generated accompanied by the both validation processes: PoW, Proof of Work (Fig. 3.9) and PoS, Proof of Stake (Fig. 3.10). The complete tracking of the chain begins from the first own transaction (refer Fig. 3.8).

What is understood in the frame the Blockchain technology under the validation processes titled PoW (Proof of Work) and PoS (Proof of Stake) correspondingly?

- PoW is the cryptographic puzzle to confirm the complexity of a particular task and service functions. This demonstrates an overuse of this service, such as a

Table 3.4 Protection goals and security mechanisms

Protection goal	Security mechanism	Example
Confidence	Encryption	AES
Compulsoriness	PKI or BC	RSA
Authorized access	Authentication	Login
Integrity	Secured data, message authentication code	CRC, Hash

Table 3.5 The most important Blockchain consensus algorithms for creating blocks

Type of consensus algorithms	Decription	Advantages	Examples
Proof-of-Work	Participants (miners) provide their computing power for a fee in order to calculate the hash values of new blocks. The difficulty of the calculation is adjusted so that blocks are created at regular time intervals	Very computationally and energy-consuming	Bitcoin
Proof-of-Authority	Only certain nodes within the network can create blocks. Row-based selection	Intolerance against a manipulation	HyperledgerFabric, Ripple
Proof-of-Stake	Participants own a network's cryptocurrency. The higher the percentage, the more likely it is that you will choose to create a block for a fee	Capital accumulation can deter other participants	Stratis, Reddcoin

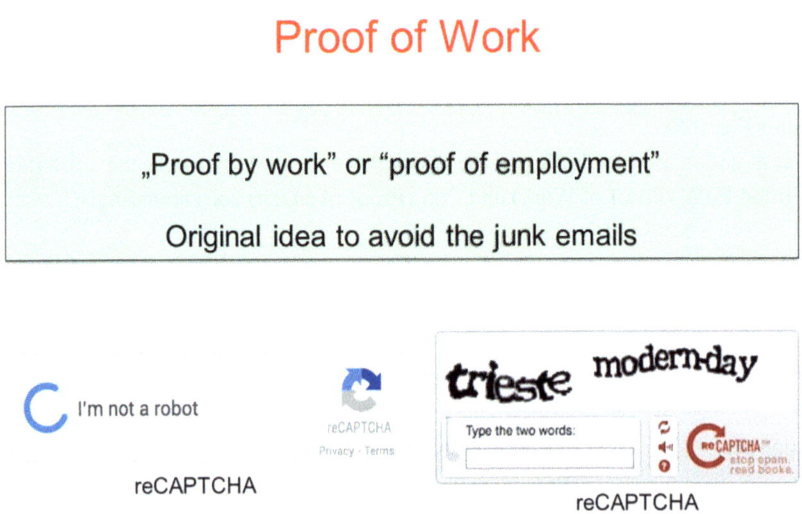

Fig. 3.9 Blockchain Operation: PoW

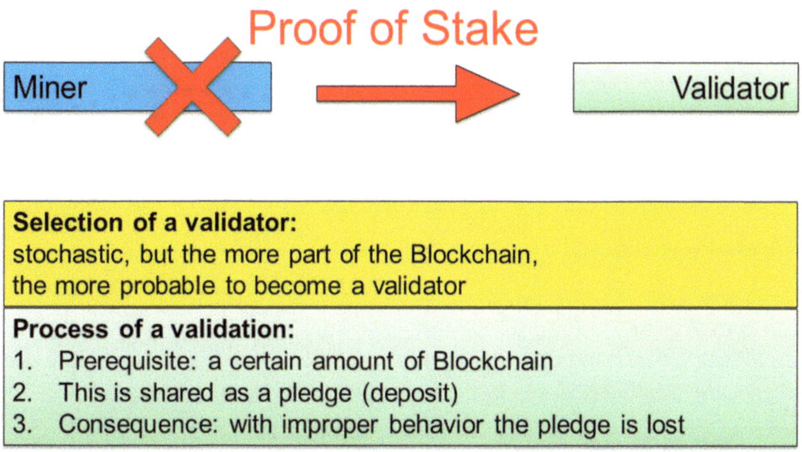

Fig. 3.10 Blockchain Operation: PoS

Denial-of-Service-Attack on an Internet host, or a spamming e-mail chain to an SMTP server. PoW should prevent such excessive use of the considered service (refer Fig. 3.9).

- Under PoS is understood a complete or partial proofing method which can create so-called 'consensus': it means which party (which participant) of the Blockchain is allowed to generate the next block (Fig. 3.10). In this case, a weighted random selection is made, in which weightings are determined for each participant, depending on the duration of the participant's participation and / or his "assets" (stake). Unlike the PoW, which uses crypto currencies Bitcoin and Ethereum, PoS does not consume time or energy like mining processing, nor can the network be "intercepted" because it has a single processing power (51% attack) (see Figs. 3.10 and 3.11). Instead of a miner, a validator is used whose selection is stochastic. The best selection options have the validators that stay in the system the longest. A specific part of the Blockchain system is reserved for verification. Such a reserved deposit or "pledge" is distributed and a check is made. In the presence of injuries/offenses, the custody is definitely lost.

3.6 BC Pro and Cons

The use of a blockchain along a supply chain offers the possibility of seamlessly tracking the condition of the goods along the entire value chain. In this way, the above-mentioned protection goals can be achieved:

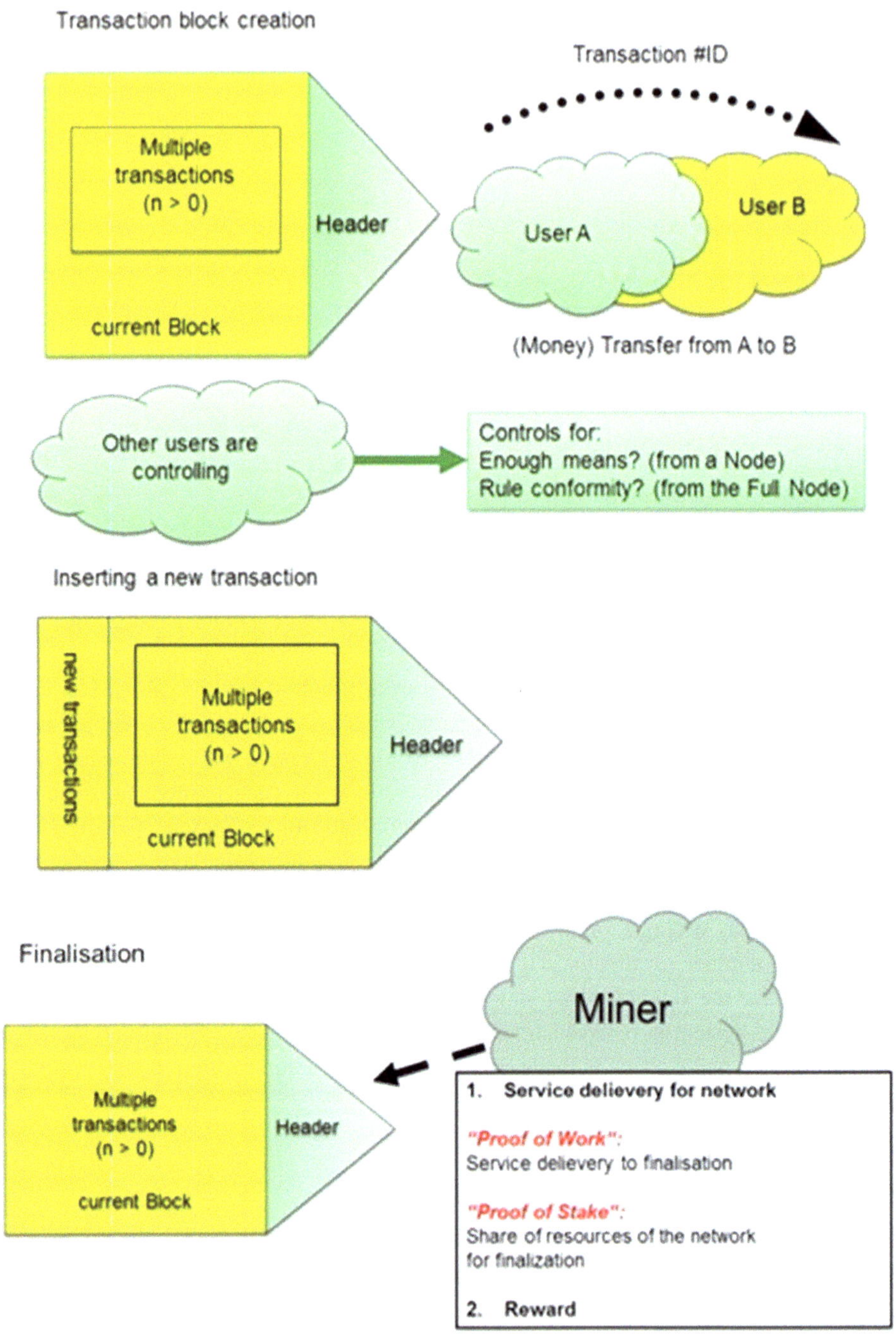

Fig. 3.11 Blockchain Operation: Miner

(a) Integrity: Sensor values recorded by IoT devices are stored in the BC in real time, where they are permanently and forgery-proof.
(b) Availability: Every company in the value chain operates its own node in the BC P2P network and thus has access to the current and complete BC at all times. This means that problems along the supply chain can be identified immediately.
(c) Confidentiality: Cross-company blockchains are usually closed systems to which only invited parties have access. There is also the option of using encryption for the transactions, for example the public key method. Programs to regulate visibility can also be stored in the blockchain so that every company within the supply chain only sees the information that is necessary.
(d) Authenticity and accountability: Another important aspect is that it can be clearly determined at any time which participant is responsible for the written data. Here, too, the blockchain offers a solution with the use of the public key procedure.

Therefore, the BC represents an actual alternative to established technologies in the given scenarios. The cross-company collaboration also fulfils all requirements for the use of a BC:

- There are multiple participants (N > 2)
- The states must be every time saved
- A third party is not always available
- All participants are known, but they are not trusted.

Specify pro and cons for in comparison to networked DB, or so-called warehouses. In summary, there are a number of advantages, but also disadvantages, compared to conventional databases, shown in the following Table 3.6.

An example of mining process under use of so-called "mining rigs" is depicted in Fig. 3.12. The miners are used within a heterogeneous P2P network. They create new transactions and obtain reward for his activity on crypto-currencies like BTC, XMR etc. What is the main difference between a Miner and a Validator? The answers are given in Table 3.7:

3.7 Practical Blockchain Applications

3.7.1 Framework MS Bletchley

In July 2016 the framework MS Bletchley was launched as a specific service BaaS (Blockchain-as-a-Service) and as a part of the Azure cloud platform. The main goal of the framework introduction [5, 6] was the acceleration of the development of the practical Blockchain applications aimed to financial institutions, manufacturing, retail,

Table 3.6 BC and their pro and cons for in comparison to networked DB, or so-called warehouses

Advantages	Disadvantages
The redundant storage of the blockchain content and unrestricted access by the individual participants results their maximal transparency	The usual ACID semantics of conventional networked databases (Atomicity, Consistency, Isolation, Durability paradigm) cannot be unfortunately applied. Atomicity is rather impossible due to the limited rollback functionality. Due to the simultaneous creation of blocks, consistency is also not always provided
By chaining the blocks, the use of hash values and distributing them in the network, the blockchain data is particularly forgery-proof, confident and secured	The creation of blocks, especially in the open blockchains, is associated with a very high amount of computing resource and energy consumption. The known CRUD functionality is rather affected (Create-Read-Update-Delete paradigm)
Through direct transactions and the given transparency, the intermediaries (inter-agents) previously required are totally superfluous our redundant. Decentralized use of the data without the leading instance	Due to the permanent storage of data in the blockchain, the later (post-factum) deletion is not possible too. This could lead to certain problems related to the GDPR (General Data Protection Regulation of EU)
Transactions in the blockchain can have any content, as well as programs and contracts can be stored within the P2P network	The sensible use of a blockchain possesses certain important restrictions. These affect the available number of participants, the type of exchanged data, the participants trust among each other as well as available third parties

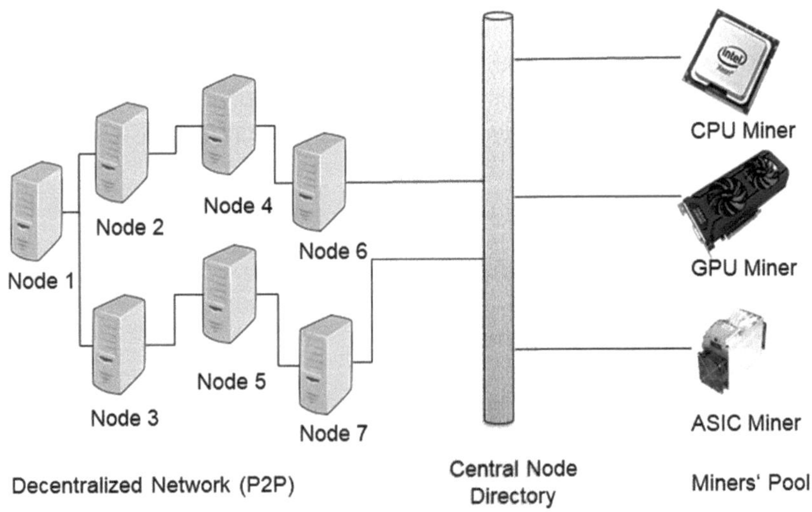

Fig. 3.12 Mining Process Example under use of so-called "mining rigs"

Table 3.7 Miners vs. Validators

Node type	Miners	Validators
Consensus algorithms	Proof of Work	Proof of Stake
Resource consumption	Higher energy expenditures	Lower (energy) expenditures
Deployment	Deployed by Bitcoin	Used by other crypto-currencies and some private Blockchains

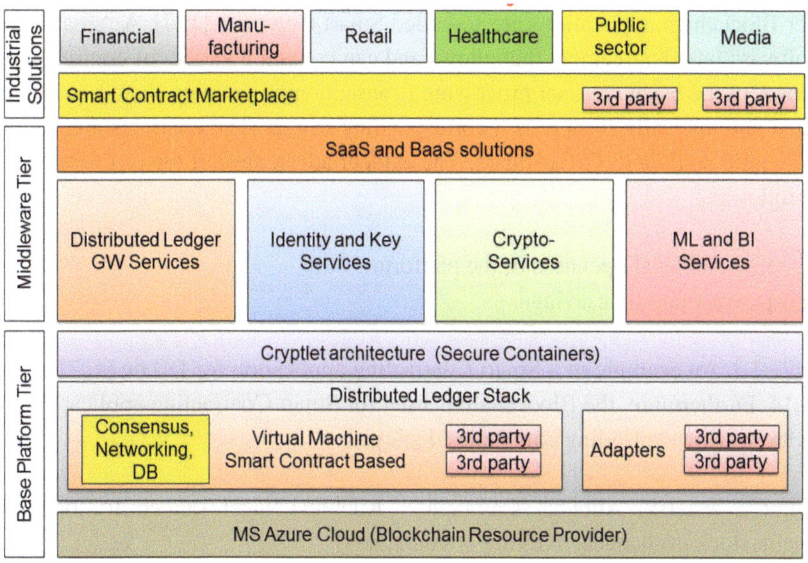

Fig. 3.13 Blockchain-Plattform MS Bletchley (source: http://www.microsoft.com)

healthcare, public sector, media on a common platform. The above-mentioned platform contains three following layers (Fig. 3.13):

1. Base platform tier
2. Middleware (MW) tier
3. Industrial solutions layer.

The framework MS Bletchley introduces two new concepts: Blockchain middleware (MW) with so-called cryptlets. To SaaS and BaaS belong i.e. so-called ML and BI functionality (Machine Learning and Business Intelligence). The Blockchain MW components support the core features of services in the clouds, such as identity management, analytics, or machine learning.

Based on the Azure cloud, these components can work with different Blockchain-based technologies aimed to creation of the practical applications. The cryptlets are the building blocks for the Blockchain technology. They are designed to ensure secure communication between the Azure cloud, the MW ecosystem and each specific customer's technology. The interoperability of the applications with Azure cloud and Azure stack is secured, as well as with the third parties, the clouds like AWS, Google and further private clouds.

3.7.2 Smart Contracting

Further Blockchain applications are so-called Smart Contracts [1, 7]. A "smart" contract is a software-based agreement that allows and can contain a variety of contract terms. In the course of the usual contract processing (transactioning), certain linked actions can be executed automatically if there is a corresponding trigger. The contracts are offered and signed within and via the Blockchain. The evident advantages of the discussed approach are as follows:

- Digitality and legal openness of the platform
- Transparency and timesavings.

Example 3.3. An example of a Smart Contracting application for DB.de [1, 7] is given in Fig. 3.14. Furthermore, the Blockchain is used for Smart Contracting applications for the following worldwide leading companies deployed:

- Walmart, Maersk, Alibaba, CartaSense, Kuehne+Nagel (aimed to logistics, sea freight, stock exchanges, marking of containers)

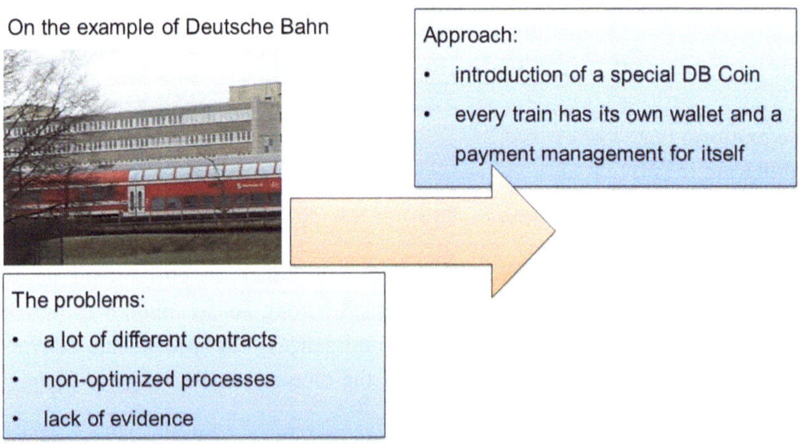

Fig. 3.14 Smart Contracting for DB.de

- Nestle, Tyson Foods, Unilever (aimed to food delivery)
- Everledger (the registers for diamond certification) and so on.

3.8 Risks and Hacking

Despite of advances in modern cryptography, viruses and Trojans are however an important threat to the Blockchain technology [18]. Sure, even the Blockchain can also store any data that is often different from financial transactions, transactions and contracts. Such malicious messages can be written to the source transaction addresses using the scripting language OP_Return. Therefore, the Blockchain can also block the extremistic or pornographic contents. As an example of such manipulation of the contents of the Blockchain, the Cryptograffiti.info website may be viewed for read and write accesses.

3.8.1 Ransomware

The next important topic is so-called Ransomware as a dangerous malware type with a blackmail background and the following features:

- Dangerous malware with extortionate action
- Virus-like operation
- Get through prepared attachments to emails
- Exploiting the vulnerabilities of web browsers
- Encryption of passwords and/or files on the hard disk with a powerful 2048-bit-long RSA key
- Prevented or completely blocked encrypted access by the victim to their own data.

Example 3.4. The ransomware prevents access to the initially encrypted user data. The victim is required to pay a ransom in crypto currency units (frequently in bitcoins), about the equivalent of $ 300–$ 500. The source of the attack or the recipient of the repayment is difficult to track through the use of cryptography. Some examples of this type of Blockchain apps have been listed in Fig. 3.15.

One of the well-known examples of ransomware was crypto trojan WannaCry (approx. 300.000 personal computers in 150 countries were affected) as well as so-called Wcrypt, WCRY, WannaCrypt, Wana Decrypt0r 2.0 etc., which are sadly known worldwide. Many international companies have suffered from these harmful applications.

The GandCrab Ransomware belongs to one of the latest recent risks. The effect of the GandCrab ransomware can be seen in the above figure. An exemplary workflow for the ‘careless users’ was depicted (Fig. 3.15). Additionally, this includes the use of the “no name” and well-unstudied development tools (kits). The ransomware also searches

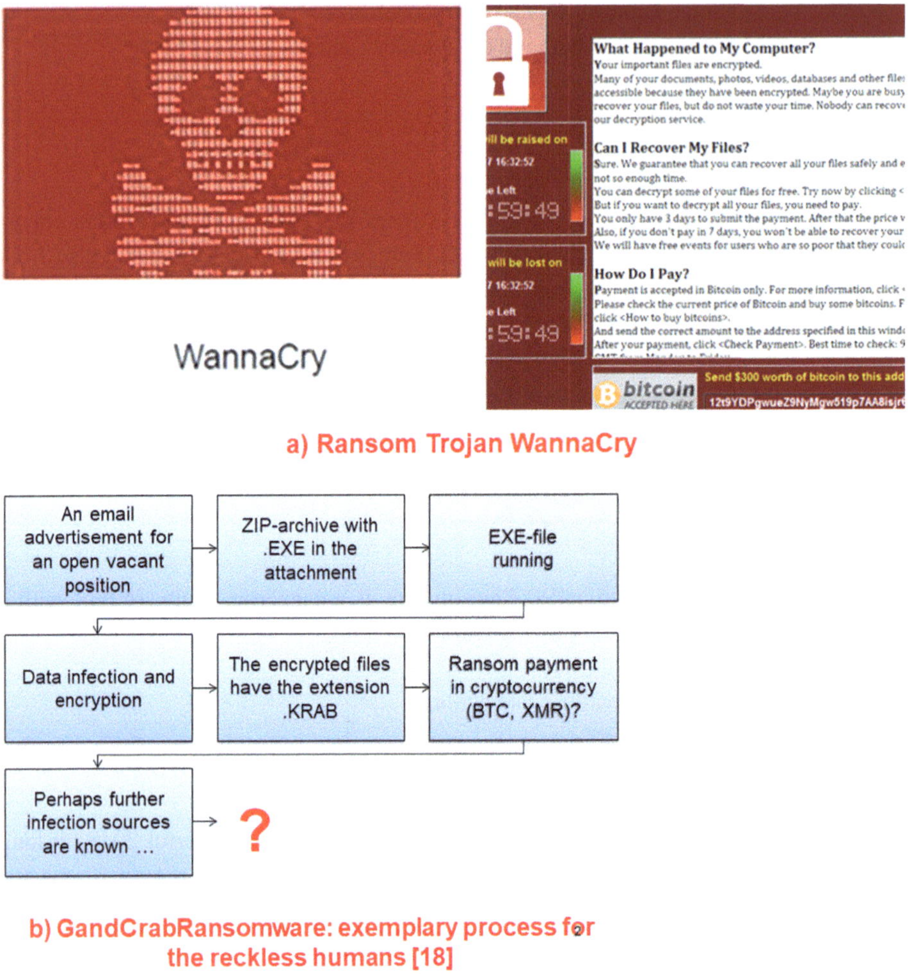

Fig. 3.15 Ransomware

for vulnerabilities in Java, IE, Adobe plugins etc. Unfortunately, the well-known anti-malware like ESET, Avira or Kaspersky do not currently have any antidotes.

Example 3.5. A Windows host receives an e-mail. The subject of the e-mail is titled "Application form to the advertised position—Nadine Bachert/Sarah Summer/Viktoria Hagen/Caroline Smith/Margareth Williams" etc. The application form contains a cover letter, a photo and a CV. The spam filters can classify these as a non-harmful. The malware is packaged as an attachment in a ZIP file. The file contains an executable file. If the user executes this file, a symmetric encryption procedure starts. For the key to restore the encrypted host data a ransom is asked. For each user, a unique URL on ananonymiser gate for hidden service is generated. Behind the gate, within the anonymised network the ransom amount and the instructions for payment can be viewed. The amount will be mostly adjusted to the victim. What is the average case, how do companies solve the problem? The distribution of primary actions for the victims is as follows:

- 78% restore the system simply with a backup
- 20% pay even blackmail money (ransom) to avoid reputation losses
- 2% even try to decrypt themselves.

The advanced firewall systems [1, 19–21] such as Cisco Umbrella and Check Point R80 contain Anti-Ransomware and can efficiently detect and prevent unauthorised access to the desktop systems, the clouds, the corporate data centres and networks under use of so-called Intrusion Detection/Prevention Systems (IDS/IPS).

3.8.2 Crypto-Currency Fraud

Some cases of large scale crypto-currency fraud in billion amounts became known to the public last time. Cybercriminals use for the robbery the anonymous Internet (so-called Darknet), which can potentially arise through a MIX cascading [1, 18–21] and is therefore untraceable. This fact effectively ruins the reputation of such MIX networks as Tor, Onion, JAP which are aimed to guarantee primarily the anonymity and privacy for the users. The anonymous Internet is accessible through special software and cannot be generally fully monitored by security analysts.

The attackers can remain anonymous for longer periods. However, the average price for mentioned malware is only $ 224, with the market value for blockchain-oriented malware now as high as $ 6.7 million with potential for growth. The attacks on the growing crypto exchange market now account for up to 25% of all Internet server attacks (according to Computer Forensic experts).

Example 3.6. According to the statistics for 2018, the following cyber currencies are recognised as the most common targets of fraudsters [18]:

- Monero, XMR (2014)—44% of hacker quota
- Ethereum, ETH (2015)—11% of hacker quota
- Bitcoin, BTC (2009)—10% of hacker quota.

The theft of cryptographic means of payment can be carried out by organised crime cartels or smaller criminal groups. However, it is often about highly skilled professionals looking for an additional criminal source of revenue on the anonymous internet. Most of these incidents related to the theft of the crypto-currency have been unfortunately recorded in USA (24), China (10) and UK (8).

3.9 DAO as Blockchained HDS

DAO means a "Digital Decentralised Autonomous Organisation" and simultaneously is signifying in Chinese the "way", "path", a holistic philosophical doctrine. DAO operates via crowd funding (group financing) by interested donors. DAO operates as a SC-based application. Such organisation works without a conventional management structure (CEO, CTO, and CIO) or board of directors and represents an instance of Ethereum Blockchain.

The above-mentioned application is related to ETH, Ethereum Classic and other Ethereum BCs as well as to recent crypto currencies.

DAO code is open-source that means the great opportunity and relevance to use at different platforms as well as by development of so-called intelligent and secured 5G Slicing (refer the previous sections). DAO can be a prospective BC Application for 5G networks.

An essential disadvantage for this solution for 5G microproviders and macroproviders can be the law conformity. DAO has, in fact, no nationality, is stateless. It means a problematic deployment because of existing EU legislation and state regulatory authorities.

3.10 Analysis: What is Stopping the Development of Blockchain?

Despite optimistic forecasts and Blockchain perspectives, development and deployment of the technology is not so fast as the optimists would like to. Certain restrictions prevent this, for example [7, 15, 18]:

- **Control and management:** Although the banks and financial institutions recognise the revolutionary character of the Blockchain applications, they also recognise the primary con: when everything is registered on the financial transactions, there is no more banking security and confidentiality! In the case of decentralised transaction processing, the lack of supervisory authorities means i.e. that even weapons and drugs can also be traded using the blocks and crypto currencies. In some countries—Romania, Kyrgizstan—a simple way was taken: the Blockchain transactions are prohibited. In opposite to them in El Salvador BTC acts a legal currency since Sept. 2021 for clearing and exchanges too like USD.
- The first steps towards the Blockchain regulation were undertaken within the EU. So, since May 2018 the General Data Protection Regulation (GDPR, in German: DSGVO) came into force, which restricts the storage and processing of (personal and privacy) information. The Blockchain has to adhere to these restrictions—at least for EU citizens.
- **Mass character:** Despite the apparent increase in crypto currency popularity, it should take several years for the crypto currency to become fully established. However, insufficient training of the population, skeptical citizens and certain unsuccessful or even criminal inset experiments (e.g. financial pyramid transactions) do not allow the Blockchain to conquer the market immediately.

- There is also negative propaganda from some financial institutions and entire states, such as India and China, where the government creates a negative image of the BTC and Blockchain and calls every transaction a criminal offense.
- **Costs:** The main algorithm for the blockchain is SHA-256. All operations are checked and executed according to this hash algorithm. However, these are very energy intensive processes. By comparison, global hashing exceeded in year 2018 the level for the power consumption of Cyprus! Moreover, these resource requirements will increase even more with a higher number of transactions [3, 4, 13].

3.11 Assessment

3.11.1 Advantages

- Blockchain is developing dynamically
- Large IT companies and financial institutions are becoming increasingly involved
- Blockchain is manipulation-proven and theoretically infinitely traceable
- Multiple applications and blockchains: crypto-currencies, Smart Contracting and in midterm automation of the domains where an intermediary (confident) is required.

3.11.2 Disadvantages

- Complete replacement of the existing centralised banking system in middle term is rather hopeless and fantastic; therefore, rather niche business as an alternative to centralised banking system and stock exchanges
- Profitability of the Bitcoin and other crypto currencies is questionable!
- One thing is evident nowadays:
- The solution to the above listed problems may be complex and resource intensive, long term mass data retrieving ("Big Data"), mass validity tests, prevention of criminal activities (ransomware, fraud) are required.

References

1. M. Staples, S. Chen et al. Risks and opportunities for systems using blockchain and smart contracts.
2. XMR: Was ist Monero? Kryptowährungen und Coins einfach erklärt (Online): https://www.cryptolist.de/monero/.
3. Monero - Secure, private, untraceable (Online): https://getmonero.org.
4. Münzwelt: die Geschichte des Geldes, MDM (Online): https://www.mdm.de/muenzwelt/einfuehrung-ins-muenzensammeln/geschichte-des-geldes/.

5. Marley Gray. Project Bletchley Whitepaper, MS Worldwide Partner Conference Toronto, July 2016.
6. MS Bletchley (Release & Roadmap). Cryplets Deep Dive (Online): https://azure.microsoft.com/.
7. Marco Liesenjohann et al. Blockchain#Banking: Ein Leitfaden zum Ansatz des Distributed Ledger und Anwendungsszenarien, Bitkom 2016 (Bundesverband Informationswirtschaft, Telekommunikation und neue Medien e. V.), 66 S. (Online): https://www.bitkom.org/noindex/Publikationen/2016/Leitfaden/Blockchain/161104-LF-Blockchain-final.pdf.
8. Kim Rixecker. Modulares Blockchain-Framework von Microsoft: Das steckt hinter Project Bletchley (Online): http://www.t3n.de/news/checked-c-microsoft-716063/.
9. Malborn Wolff. Blockchain als Open Source: das Hyperledger-Projekt (Online): https://www.maibornwolff.de/blog/blockchain-als-open-source-das-hyperledger-projekt.
10. Blockexplorer (Online): https://btc.com/3fb5f4fd6d1129b9797d66be1750b208a10ac2d551125990305211834be94cd9.
11. Bundesanstalt für Finanzdienstleistungsaufsicht; Artikel: Blockchain Technologie; (Online): https://www.bafin.de/DE/Aufsicht/FinTech/Blockchain/blockchain_node.html.
12. Welt-Bruttosozialprodukt, Bundeszentrale für Politische Bildung (Online): https://www.bpb.de/nachsc-fakten/globalisierung/52655/welt-bruttoinlandsprodukt.
13. Christopher Malmo. Bitcoin hat ein großes Problem: Die Kryptowährung ist einfach nicht nachhaltig (Online): https://motherboard.vice.com/de/article/4xa9ed/das-oeko-problem-von-bitcoin-darum-ist-die-krypto-waehrung-nicht-nachhaltig-3920.
14. Elisabeth Oberndorfer. Blockchain as a Service: Microsoft baut einen Blockchain-Marktplatz (Online): http://www.t3n.de/news/microsoft-blockchain-marktplatz-680639/.
15. Phillipp Giese, Maximilian Kops, Sven Wagenknecht, Danny de Boer, Mark Preus: Die Bitcoin Bibel - Das Buch zur digitalen Währung, Verlag BTC-Echo, 2016.
16. Simon Huelsboemer. Was ist Blockchain? (Online): https://www.computerwoche.de/a/blockchain-was-ist-das.
17. Secure Hash Standards (SHA-1, SHA-224, SHA-256, SHA-384, SHA-512), in Federal Information Processing Standards, NIST, Release Oct. 2008, 33 p. (Online): http://csrc.nist.gov/publications/PubsFIPS.html#fips180-4.
18. Ransomware GandCrab (Online): https://www.searchsecurity.de/news/252448017/Ransomware-GandCrab-kommt-als-Bewerbung-getarnt.
19. Andriy Luntovskyy, M. Klymash. Examination of Modern Concepts for Firewalls and Collaborative Intrusion Detection, International Research Journal Telecommunication Sciences, Kiev (Registration Certificate KB No. 15064-3636P, ISSN 2219-9454, UDC 621.39).-2016. - Issue 14, No.2, pp. 44–49.
20. Luntovskyy, Andriy, Spillner, Josef. Architectural Transformations in Network Services and Distributed Systems: Current technologies, standards and research results in advanced (mobile) networks, Springer Vieweg Wiesbaden, 2017, 344p., ISBN: 9783658148409 (https://www.springer.com/gp/book/9783658148409#otherversion=9783658148423).
21. Luntovskyy, Andriy, Gütter, Dietbert. Moderne Rechnernetze: Protokolle, Standards und Apps in kombinierten drahtgebundenen, mobilen und drahtlosen Netzwerken, 500 Seiten, 263 Abb., Springer Nature, Juli 2020, ISBN: 9783658256166 (https://www.springer.com/gp/book/9783658256166).

From Big Data to Smart Data: Best Practices for Data Analytics

4

Cyber-PHY, IoT, sensor networks, Robotics (thick and serverless mobile applications), real time network applications (thin clouds clients) can generate large arrays of unmanaged, weakly structured, and non-configured data of various types, known as "Big Data". With the acceleration of industrial development "Industry 4.0" processing of such data became considerably more complicated. However, so-called problem "Big Data" is hard to solve or resist nowadays. The given chapter discusses the Best Practises and Case Studies for Data Analytics aimed to overcoming of the Big Data problematics under a slogan: "From Big Data to Smart Data!". Conventional and new solution paradigms are overviewed (R Tools, SPSS, Grafana vs. Hadoop, Hbase). So-called "light-weighting" approaches based on cloud analytics as well as ML deployment are examined and favorised. The Big data shortcomings for important areas were discussed.

4.1 Motivation

Big Data accumulation (Fig. 4.1) is nowadays typical for trading and marketing, electronic payments, production process, for the traffic from mobile providers, international justice and forensics, for public fiscal authorities, pharceutical and advertising industry. A large number of research institutes, organisations and universities accumulate, store and process large amounts of technical and scientific information.

The sources can be primarily classified as follows [1–4]:

- Data from sensors and production capacities (Smart Manufacturing)
- Navigation data from GPS and other navigation systems, e.g. Galileo
- Data from GIS (Geographic Information Systems)
- Traffic and telematics data (IoT)

A. Luntovskyy and D. Gütter, *Highly-Distributed Systems*,
https://doi.org/10.1007/978-3-030-92829-2_4

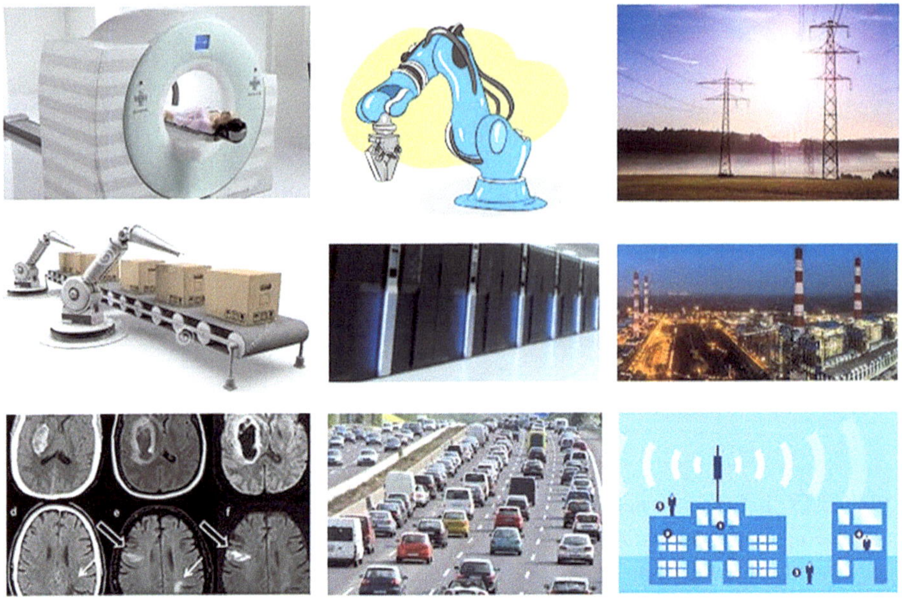

Fig. 4.1 Sources of unmanaged and unstructured data under Industry 4.0

- Robotics data (navigation, sensing, measurements)
- QoS management data and traffic data from mobile providers
- Power and renewable energy stations (data for smart metering)
- Data from scientific simulations and parallel computing (clusters, grids)
- Imaging data of patients from CT and MRT (Computer and Magnetic Resonance Tomography) in healthcare
- Marketing information from cookies, social networks and Web2.0 semantic networks
- Collected personal data of citizens and much more.

In the conditions of modern industrial development, under so-called "Industry 4.0", there are even more Big Data sources: home automation, patient health data, M2M and robotic data, business intelligence, pharmacological research, networking and experimental data.

The mobile networks and aps for the 5G will definitely take an active part in the process of receiving and processing of large data amounts [2, 5–7] too. By year 2020, the new 5G networks will use more than 50–100 billion sensors to download comprehensive information about how we interact with things that surround us or even are inside of us?!

As one of the most interesting further topics the Blockchain technology occurs. This is nowadays exponentially increasing and enables modern cryptocurrencies: e.g., Bitcoin, Monero, Ethereumetc. The technology of Blockchain and its associated applications like cryptocurrencies are so-called "resource eaters" due to their enormous energy and memory consumption. Large amounts of chained crypto blocks are causing surely "Big Data problematics".

Long-term mass data retrieving and processing requires large amounts of memory: decentralised, i.e. on private hosts and even on mobile devices, as well as centralised, in data centres and clouds. This means Big Data shortcomings too.

Another general problem is the economic and technological recognition of crypto technologies, processing of mass tests for the validity of blocks, the prevention of criminal activities (extortion, fraud) based on Blockchain. The further chapter content was built as follows:

- State-of-the-art approaches are analysed.
- Big data shortcomings for important areas (best practices/case studies) are discussed like WSN, health care, IoT and robotics.
- Analytics placement options to overcome Big Data are offered (cloud and fog based).
- Regular paradigms for Big Data are overviewed as well as empirical Data Analytics for Big Data are discussed.
- Finalising, conclusions and outlook have been done.

4.2 State-of-the-Art

Huge as well as heterogeneous data volumes (approx. 100 PB to 100 EByte) are nowadays so immobile that their access and management becomes only very efficient last time without use of special techniques [1–3]. Big Data are nowadays too complex for their efficient processing with classical manual methods for data structuring. Herewith a recognition of the problem is given by one of the world's leading research and consulting company as very serious one: the company Gartner Inc. notes "the problem of Big Data as one of the most important trends of IT-infrastructure development along with virtualisation and energy efficiency of IT". Under use of convenient retrieving, archiving and analytics methods for Big Data from [1–3] like databases and data warehouses, electronic sheets, formatted texts, CSV-data, Web- and XML-documents, graphical, audio and video documents data become faster suitable for nothing and inadequate. To solve this problem the computational techniques of so called "Data Mining" inter alia with ontologies, data pattern recognition, fuzzy logic etc. are used. These techniques can correspondingly be based on cloud analytics and disassemble the complexity and heterogeneity [1–10]:

- Taxonomies and predictive classification.
- Preliminary clustering and compression.
- Predictive deviation detection (based on regular statistics).
- Descriptive association rule discovery (fuzzy based).
- Ontologies and knowledge data bases (KDB).
- Regression Models.
- Sequential Pattern Discovery (Machine Learning).

Otherwise, such a large amount of data remains unstructured and will be in future characterised by the over-proportional management complexity. Additionally, significant increase of network traffic can be forecasted during the data processing under use of typical today solutions like Cloud Computing with heterogeneity of networks, replication across multiple geographically distributed computing nodes [14–21].

4.3 The 6V Prevention Factors

However, there are six factors, which affect the complication or prevention of the use of the regular and/or (cloud-based) analytical methods (Fig. 4.2). They build so-called 6V paradigm which distinguishes from known 5V or even 10V-paradigms [1, 3, 14, 18]:

- Volume: the amount of generated and stored data is moving from 100 PB up to 100 EByte nowadays.
- Velocity: it means the frequency and speed of data acquisition and processing (from slow batch techniques to real time with strong limitation on reaction time or latencies).
- Variety: multiple codec types of multimedia data. This helps people who analyse it to effectively use the resulting insight.
- Veracity: The quality of captured data can vary greatly, affecting accurate analysis.
- Violation: The discussed data are often scattered, without any clustering or any structuring.
- Value: Inconsistency of the data sets can negatively affect the processes of data handling and management.

Unfortunately, the depicted V-factors for Big Data are growing faster than the performance of their analysis (in Mbyte/s or in GFLOPS) via classical computational techniques. Therefore, for Industry 4.0, IoT systems and M2M not only improvements

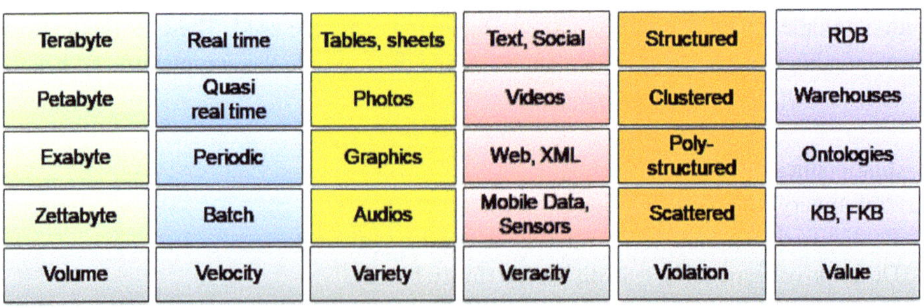

Fig. 4.2 Own 6V paradigm

in networking but also in analytic blocks became very significant [1–7, 14, 18], which aimed to decreasing of data streams and their algorithmic complexity.

4.4 Unmanaged and Unstructured Data in Industry 4.0

The goal of clustering and further processing of unmanaged and unstructured data from Industry 4.0 is the reducing of costs for their storage, better traceability, forecasting in process trends, workflow and production process optimisation, data security and human private sphere protection. The chapter discusses the Best Practises and Case Studies aimed to overcoming of the Big Data problematics if possible under use of "data compression" via their transformations (Fig. 4.3).

The "heavy-weighting approaches" involving precise analytical techniques and expensive specialised software are used for this aim. On the other hand, there is the opportunity to solve the Big Data problem under use of some "light-weighting approaches" based on agility: freeware, multipurpose techniques, minimal challenges on the personnel training and competencies. The chapter examines the techniques and case studies of both topics. The "heavy weighting approaches" (ontologies, knowledge

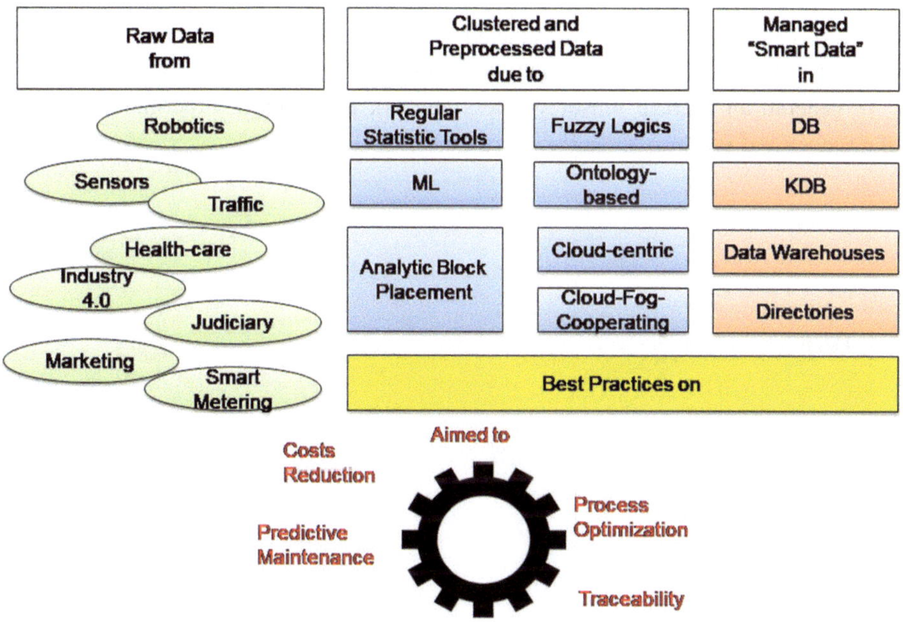

Fig. 4.3 Method classification: what is the aim of the clustering and further processing of unmanaged and unstructured data?

bases, fuzzy logic and fuzzy knowledge bases) are compared to "light-weighting" one. The existing reference solutions are discussed below.

So-called "Data Scientists" need to answer the following questions [8–10]:

- How is the acquired data available?
- When and with which periodicity the data survey acquisition and accumulation must be done?
- In which way must data sampling be done?
- How can the data be accessed?
- Which formats and codecs are required to apply within data evaluation system or for processing in cloud analytics blocks?

The typical sampling, survey and accumulation models are shown in Fig. 4.4.

Mostly, the collected data is large unstructured and unmanaged.

Example 4.1: Data Analytics for "Smart Data". Depending on the event frequency, the event-driven, periodically, permanent, and the behaviour of the sensors (push, pull), a sensor survey and data accumulation are performed. Let's a WSN consists of 15,000 sensors.

Each sensor can transfer a short telegram up to 100 Bits. Thereby:

- The survey for each sensor is conducted 20 times per hour: 2000 Bits/h
- x 24 h = 48,000 Bits/daily = 6000 Bytes/daily for each sensor
- In general, an average sensor wireless network accumulates experimental data for 3 years × 365 days:

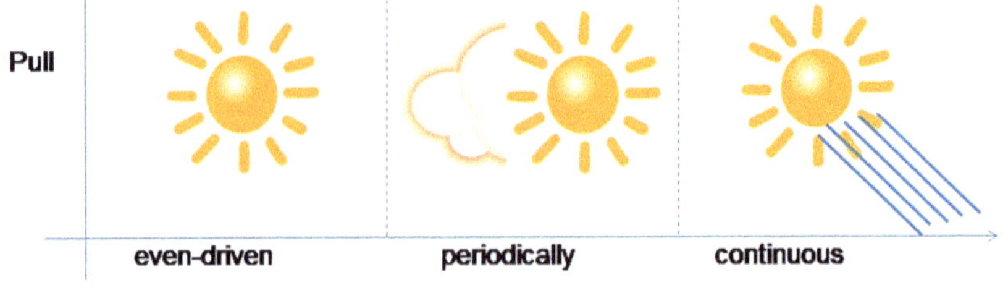

Fig. 4.4 Typical data acquisition and evaluation model

- $\times 365 \times 3 = 6.57$ MB for each of the sensors.
  The overall data for the mentioned network:
- 6.57 MBytes × 15,000 sensors = 100 GB of raw data!

Further research of the data accumulated in this way can be carried out by manual and automated methods both (Fig. 4.5). As a result of such research data evaluation the following common documents can be issued [9, 10]:

- Management Template
- Production Report
- Error Report
- Fault Forecast
- Feedback Recognition.

Often the data research cannot be completely studied or their trends recognised due to the complexity and heterogeneity. The future transition "Big Data to Smart Data" is given in Fig. 4.6.

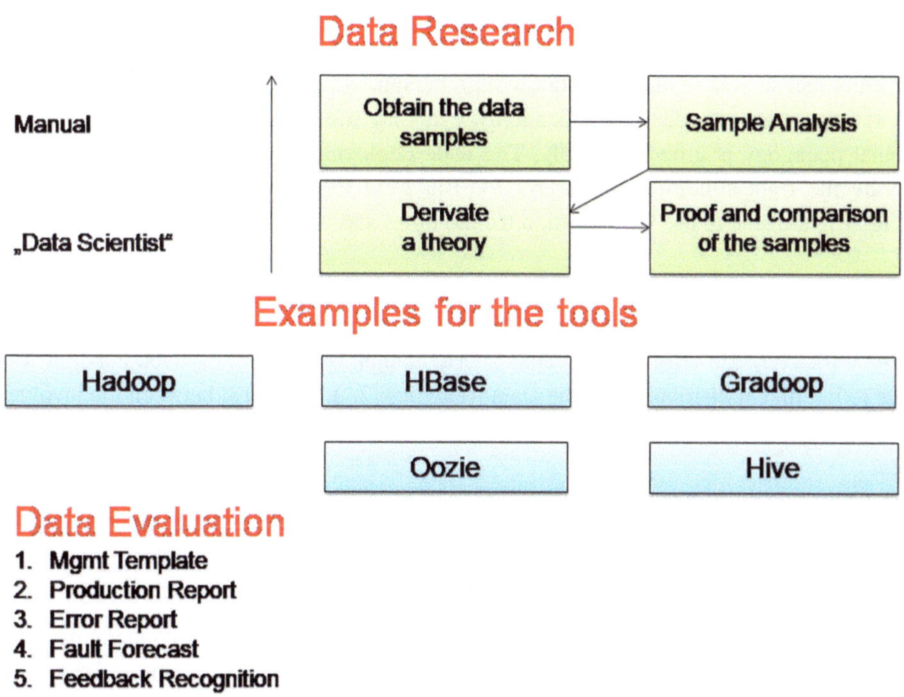

Fig. 4.5 Data Analytics: Manual and Automated

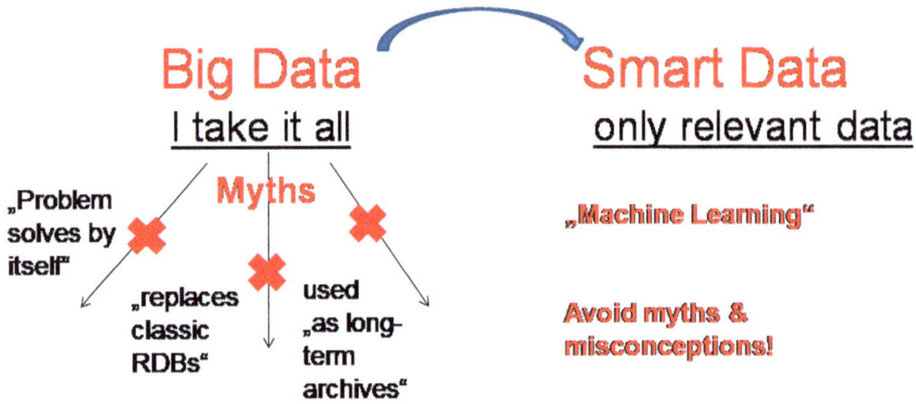

Fig. 4.6 Transition "Big Data to Smart Data"

The overcoming of myths and misconceptions is possible due to use of Machine Learning (ML).

4.5 Big Data Problematics for IoT and Robotics

The number of robots that cooperate with a person, with machines and with each other in various fields of industry and economy, everyday life and the entertainment industry, social pedagogy is growing rapidly. The wide deployment of robots and robotics data create the right conditions nowadays for Big Data too. Up-to-date robotic platforms (Fig. 4.7) are able to use virtualisation technologies and virtual robots over the network.

Example 4.2: The possible analytics placement options to overcome Big Data are given below. The up-to-date IoT and robotic applications can be classified into three groups (Fig. 4.8). They can affect the Big Data shortcomings, which can be solved in better way due to intelligent distribution of the Data Analytics [2, 3, 6, 14, 18] between the involved parties:

(1) Conventional Robots.
(2) Cloud-Centric Robots.
(3) Distributed (Fog-Cloud-cooperating) Robots.

Only the cloud centric solution (2) and further distributed, fog-cloud-cooperating solution (3) both are able to overcome the discussed problem nowadays and in the future.

The analytic blocks, migration agents as well as further adaptive interfaces are delegated to the clouds and, possibly, after preprocessing and clustering backwards to the so-called 'fog' (refer Fig. 4.8) under use of the mentioned solutions and protocols. The

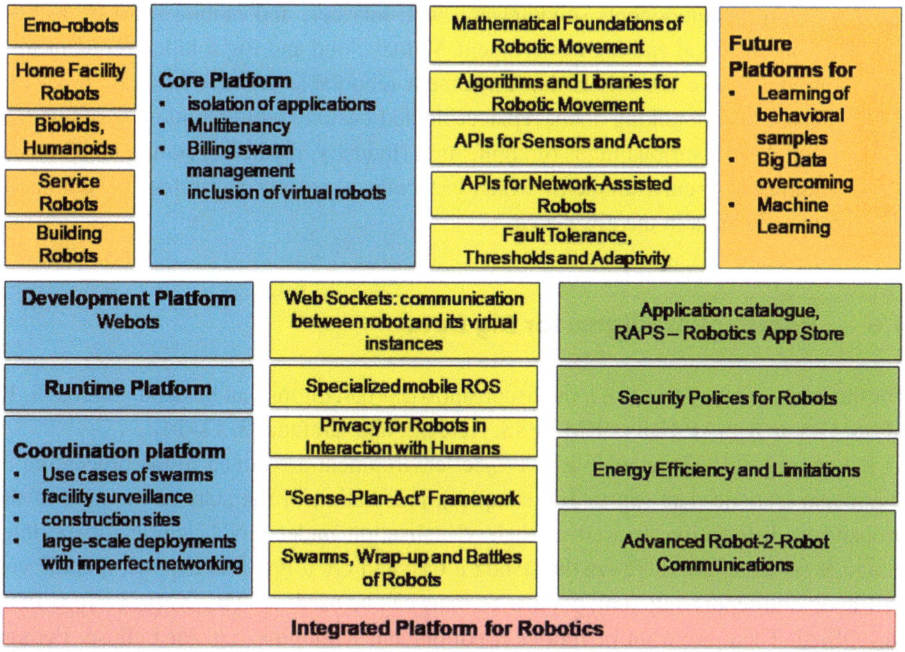

Fig. 4.7 The properties of the future robotics platforms

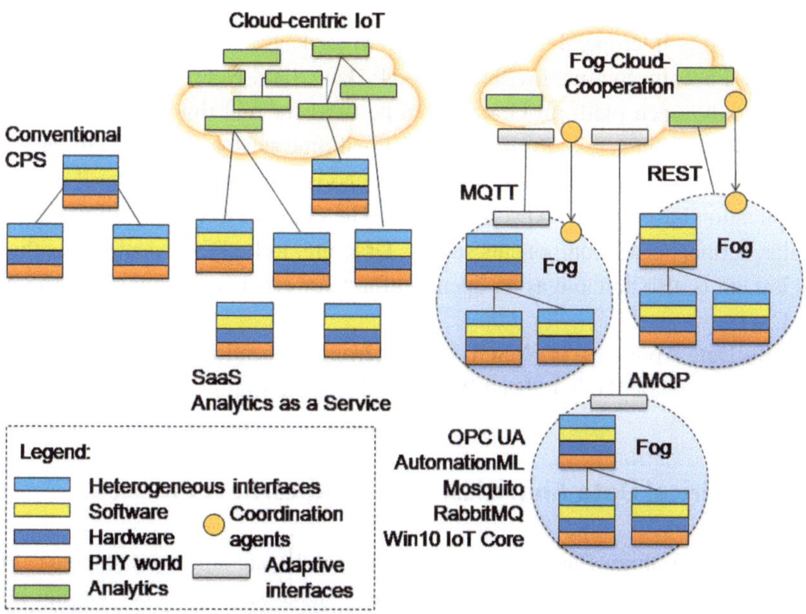

Fig. 4.8 Three communication ways: Analytics Placement Options to overcome Big Data

virtual analytical components (middleware, web services, and mobile software agents) are placed in the cloud and fog environment. Virtual cloud and fog solutions contain software components for the robots that implement rebootable (virtual) business processes [3, 14]. Thus, we are talking about replaceable and customizable robotic algorithms (cp. Fig. 4.7 and 4.8) in various fields of application (industry, medicine, communication and telecommunication, entertainment) which can collect, process and retrieve voluminous heterogenic Big Data in the given area.

4.6 Regular Paradigms for Big Data

For the processing of Big Data, the usual statistical paradigms can be deployed like S, R, SPSS, Oracle R, SAP Hana, IBM SPSS, Netezza, and Grafana [3, 11–14].

R Language (1992) as well as R Project environment are aimed to the processing of significant statistic data (up to 106 samples), calculation of the statistic moment of the probability distribution as well as their visualisation under GNU license. In the integrated R environment there are the standard CLI and GUI. For data exports, graphic visualisation and report output all popular formats can be used: HTML, XML, LaTeX, PDF, MS Office. There are a lot of further R-editors and R-plugins, e.g. for Eclipse, Python, Gretl for all significant OS.

SPSS (IBM 1968) is a trademark of IBM, which develops and distributes software for statistics and data analysis (IBM SPSS Statistics v25, PASW—Predictive Analysis SoftWare for MS Win, GNU/Linux, Mac OS X).

R Language is interoperable to many modern packages for Big Data processing like SPSS, SAS, Platform Symphony, Tableau, SAP Hana, IBM Netezza.

Recently, the open platform Grafana can be used for real time visualisations of fast processes of inter alia network traffic data and real time sensor data. The Grafana system is designed to construct dashboards and graphs and is a complex multi level open source web application based on the Apache License. The Grafana web application supports, in particular, further server components such as Graphite, InfluxDB, Opentsdb (as the backend). Grafana is also suitable as an open platform for efficient analytics and monitoring of Big Data.

The Big Data Appliance is NoSQL-Cluster from application servers for massive parallel analysis based on the integrated R tools and Apache Hadoop.

4.7 Empirical Data Analytics: Case Studies/Best Practices

Example 4.3: Ontology Based Medical Imaging and Data Analytics. Medical formats for CT and MRT applications are based on the open standard DICOM (Digital Imaging and Communications in Medicine) founded by Siemens and Philips. DICOM stores, manages or transmits the compressed patient Big Data. DICOM standardizes both the data storage format and communication protocol, which is represented by digital medical images, data for their segmentation, surface definition or registration. The DICOM format is also the basis for archiving (Picture Archiving and Communication System, PACS) in the clinics and hospitals. The ontology based framework for data analysis [3, 14] process enormous mass of patient data, which is stored in a variety of formats: handwritten texts, body imaging (CT, MRT), tests, forms with laboratory results, genomic data. Often such data should be processed in a few steps. Unstructured and unmanaged data must be converted to structured formats through their classification (taxonomies) and ontology creation The given method is so-called "heavy-weighting" [3, 8–10, 14–20].

Example 4.4: A Light Weighting Approach. Let us to examine a system example for overcoming of Big Data complexity which was developed via TIQ Solutions in Leipzig [3, 15, 16]. The mentioned reference solution of TIQ Leipzig is represented below. The system is based on 'a light-weighting approach' and is oriented to processing of Big Data in the domain of Business Analytics. The system architecture is depicted in Fig. 4.9. The most important quality criteria of the referred solution are wide scalability and suitability for the Linux clusters. The system allows the connectors to the conventional DB. The complex polystructured and redundant retrieved data can be processed with higher performance within an enterprise or institution data centre or a cluster. Some in Java implemented modules allow even real time control. The expenditures in the form of investments CAPEX (Capital Expenditures for hardware, cable infrastructure, premises) and OPEX (Operational Expenditures for licenses, personnel, electricity, ongoing maintenance) are significantly reduced based on proven freeware components. Therefore, the solution is relevant for so-called middle range enterprises.

The discussed architecture is characterised via agility, possesses no rigidity and offers only small license costs by the use of proven freeware.

The components of the architecture [3, 8–10, 14] are as follows:

- Apache Hadoop (2008) was implemented in Java and represents a freeware framework for scalable distributed applications under use of Google similar algorithms and GFS (Google File System), which supports computing with significant data (Petabytes-area) on the computer clusters.
- Apache HBASE is a scalable DB for management of Big Data within the Hadoop clusters.
- Apache Phoenix is a massively paralleled RDB-Engine with OLTP concept for Hadoop with Apache HBase as back-store. SQL queries are processed, then compiled into the series of HBase scans and are orchestrated to produce the regular JDBC results.
- Apache Hive: a DBMS for Hadoop, aimed to data aggregation, queries and analysis with a SQL-similar interface. The data can be retrieved in diverse DB and file systems.

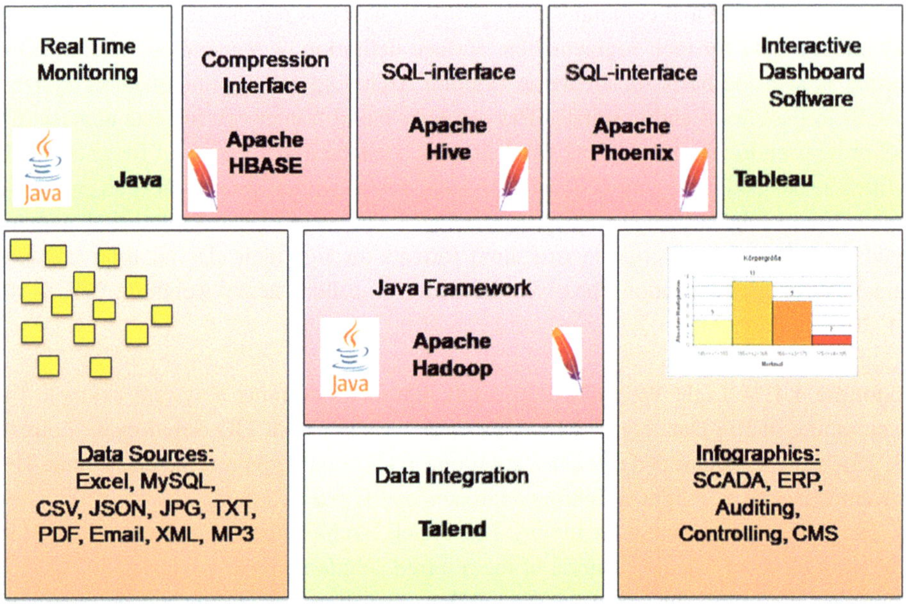

Fig. 4.9 Architecture Example for Big Data processing

- Tableau (2003): software for data visualisation and reporting.
- Talend: cloud based software for Big Data integration.
- SCADA (Supervisory Control and Data Acquisition): the computer assisted monitoring and control of technical processes.

4.8 Deployment of ML and AI

Nowadays, artificial intelligence is widely used in all spheres of life without stopping in its development. In midterm, the standards for ML and AI will accompany the industries, digital economy and everyday life over the world and for each institution. Surely, they will find it deployment in the area of Big Data. Therefore, let us to give a more detailed overview below. Frequently, to overcoming of misconceptions on Big Data the novel AI and ML techniques are used. The intelligent algorithms can provide clusterisation and data preprocessing too. When talking about artificial intelligence (AI), we should also mention ML as an important basis. To begin with, we would like to represent what the difference between the convenient deterministic algorithms (cp. classical flowcharts, like SSADM or PAP by DIN 66001) and Machine Learning [3, 14, 21]. Figure 4.10 shows that in classic algorithms we have Input Data and an algorithm that allow s us to get results. In ML, we have Input and Output Data, which help us to get a neural network a learning algorithm and it will help in the future to make the neural network more powerful via training.

ML systems, based on cloud concentrated knowledge (knowledge data bases), create statistics and regressions on obtained voluminous experimental data in background mode. An artificial system is 'learned' from samples and examples and can summarize them after the completion of the study and evaluation (training) phase. The ML system recognizes templates and trends in research data.

Thus, the ML systems within modern highly-distributed applications (Fig. 4.11) can also evaluate data on representativeness and compliance. Three mostly used types of ML algorithms are depicted below. The main purpose of machine learning (ML) is to look for learning algorithms, performed without human intervention, where the time and amount of data required for training is one of the most important indicators of performance. The main ML approaches are traditionally divided into Supervised Learning (SL), Unsupervised Learning (UL), and Reinforcement Learning (RL). The third approach is mostly important for the presented work. It means the mostly appropriate deployment chances for Big Data and eco-system (refer Fig. 4.11).

Supervised Learning (SL) is an ML method that takes training data (organised into an input vector (x) and a desired output value (y)) to develop a predictive model for the traffic classes. It is easy for the humans, e.g. to understand, where in some types of data we have video, music or pictures, but for a robot it becomes a more difficult task. So, we translate these data into a robot understandable language, and indicate, were we have

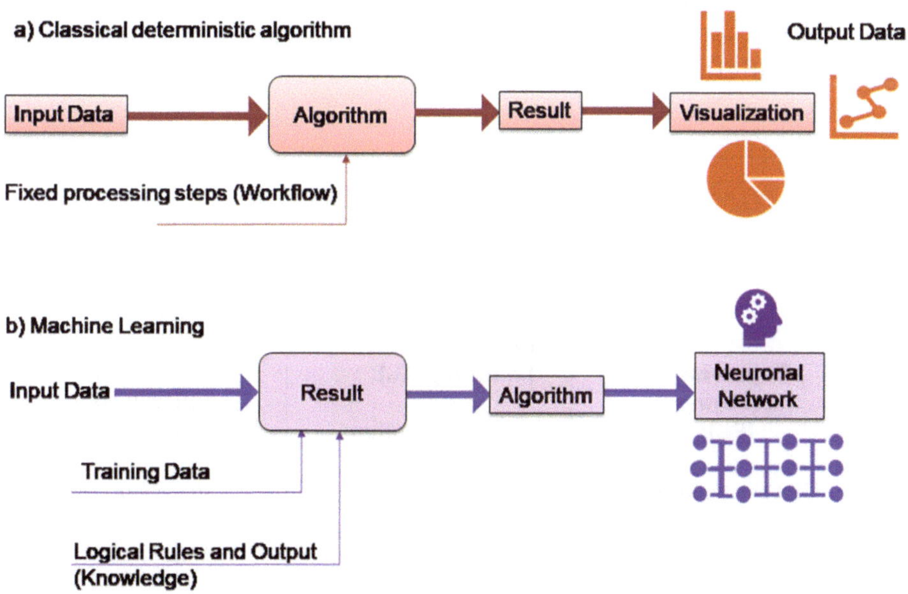

Fig. 4.10 Comparison of ML to deterministic algorithms

each kind of data. This will be our training data, that the algorithm memorizes and can already analyze the data flow. Knowing what type of data the user applies, we will give the algorithm the ability to classify them correctly (refer Fig. 4.12).

By UL (Unsupervised Learning), we allow the robot to learn on its own manner without giving the correct answers to the problem we want to solve. The aim is as follows: without knowing what kind of data we have, the data could be divided via the algorithm into several classes (Fig. 4.13).

Reinforcement Learning (RL) algorithm try to understand how to correctly configure parameters to achieve a specific goal (Fig. 4.14). Many problems and sequential tasks cannot provide a clear answer to a problem, and when it performs poorly, RL has already proven effective in many real world applications such as robotics, stand-alone helicopters and drones, fixed network routing, automated and hazardous works etc.

The (cloud based) ML system enables artificial creation of knowledge from the obtained voluminous experimental data in background mode. An artificial system is 'learned' from samples and examples and can summarize them after the completion of the study and evaluation (training) phase. The ML system recognizes templates and trends in research data. Thus, the ML (cloud) system can also evaluate data on representativeness and compliance. Evidently, at this time, we can propose only one way to improve this approach [3, 21]. We can use the cloud centric solution, i.e. the stored behavioral data from offered knowledge bases, which provide real time monitoring and learning. We want to introduce the stored behavioral data from offered knowledge bases. In this case, we will collect the data of each sensor or robots and send them to a common knowledge database. This will be an analogy as a common knowledge base for Tesla's autopilot, which can help us to make "studying models" for robots or sensors. Then our devices such as sensors can automatically predict the mass movement of data into specific places [20, 21].

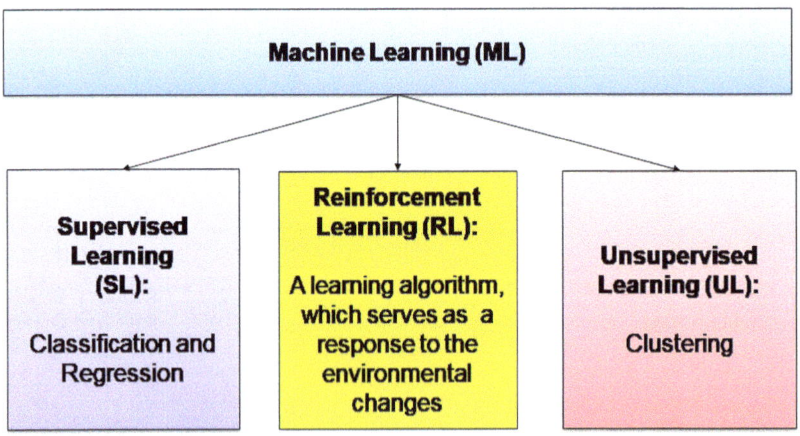

Fig. 4.11 Three mostly used types of ML algorithms

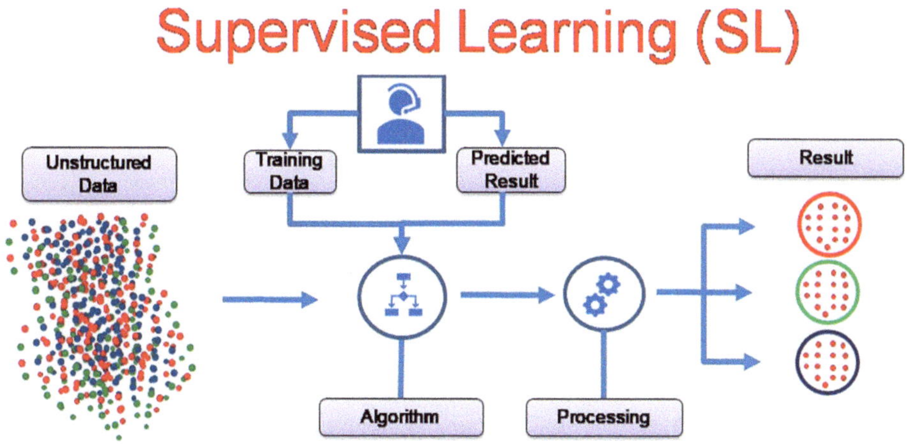

Fig. 4.12 Supervised Learning for Big Data Clusterisation

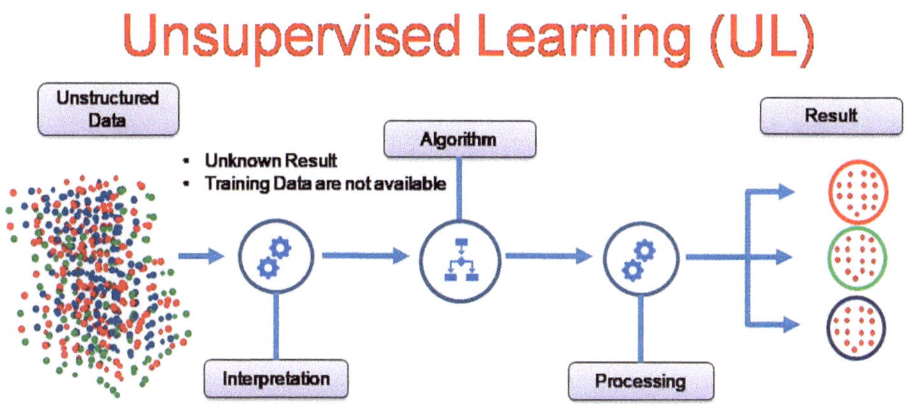

Fig. 4.13 Unsupervised Learning for Big Data Clusterisation

Example 4.5: An Approach of Big Data complexity overcoming for Telecom Operator. The telecom companies process huge multimedia data amounts. Unfortunately, such highly dynamic sphere is characterised nowadays by an essential lag in the discussed problematics. Telecom data require involving of special approaches, methods, algorithms and tools for Big Data analytics and processing (Fig. 4.15):

- call options (different parameters of customer calls)
- customer profiles (personal customer data, duration of customer calls and other used services, billing records)
- data from financial reports, questionnaires, advertising, company plans, users' applications etc.

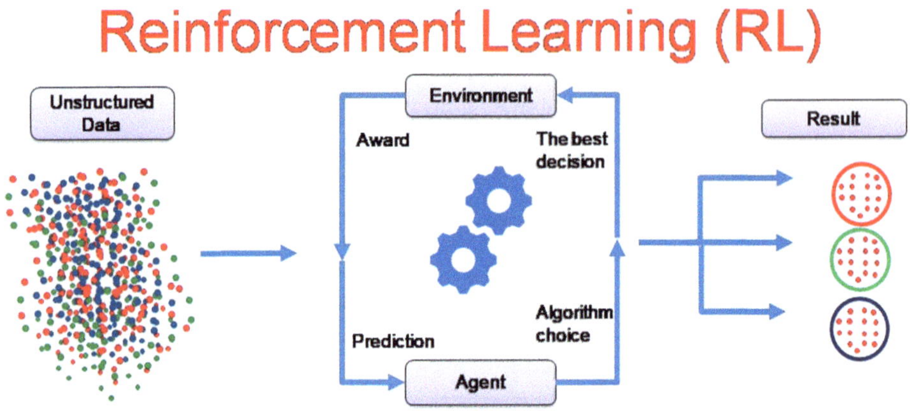

Fig. 4.14 Reinforcement Learning for Big Data Clusterisation

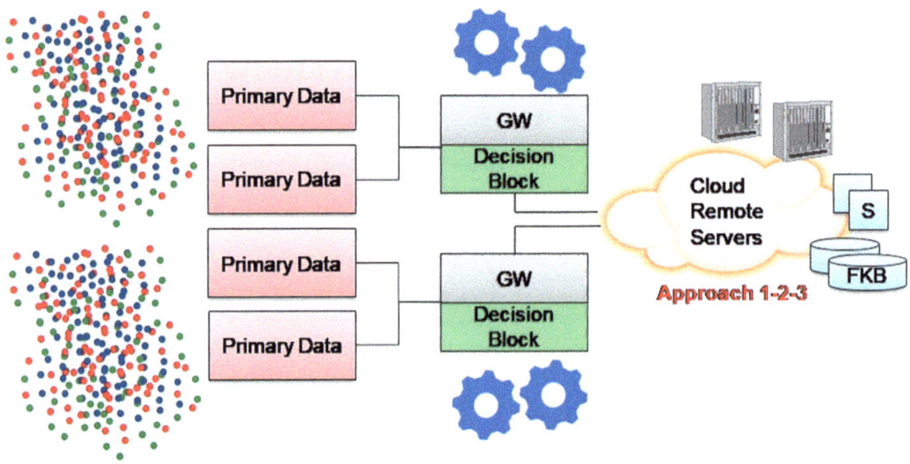

Fig. 4.15 Cloud architecture with the Decision Supporting Blocks used for primary telco data: so-called Approach 1-2-3 [3, 20]

Therefore, the following data taxonomies and data scales should be considered:

- amount of records
- amount of calling (and called) customers
- diversity in calling behaviour
- amount of used services and services delivery options (characteristics)
- transmitted video-volumes.

However, every analytic request uses only part of all these data. In this regard, there is a need to systematize and classify data from various sources. Ontological models are used as a problem solution [17]. This is the first step for "Big Data" overcoming. The second step is loading the data to execute the query via the GW (gateways) with available Analytics (Decision Blocks). This requires a dynamic workflow generation mechanism based on the ontology of data, knowledge and computational components. One of the approaches to designing such kind of mechanism is discussed in [15–17].

The third important step is fuzzy knowledge base (FKB) design for possibility to use not very hard computational process during the telco operation.

The huge amount of statistics can be classified and clustered using the algorithms considered in [19]. However, every cluster is only one fuzzy rule. The rule set is an FKB, but to check it correctness it is possible using so-called metagraphs. FKB correctness checking is discussed in [3, 14–20]. The obtained FKB can be used as the basis for a decision block for IoT or others telecoms platforms (refer Fig. 4.15).

4.9 Conclusions and Outlook

The given work is dedicated to Big Data sources in such modern issues like Industry 4.0, Robotics, IoT, 5G, Blockchain and best practices for analytics. The chapter discusses the Best Practises and Case Studies aimed to overcoming of the Big Data problematics if possible under use of "data compression" via their transformations:

- Own 6V-paradigm on Big Data distinguishing is formulated.
- Primarily the state-of-the-art approaches are analysed. The mostly common techniques classification and overview is given.

So-called "heavy-weighting" (ontologies, FKB, meta-graphs) and "light-weighting" approaches are examined.

- The Big data shortcomings for important areas (examples 1–5) were discussed like WSN, healthcare, telecom companies and robotics.
- Regular solution paradigms for Big Data are overviewed (e.g. R Tools, SPSS, Grafana).
- The (cloud and fog based) analytics blocks placement options to overcome Big Data are offered.
- Best Practices for Empirical Data Analytics for Big Data are discussed.

This chapter can be qualified as a "work-in-progress". The authors try to find new efficient method combinations to overcome the above-mentioned 6V-paradigm.

References

1. M.Ulema. Big Data and Telecommunications Telecom Analytics, Tutorial at Int. IEEE Conf. BlackSeaCom-2016, Varna, Bulgaria.
2. A. Konys, W. Rogoza. Big Data and Ontologies. Talk at ACS Int. Conf. 2016 in Miedzyzdroje, Oct. 2016, 3p.
3. A. Luntovskyy, L. Globa, B. Shubyn. From Big Data to Smart Data: the Most Effective Approaches for Data Analytics, in Advances in Information and Communication Technology and Systems, Springer Nature, Part of the "Lecture Notes in Networks and Systems Series", MCT 2019, Vol. LNNS 152, Chapter 2, pp. 23–40, Sept. 2020 (ISBN: 978-3-030-58359-0, DOI: https://doi.org/10.1007/978-3-030-58359-0_2).
4. E. Kuiler. From Big Data to Knowledge: An Ontological Approach to Big Data Analytics, Review of Policy Research, Vol.31, No.4 (2014).
5. S. Russell, P. Norvig. Artificial Intelligence: A modern approach, New Jersey, Upper Saddle River, 2010.
6. M. Jones. Artificial Intelligence: A Systems Approach, Hingham, Massachusetts, New Delhi, Infinity Sci. Press LLC, 2008.
7. B. Marr, J. Wiley. Big Data: Using SMART Big Data, Analytics and Metrics to Make Better Decisions and Improve Performance, Sons Ltd, 2015.
8. IBH Reports, IBH Dresden Workshop on 23.3.2017.
9. D.Blankenberg. Big Data in der Industrie 4.0, TIQ Solutions Leipzig, Workshop an der IBH Dresden, 13.3.2018.
10. TIQ Solutions Leipzig: https://www.tiq-solutions.de/.
11. R-Project: http://www.R-project.org/.
12. SPSS: http://www.ibm.com/analytics/de/de/technology/spss/.
13. Grafana: http://docs.grafana.org/.
14. L. Globa, I. Svetsynska, A. Luntovskyy. Case Studies on Big Data, Journal of Theoretical and Applied Computer Science, JTACS, Polish Academy of Science, Gdansk, No. 2, 2016, ISSN 2299-2634, 10p.
15. L. Globa, R. Novogrudska, A. Schill. The approach to engineering tasks composition on knowledge portals, Proc. of SPIE, Vol. 10445, 2017: https://www.scopus.com/record/display.uri?eid=2-s2.0-85058990971&origin=results.
16. L. Globa, M. Ternovoy, O. Shtogrina, O. Kryvenko. Based on force-directed algorithms method for metagraph visualisation, Advances in Intelligent Systems and Computing, vol.342, 2015, pp. 359–369.
17. L. Globa, R. Novogrudska, A. Koval. Ontology Model of Telecom Operator Big Data, Int. IEEEConf. BlackSeaCom-2018: https://ieeexplore.ieee.org/xpl/mostRecentIssue.jsp?punumber=8419227.
18. A. Luntovskyy, D.Guetter. Allgegenwaertige Vernetzung: Industrie 4.0, Internet der Dinge, Fog Computing u.v.m!, in BA Saxony Magazine "Wissen im Markt", Vol. 2/2018, Glauchau, 8p.
19. L. Globa, V. Kurdecha, I. Ishchenko, A. Zakharchuk. Fuzzy Logic Usage for the Data Processing in the IoT Networks, Proc. of OSTIS-2018, Minsk, Belarus, p. 287–292.
20. L. Globa, V. Kurdecha, I. Ishchenko, A. Zakharchuk, N. Kunieva. The Intellectual IoT-System for Monitoring the Base Station OoS, Int. IEEE Conf. BlackSeaCom-2018: https://ieeexplore.ieee.org/.
21. T. Maksymyuk, L. Han, S. Larionov, B. Shubyn, Luntovskyy, M. Klymash. Intelligent Spectrum Management in 5G Mobile Networks based on Recurrent Neural Networks, IEEE CADSM 2019, IEEE Xplore: https://ieeexplore.ieee.org/document/877930.

Green IT: Energy Efficient Constructions and Applications for Data Centres and Clusters

5

The presented chapter is devoted to an actual concept of "GreenIT" with the background of the "hardcore" resource consuming applications. The given work discusses the energy efficient constructions for computing centers and clusters, which provide optimization of performance as well as PW, PUE and ERE factors [1–6].

The following both approaches are examined:

- Workload consolidation and intelligent VM migration
- Optimization of the cooling process (guaranteeing PUE) and energy recycling (guaranteeing ERE and CO_2 footprint reduction).

The optimal ERE within the data centers and computing clusters provides advanced resource-consuming distributed applications like pharmacology simulations, Big Data analytics, Blockchain, crypto-currency mining.

5.1 Green IT problems and Challenges: Motivation

So-called "Green IT" concept stems from topical problems [1–4]:

1) Resources are limited and renewable energy is too expensive,
2) Electricity costs are rising permanently – partly because of IT – developing into a significant cost factor; individual devices indeed consume less energy; however, the overall count of consuming devices steadily increases due to clustering, cloud computing and IoT
3) Growth of "greenhouse gas" emissions and, especially, CO_2 footprints for IT, are to be mentioned.

A. Luntovskyy and D. Gütter, *Highly-Distributed Systems*,
https://doi.org/10.1007/978-3-030-92829-2_5

Therefore, providing a reasonable and sustainable Green IT concept is only possible under consideration of the following two groups of aspects [5–7]:

1. economic aspects, including
   - efficient use of hardware resources, energy efficient hardware, software and network protocols, energy efficient cooling processes in IT,
   - recycling of waste heat, as well as
2. environmental and social aspects, including
   - increased people awareness, reduction of electronic waste and electro-smog,
   - UN and EU climate programs and environment-friendly IT placement.

Surely, there are many politically motivated statements, too, such as the following:

- *"Not the CO2 emissions of factories, aeroplanes or cars, but the Internet is the climate killer " (translated from a 2017 article in Die Welt),*
- *"A single Google search query uses just as much electricity as a 10WCFL needs to light up for 108 s" (according to Google*[1]*), and*
- *"The data centres worldwide can generate so much CO_2 emissions as international air traffic" (claimed in a 2015 article in The Guardian).*

Besides these publicity claims and whether one believes them or not, the statistics are inexorable [5–13]:

- in 2017, IT services consume reached 3% of world's energy overall consumption,
- in 2025, IT services will consume 20% of the world's available energy!

IBH IT-Service GmbH, as an Internet service provider and system integrator, has many years of experience in the professional equipment and service-oriented operation of data centres. The following questions are often discussed between the provider and users [8]:

1. What standards must be applied to the construction of high-performance data centres?
2. How can standardised 19″-racks be optimally used?
3. What role do lightning and fire protection as well as energy efficiency play?

IBH represents the most modern design principles and important technical details on fail safe power supply, expansion of redundant communication channels and highly secure services [8]. Nowadays however, massively energy efficient computing power is required

[1]The data was provided by Google in 2009. Since then, the efficiency of data centres surely has increased. Nevertheless, it definitely debunks a often stated '1 hour per search' - claim to be found in the Internet and many publishing media.

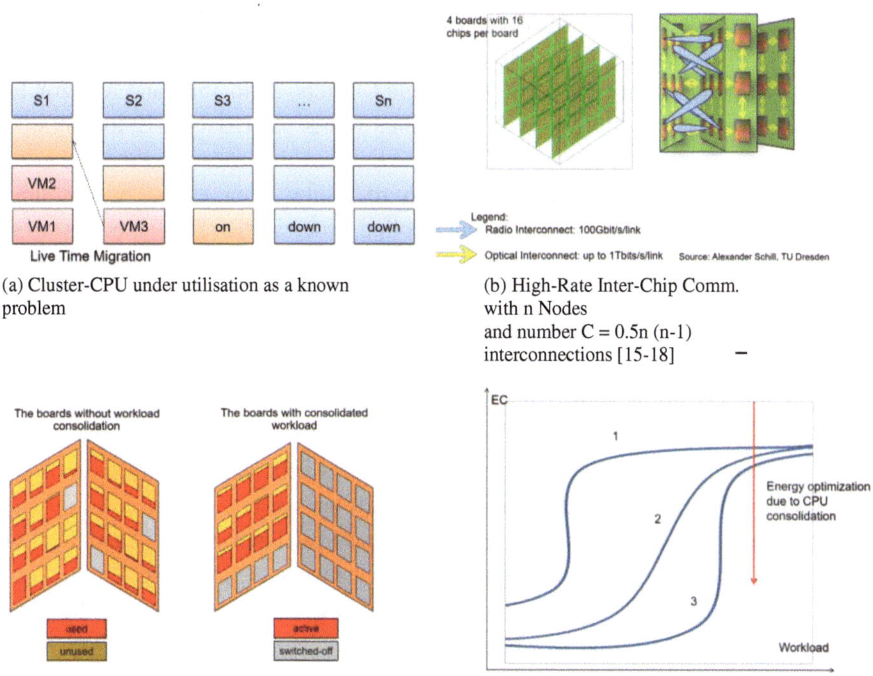

(a) Cluster-CPU under utilisation as a known problem

(b) High-Rate Inter-Chip Comm. with n Nodes and number C = 0.5n (n-1) interconnections [15-18]

(c) Workload Consolidation by Distributed Computing [15-18]

(d) Energy Optimisation due to CPU Consolidation by a growing Workload

Fig. 5.1 Migration and Clustering Effects (**a**) Cluster-CPU under utilisation as a known problem (**b**) High-Rate Inter-Chip Comm. with n Nodes and number C=0.5n (n-1) interconnections [15–18] (**c**) Workload Consolidation by Distributed Computing [15–18] (**d**) Energy Optimisation due to CPU Consolidation by a growing Workload

due to the deployment of new applications [1–5] such as data mining, AI, or Blockchain. This can be illustrated under use of the example of CLOUD&HEAT Technologies GmbH [14]:

1. What is the PUE (Power Usage Effectiveness) factor and how is it calculated?
2. What is the ERE (Energy Reuse Efficiency) factor and how is it calculated?
3. How do you achieve a higher level of CPU and GPU integration?
4. How is the server waste heat recycled to useful heat, e.g., in a building's hot water and heating supply?

CLOUD&HEAT delivers fully integrated solutions for secure data centres and computing infrastructures with higher energy efficiency. The projects enable real transition to a digital future through the deployment of powerful racks and highly integrated CPU containers with efficient water and air cooling as well as energy recycling.

There is a further well-known fact, too: cluster CPU under utilisation is a common problem in IT [1–5] (cf. Fig. 5.1a). Of their available CPU time, under utilisation can be quantified as Twitter Clusters: 40 to 50%, Google Clusters: 25 to 35%, and Amazon EC2 Clusters: approx. 20%. It often leads to suboptimal values of PUE: ca 1.12 to 1.89 in this area.

Many solutions have no predesigned energy recycling; often, excess heat is simply cooled away, however at least an efficient cooling process is used. Fig. 5.1a provides a scenario for VM migration[2] with three virtual machines: VM1, VM2, and VM3. It is aimed at load balancing, dynamic deployment, and efficient utilisation of the physical servers S1, S2, till Sn via live time migration[2].

Therefore, the next question might be formulated like this: What is energy efficient service execution? Current approaches to Green IT are herewith as follows [15–21]:

- design of multi dimensional, highly scalable, energy efficient multiprocessor servers with multiple CPUs per board and multiple boards per server,
- efficient inter processor communication via powerful links in radio and light frequency ranges for interaction and load balancing between the jobs, and
- flexible virtualisation and VM migration processes taking into account the power, load, network, storage usage as well as jobs structure in order to increase QoS.

Furthermore, the following challenges for Green IT can now be expressed:

1. Advanced efficient constructions for computing centres are required.
2. Workload consolidation and intelligent VM migration are necessary.
3. Optimisation of the cooling processes (guaranteeing PUE) and energy recycling (guaranteeing ERE and CO_2 footprint reduction) gain on the meaning.
4. The optimal ERE within the data centres and computing clusters provides advanced and – simultaneously – resource consuming distributed applications such as pharmacology simulations, Big Data analytics, Blockchain, or crypto currencies mining.

The remainder of this chapter's content follows as:

- State of the art and energy efficient constructions and approaches to performance and energy consumption, as well as QoS optimisation are discussed.
- Performance parameters and a generalised law for clusters speedup are examined under considering of the amount of the available processors or VMs and the workload.
- CLOUD&HEAT efficiency parameters are discussed.

[2] Live Time Migration: the VM remains active while it is migrated during run time. Service interruptions are minimised, ideally not occurring at all.

- The evaluation of the solutions is offered and proposed approaches are given.
- We investigate Blockchain and mining of crypto currencies implemented via powerful CPU + GPU clusters, which are operated under use of free of charge power or gained power via waste recycling due to higher ERE.

5.2 Best Practices for Energy-Efficient Constructions

In order to better understand the underlying principles and problem, the following example shall be considered.

Example 5.1:
Within the HAEC project at TU Dresden [15–18, 37] the best practices for energy efficient constructions were investigated. The cluster architecture examined by the HAEC team consists of a large set of CPUs with multiple radio interconnections (each link of 100 Gbps) as well as of multiple optical interconnections (each link of 1 Tbps) (cf. Fig. 5.1b). Due to the large count of participating nodes (n) the quantity of interconnections (C) grows quadratic. The given architecture solves the problem of high rate inter chip communication. However, the computing resources consume a large amount of energy even when idling (cf. Fig. 5.1c). The more complex the servers, the higher their idle power consumption. With the aim of achieving energy efficiency, switching off idle or under-utilised resources are of a big interest. HAEC's proposed algorithm of the workload consolidation by distributed computing is as follows:

1. Estimation the incoming workload.
2. Live time VM migration (migration of containers).
3. Workload consolidation enables the downtime for the under-utilised physical servers.
4. This migration involves image encoding and transfer from a source server to a destination server.

Surely, the energy gain depends on migration latency (MT) and energy consumption (MC). These parameters are dependable on the VM size and the available network bandwidth (BW). The energy costs of a VM migration is a function of MT and MC. Interestingly, in some their publications members of the HAEC team claim MC increases superproportionally as the available BW decreases linearly. Design of scalable multiprocessor servers with multiple CPUs and GPUs per board and multiple boards per server requires efficient interprocessor communication for interaction and load balancing between the executed jobs. In such cases, flexible virtualisation of execution engines is required. Migration policies taking energy, load, service quality and other parameters (such as memory or networking) as well as jobs structure into account [17–22].

As a future work, the scheduling mechanisms are necessary for detailed control and decision making concerning VM placement and migration to achieve load proportional

Table 5.1 Availability Boundaries and Classes [23]

#	Availability and annual downtimes			
	Yearly availability, %	Downtime yearly, Tdown	Units	Availability classes
1	90	36,52	days	1 – reliable up to 90%
2	95	18,26	days	
3	98	7,30	days	
4	99	3,65	days	2 – stable
5	99,5	1,83	days	
6	99,8	17,53	h	
7	99,9	8,77	h	3 – available
8	99,95	4,38	h	
9	99,99	52,59	min	4 – highly available
10	99,999	5,26	min	5 – insensitive to errors
11	99,9999	31,56	sec	6 – fault tolerant
12	99,99,999	3,16	sec	7 – error resistant

Average sidereal year with 365.256 days (with leap days)

energy consumption and to reduce absolute energy consumption and CO_2 footprint in total.

5.3 Performance Parameters and Energy Optimisation due to Load Consolidation

Within this Section the mostly important QoS-parameters of the clusters and data centers are discussed, among them the parameters are as follows [1–5, 8, 22–24]:

- Cluster availability classes 1–7,
- Cluster performance Pcluster,
- Cluster PW (Performance-2-Power, or Performance-2-W),
- Cluster average downtime T_{down},
- Cluster PUE (Power Usage Effectiveness),
- Cluster ERE (Energy Reuse Efficiency).

An explanation of such important computing cluster and data center parameters [8] like average downtime Tdown is summarized in Table 5.1.

Notice: A sidereal year has 365.256 days. The Gregorian leap day regulation is as follows: a leap year can be divided by: /4,/400; exceptions for divisibility are: /100 and 1000. The following classes 1–7 of service availability are given based on the sidereal yearunder consideration of the leap years. The given representation is based on [8–11] and depicts the services from the 90%-reliable clusters and data centers up to the error resistant systems. The Cluster Speedup shows acceleration and performance increasing with n available CPUs in relation to normal performance of 1 CPU:

$$\mathrm{An} = \mathrm{Pcluster}/\mathrm{P1} \tag{5.1}$$

where P1 is the performance on single CPU (ca. 100 GFLOPS for a convenient desktop PC). The parameter PW (Performance-2-W) represents the cluster efficiency by reaching of the specified performance:

$$\mathrm{PW} = \mathrm{Pcluster}/\mathrm{PIT} \tag{5.2}$$

where PIT is the nominal electrical power of the cluster.

5.4 Generalized Law for Cluster Speedup

The Sun-Ni's Law as the Generalized Law for Cluster Speedup [26–28] was developed in the early 1990s. It refers the memory bounds and increasing workload within a cluster and then leads to a generalization of both, Amdahl's and Barsis-Gustafson's models [10, 11]. Considering that:

$$1 = \mathrm{e} + \mathrm{p}, \qquad \mathrm{p} = 1 - \mathrm{e} \tag{5.3}$$

where p is the parallelism grade.

If n CPUs are available within a cluster for a workload W, then the parallel portion of the workload Wnew will be increased in G(n) times as a reflection of the increasing of the memory consumption:

$$\mathrm{Wnew} = \mathrm{eW} + (1 - \mathrm{e}) \times \mathrm{G(n)W} \tag{5.4}$$

where Wnew is the parallel portion of the workload W. Then the speedup which considers the memory bounds can be calculated as follows:

$\mathrm{An} = [\mathrm{eW} + (1 - \mathrm{e}) \times \mathrm{G(n)} \times \mathrm{W}] / [\mathrm{eW}/1 + (1 - \mathrm{e}) \times \mathrm{G(n)} \times \mathrm{W/n}]$, and thus:

$$\mathrm{An} = [\mathrm{e} + (1 - \mathrm{e}) \times \mathrm{G(n)}]/[\mathrm{e} + (1 - \mathrm{e}) \times \mathrm{G(n)/n}] \tag{5.5}$$

where p is the parallelism grade and $\mathrm{p} = 1 - \mathrm{e}$.

Depending on the growth factor G(n), there are three possible cases:

1. $\mathrm{G(n)} = 1$, like Amdahl's (cf. [1, 2, 10–12]):

$$\mathrm{An} = [\mathrm{e} + (1 - \mathrm{e})]/[\mathrm{e} + (1 - \mathrm{e})/\mathrm{n}] = 1/(\mathrm{e} + (1 - \mathrm{e})/\mathrm{n}) = 1/((1 - \mathrm{p}) + \mathrm{p/n}) \tag{5.6}$$

2. G(n) = n, like Barsis-Gustafson's (cf.[1, 2, 10–12]):

$$\begin{aligned} An &= [e + (1 - e)n]/[e + (1 - e)n/n] = e + (1 - e)n = (1 - p) + pn \\ &= 1 + (n - 1)p \end{aligned} \tag{5.7}$$

3. G(n) = m > n, for Sun-Ni's Law:

$$\begin{aligned} An &= [e + (1 - e) \cdot m]/[e + (1 - e) \cdot m/n] = [e + (1 - e) \cdot m]/[e + (1 - e) \cdot m/n] \\ &= [1 - p + pm]/[1 - p + p \cdot m/n] \end{aligned} \tag{5.8}$$

Sun-Ni's Law's formula is used for the case that the workload grows faster than memory. The speedup is slightly higher than Barsis-Gustafson's, i.e. the ratio (m/n) = 1.1 … 1.5.

Data center virtualization technologies have attracted a lot of attention to enable various cloud computing services and to facilitate VM migration.VM migration can help service providers to achieve the goals of saving energy, enhancing resource efficiency and quality of service (QoS).

In order to ensure the QoS, the migration time and the downtime of VM should be considered while implementing the VM migration.

5.5 Workload Consolidation and VM Migration

Computing resources consume a large amount of energy even when they are idle. The workload consolidation and VM migration leads [15–21] to the decreasing of idle times, overall costs and energy overhead (waste) but optimization of the Migration Time (MT), Migration Power (MP), Migration Energy Consumption (EC) must be also taken into account. The following VM migration models are examined and represented in the literature; refer (5.9) and (5.10). Linear regression is used as a rule [17–19]:

$$EC = MP \cdot MT \,|\, EC = a \cdot Vmig + c \,|\, EC = a + b \cdot Vload + c \cdot BW \tag{5.9}$$

where {a, b, c} are constants, Vmig is the volume for a migrating VM, Vload is the workload volume, and BW is the bandwidth of the inter-CPU-links (e.g. Fibre Channel with 16 Gbps).

$$MT = a + b \cdot INST + c \cdot VL3\text{-}cache \tag{5.10}$$

where {a, b, c} are constants, INST is the volume of the processed instructions during a VM migration, and VL3-cache is the volume of the processed lines of L3-cache during a VMmigration. Therefore, flexible migration is possible when considering the power, load, network, and storage usage as well as jobs structure. Only under such conditions QoS increasing by a growing workload within the cluster can be reached.

The energy optimization due to CPU consolidation is depicted in Fig. 5.1 (a–d). The depicted curves 1, 2, and 3 represent in Fig. 5.1(d): 1 – without VM migration, 2 and 3 – with (partial) migration (less energy by more workload).

5.6 Energy Efficiency Parameters and Energy Recycling

5.6.1 Energy Efficient Computing and Applications

In addition to the performance parameters such as TFLOPS and PFLOPS, the experts speak more and more about energy efficiency in Cluster Computing, namely about the values [1–13], such as PUE (Power Usage Effectiveness) andERE (Energy Reuse Efficiency), which are defined as follows:

$$\mathrm{PUE} = (\mathrm{PIT} + \mathrm{PA})/\mathrm{PIT} \tag{5.11}$$

$$\mathrm{ERE} = (\mathrm{PIT} + \mathrm{PA} - \mathrm{PR})/\mathrm{PIT} \tag{5.12}$$

where PIT is the necessary power for IT and computing in the cluster, PA is the waste heat power, and PR is the recycled power. The first one describes effectiveness of the power usage within a data center cluster under considering of the waste heat PA. The second represents the efficiency of the energy recycling PR under considering of the PA value. The branch's best [3–6] computing centers and clouds delivers PUE = 1.02…1.12. By significant improvement [5] via cooling and energy recycling ERE can be estimated in the vicinity of ERE = 1.02…1.06.

5.6.2 Best Practices for Energy Efficiency: Hot Water Cooling

The solutions for Green IT [8–15], energy efficiency for computing clusters and data centers have been in the press since more than a decade, for example: Google's Data Centers in The Dalles, OR and Douglas County, GA in 2010 (cp. Fig. 5.2 (a) and (b)), CoolMUC-2 in Munich in 2016, Mäntsälä in Finland in 2017, or hot water cooling at TU Dresden's Centre for High Performance Computing in 2017.

Currently, CLOUD&HEAT [6, 14] belongs to the few Cloud providers following the Green IT best practices in Germany, and maybe worldwide, too:

(a) Cooling Towers at Google's Data Center in The Dalles, OR (Photo: Google) [4,5]

(b) Central cooling plant in Google's Data Center in Douglas County, GA (Photo: Google/Connie Zhou) [4,5]

Fig. 5.2 Best Practices for Energy Efficiency: Cooling Processes (**a**) Cooling Towers at Google's Data Center in The Dalles, OR (Photo: Google) [4, 5] (**b**) Central cooling plant in Google's Data Center in Douglas County, GA (Photo: Google/Connie Zhou) [4, 5]

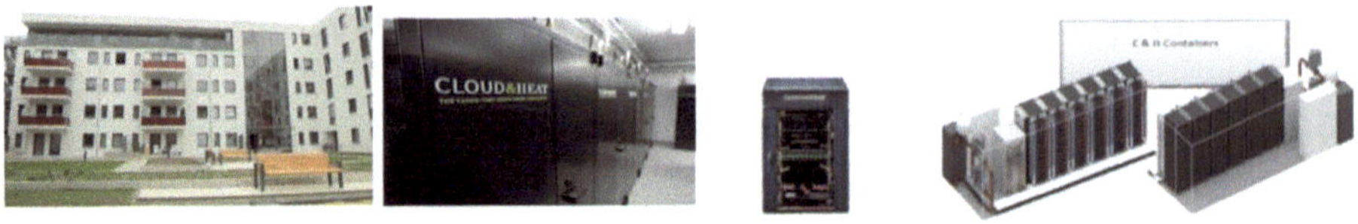

CLOUD&HEAT Datacenter within the basement of a regular civic building in Dresden (Photo courtesy of CLOUT&HEAT Technologies GmbH, Dresden, Germany) [6,14]

a) CLOUD&HEATwater-cooled rack
b) CLOUD&HEAT water-cooled datacenter containers in two constructive variants

CLOUD&HEAT energy efficient datacenter hardware [6,14]

Fig. 5.3 Best Practices of CLOUD & HEAT CLOUD&HEAT Datacenter within the basement of a regular civic building in Dresden (Photo courtesy of CLOUT&HEAT Technologies GmbH, Dresden, Germany) [6, 14]. **a** CLOUD&HEAT water-cooled rack. **b** CLOUD&HEAT water-cooled datacenter containers in two constructive variants CLOUD&HEAT energy efficient datacenter hardware [6, 14]

- innovative and highly efficient hot-water cooling system,
- fully water-cooled hardware with waste heat reuse (to heat of buildings directly with servers running in the basement instead of a district heating connection), and
- integrated hydraulic system to reuse server waste heat by approx. 55 °C on the processor surface and heat disposal.

Whole datacenters can be collapsed to an energy-efficient box, or specialized water-cooled 19″-rack, or container for a lot kW of recycled waste heat (refer Fig. 5.3ab). The mentioned hardware solution consists of a freely scalable number of server blades and storage units: Intel Xeon technology with up to 400 cores per rack, up to 360 TB storage with redundant storage servers etc. The systems are redundantly dimensioned for all essential components for failure-tolerant operation.

5.7 Blockchain: Energy-Efficient Mining of Crypto-Currencies is available?

Blockchain is a cryptographical concept supporting decentralized applications with demand of consistency, integrity, and accountability assurances. Such decentralized applications include, but are not limited to, payment systems, financial online transactions, or crypto-currencies. By design, the Blockchain is a peer-to-peer (P2P) concept. The economic success of this cryptotechnology will be evident in the next 10 up to 20 years. The deployment of Blockchain technology speaks mainly for a decentralized financial system. The advantages of such a solution seem obvious [6, 14, 29–32]:

- sustainability, general transparency and commitment,
- accelerated economic workflows and digitization processes (so-called "IT in the digital age"), and
- it is also well suited to supporting current crypto-currencies (such as Bitcoin, Ethereum, Ripple, Litecoin, ZCash, Monero, Stellar, etc.).

5.7.1 Blockchain Architecture at Glance

Decentralized, cryptographically secured and unified blocks, their chains and transactions are grouped under a general, global public ledger (account), the structure of which is as follows: The Blockchain, as a networked Public Ledger, consists of participating nodes that represent an efficient P2P communication model. Typical features of the Blockchain are [33–36]:

- redundancy and synchronization,
- cryptographic hash procedures for integrity assurance and attack safety,
- decentralized management and control of the Blockchain,
- network subscribers are also referred to as Nodes (Full-Nodes, Miners, Validators [29–36]) and run redundantly with mutual synchronization, and
- often also considered a (anti-)feature: large block volumes can cause "Big Data" problem.

The different Nodes are defined as:

- *(Generic) Node*: All Blockchain participants (computers, smartphones, tablets, clusters, …) are qualified as Node, if they have installed the corresponding software.
- *Full-Node*: A Node with a full (local) copy of the Blockchain; it checks for so-called "consensus rules."
- *Miner* and *Validator*: The individual participants or mining pool (with high resource requirements regarding hardware and energy consumption); they finalize blocks, where in a Miner "generates" a new block by calculating valid hashes, and a Validator proves the validity of the new block. Externally, they each act like a large participant; however, in fact many small blocks are generated for fractions of the system units (e.g., for small payments within a crypto-currency).

However, the following problems may occur during Blockchain operations:

- enormous energy consumption due to "mining" of valid blocks (processing of the hash blocks via its algorithmic complexity),
- exponential memory growth (including capacity migration between USB media, smartphones, PC, storage media such as SAN/NAS, as well as cloud storages), and
- privacy issues may arise even though cryptographic data security is guaranteed.

Table 5.2 Features of a Mining Device (source: amazon.de)

Features of AntMiner S15	High-Performance Mode	Low-Energy Mode
Hash rate	28 TH/s	18TH/s
El. power	1596 W	900 W
Power Efficiency	57 J/TH	50 J/TH
NW connection	Ethernet	Ethernet
Weight and dimensions	7 kg, 240 mm × 178 mm × 296 mm	

A solution may be not to process the complete Blockchain with all the transactions, but only use of excerpts of the Blockchain without a prehistory.

5.7.2 Mining of Crypto-Currency and Resource Consumption

At least eight factors are necessary and must be considered when calculating the profitability of the mining [29–36]:

1. *Investment* (basis device costs): hardware investment for a mining rig or cluster.
2. *Electricity Costs*: in real life currency per kWh (e.g. €/kWh).
3. *Energy Consumption*: electrical power of Mining Rig in KW.
4. *Hash Rate*: hash values computed each second.
5. *Mining Difficulty*: factor has always a new actual value.
6. *Pool Fees*: how much in % belongs to the joined pool.
7. *Block Reward*: monetary or ideological compensation for each new computed block (or for other units).
8. *Unit Price*: exchange course for the BTC (or other units like XMR, ETH, ZEC, or LTC).

There are different multi-purpose and specialized devices for mining available, so-called Mining Rigs. Standard PCs or smartphones can also be used, but considering the energy consumption problems, they provide rather more disadvantages than justifiable use. The following types of devices can be deployed to generate the hash values in the mining process:

- CPU mining: powerful processor is required,
- GPU mining: powerful graphics card is required, or
- ASIC mining: using *A*pplication-*S*pecific *I*ntegrated *C*ircuits (cf. Table 5.2).

Example 5.2:
Using AntMiner as mining rig [37], which can calculate 28 TH/s with a power consumption of 1596 W: The average daily gain from this performance is 0.00115956 Bitcoin. The contribution to the joined mining pool is 1.5%. The current value of a Bitcoin is

6,640 €.[3] In Germany, the average electricity price is 0.3137 €/kWh.[4]When looking for the expectable investment to conduct a week's mining one shall consider:

- the expenditures E for the used electricity,
- gain/proceeds from the BitcoinsGB,
- efficiencyEff (in J/TH), and
- gain or losses (in €) GL per week of such mining.

The purchase costs are negligible. As a result, one can calculate:

$$\begin{aligned}
\Rightarrow \mathrm{E} &= 1.596\ \ \mathrm{kW} \times 168\ \mathrm{h/week} \times 0.3137\ €/\mathrm{kWh}\\
&= 268.128\ \mathrm{kWh/week} \times 0.3137\ €/\mathrm{kWh}\\
&= 84.1117\ €/\mathrm{week}\\
\Rightarrow \mathrm{GB} &= 0.00115956\ \mathrm{Bitcoin/d} \times 7\ \mathrm{d/week}\\
&\qquad \times 6640\ €/\mathrm{Bitcoin}\\
&= 53.8963\ €/\mathrm{week}\\
\Rightarrow \mathrm{GL} &= 53.8963\ €/\mathrm{week} - 84.1117\ €/\mathrm{week}\\
&= -30.2154\ €/\mathrm{week}\\
&\text{(losses due to negative value)}\\
\Rightarrow \mathrm{Eff} &= 1596\ \mathrm{W}/28\ \mathrm{TH/s} = 57\ \mathrm{J/TH}
\end{aligned}$$

A simple cross-multiplication of Eff shows that AntMiner rig would need to at least calculate 43,7 TH/s in order to cover the cost and turn a marginal profit.

The AntMiner rig has an extra energy saving mode. In that mode, the rig calculates only 18 TH/s with a reduced power consumption of 900 W. The process results in gain of 0.00074543 Bitcoin per day. All other parameters remain unchanged. Then the new calculation derives:

$$\begin{aligned}
\Rightarrow \mathrm{E} &= 0.9\ \mathrm{kW} \times 168\ \mathrm{h/week} \times 0.3137\ €/\mathrm{kWh}\\
&= 47.4314\ €/\mathrm{week}\\
\Rightarrow \mathrm{GB} &= \ 0.00074543\ \mathrm{Bitcoin/d}\ \times 7\ \mathrm{d/week}\\
&\qquad \times 6640\ €/\mathrm{Bitcoin} = 34.6476\ €/\mathrm{week}\\
\Rightarrow \mathrm{GL} &= 34.6476\ €/\mathrm{week} - 47.4314\ €/\mathrm{week}\\
&= -12.7838\ €/\mathrm{week}\\
\Rightarrow \mathrm{Eff} &= 900\ \mathrm{W}/18\ \mathrm{TH/s} = 50\ \mathrm{J/TH}
\end{aligned}$$

[3]As of 18 April 2020.

[4]National average calculated by the German Federal Grid Agency. Given price represents average as of 17 April 2020.

Once again, the rig produces a net-loss.

The interesting question fora private individual is as follows: How high does the price PKWH for 1kWh have to be, so that mining is rentable for a given situation?

- High-Performance mode:

$$\begin{aligned} \Rightarrow \text{GB} &= 53.8963 \text{ €/week} \\ &= 268.128 \text{ kWh/week} \times \text{PKWH} \\ \text{PKWH} &= 0.2010 \text{ €/kWh} \end{aligned}$$

- Low-Power mode:

$$\begin{aligned} \Rightarrow \text{GB} &= 34.6476 \text{ €/week} \\ &= 151.2 \text{ kWh/week} \times \text{PKWH} \\ \text{PKWH} &= 0.2291 \text{ €/kWh} \end{aligned}$$

This practical example shows that in a lot of cases mining of the crypto-currencies like BTC, ETH, ZEC, XMR, etc. leads to "no reward"-scenarios due to the large energy consumptions in combination with the unsustainable electricity price. Additionally, the initial purchase costs for the rigs cannot be considered negligible. In contrast, positive results can be reached if the powerful CPU+GPU or ASIC chip clusters are used with access to low-cost renewable energy,[5] or can gain energy via waste recycling (both with minimal PKWH). A cluster with an optimal ERE value can use such recycled energy. The gained PR can then be targeted for crypto-currency mining.

Example 5.3:
Considering a no-name XMR mining operation using a similar container concept [37] like it was described earlier [6, 14, 32–34]. The mostly important container features are—as usual—the electrical power and water proximity. E.g., PIT+PA=2.4 MW for overall electrical input, PA=0.25 MW for waste heat, as well as PR=0.15 MW can be recycled additionally. Therefore: PUE=2.4 MW / 2.15 MW=1.116.

$$\text{ERE} = (2.4 \text{ MW} - 0.15 \text{ MW})/2.15 \text{ MW} = 2.25 \text{ MW}/2.15 \text{ MW} = 1.047$$

[5] Such as hydro-electric power in Norway, or geo-thermal power in Iceland. Both locations are favored by many crypto currency mining operations.

For 1 ASIC chip:

$$\Rightarrow \text{E} = 47.4314 \text{ €/week}$$

$$\Rightarrow \text{GB} = 0.00521801 \text{ Bitcoin/d} \times 7 \text{ d/week} \times 6640 \text{ €/Bitcoin} = 242.5331 \text{ €/week}$$

$$\Rightarrow \text{GL} = 195.1017 \text{ €/week} \quad \text{(profit)}$$

For a heat waste recycling cluster with

$$\text{PR} = 0.15 \text{ MW}, \text{Pchip} = 900 \text{ W}, \text{Nchips} = \text{PR/Pchip} = 166{,}666.7$$

Hence, with 166,667 ASIC chips and reutilized waste heat:

$$\Rightarrow \text{GB} = 242.5331 \text{ €/week} \times \text{Nchips} = 40,422,264.1777 \text{ €/week}$$

$$\Rightarrow \text{GL} =\sim 40.4 \text{ million €/week} \quad \text{(pure profit)}$$

The calculation results in a pure profit as the expenditure E can be set as E = 0 since the investment into the power costs is calculated against the primary task of the cluster rather than the "waste-product" of Bitcoin mining.

A 40'-standard container can include ~100,000 ASIC chips, which are divided into multiple rows of 19"-racks including liquid cooled servers, pump groups for high and low temperature levels, heat-exchangers and controlling units, as well as the hydraulic pipe systems [6, 14] (cf. Fig. 5.3). Such CLOUD&HEAT data centers and computing clusters are deployed no further than 30 km away from the cloud users [4, 5] because of a targeted latency of $\Delta < 1$ ms. These energy-efficient containers with data centers and computing clusters can be implemented in two variants:

- variant 1: PIT = 500 kW in a 20' shipping container;
- variant 2: PIT = 1 MW in a 40' shipping container [6, 14].

5.8 Conclusions and Outlook

1. Advanced efficient constructions for computing centres are required.
2. Workload consolidation and intelligent VM migration are necessary.
3. Optimisation of the cooling processes (guaranteeing PUE) and energy recycling (guaranteeing ERE and reduction of the CO_2 footprint) are increasingly important.
4. Optimised ERE within data centres and computing clusters provides advanced and, simultaneously, resource consuming distributed applications, such as pharmacology simulations, Big Data analytics, Blockchain, or crypto currencies mining.

References

1. A. Luntovskyy, J. Spillner. Architectural Transformations in Network Services and Distributed Systems: Service Vision. Case Studies, Springer Nature, 2017, 344p. (ISBN: 9-783-6581-484-09).
2. A. Luntovskyy, M. Klymash. Software Technologies for Mobile Apps, Apps for Fog Computing, Robotics and Cryptoaps, Lviv, 2019, 247 p. (Monograph, in Ukrainian, ISBN 978-617-642-399-7).
3. A. Luntovskyy, D. Guetter. Moderne Rechnernetze - Theoriebuch: Protokolle, Standards und Apps in kombinierten drahtgebundenen, mobilen und drahtlosen Netzwerken, Springer Fachmedien Wiesbaden GmbH, 15 July 2020, ISBN: 978-3-658-256173 (in German, Vorwort Alexander Schill), 419 p.
4. 2020 Annual Data Center Industry Survey, Uptime Institute (Online): https://uptimeinstitute.com/2020-data-center-industry-survey-results/.
5. Is PUE actually going up? Uptime Institute, May 2019 (Online): https://journal.uptimeinstitute.com/is-pue-actually-going-up/.
6. Green IT workshop at BA Dresden, on 26.10.2018 (Online): https://www.ba-dresden.de/.
7. Lutz Zipfel. Green IT Problems: Welcome Word, Workshop Green IT at BA Dresden, Oct. 2018 (in German).
8. T. Horn. Energieeffiziente RZ in Dresden und Sachsen, IBH IT Service GmbH Dresden, Workshop Green IT at BA Dresden, Oct. 2018 (in German).
9. A. Schill. Energy Efficient Service Execution, TU Dresden Workshop Green IT at BA Dresden, Oct. 2018 (in German).
10. J. Struckmeier. Cloud&Heat Data center for AI, AR and Big Data: Secure, scalable and sustainable, CLOUD&HEAT Dresden, Workshop Green IT at BA Dresden, Oct. 2018 (in German).
11. D.Guetter. Green IT: Versorgungssicherheit und Effizienz in Stromnetzen, Workshop Green IT at BA Dresden, Oct. 2018 (in German).
12. IBH IT-Services (Online): http://www.ibh.de/.
13. HPC Clusters in Dresden, ZIH@TUD (Online): http://zih.tu-dresden.de/.
14. CLOUD&HEAT Technologies (Online): https://www.cloudandheat.com/.
15. Alexander Schill. Manuscripts of the Chair of Computer Networks (Online): http://www.rn.inf.tu-dresden.de.
16. HAEC Project with TU Dresden, SFB 912 for Highly Adaptive Energy- Efficient ComputingOnline): https://tu-dresden.de/ing/forschung/sonderforschungsbereiche/sfb912.
17. K. Rybina, W. Dargie, S. Umashankar, A. Schill. Modelling the Live Migration Time of Virtual Machines; OTM Conferences 2015, in Springer International Publishing Switzerland, 2015.
18. K. Rybina, A. Patni, A. Schill. Analysing the Migration Time of Live Migration of Multiple Virtual Machines, in Int. Conference on Cloud Computing and Services Science (CLOSER 2014), April 3–5, Barcelona, Spain, 2014.
19. Gang Sun et al. A new technique for efficient live migration of multiple virtual machines, in Future Generation Computer Systems, vol.55, 2016, pp. 74–86.
20. Kejiang Ye et al. Live Migration of Multiple Virtual Machines with Resource Reservation in Cloud Computing Environments, in IEEE Int. Conf. Cloud Computing (CLOUD 2011), ISBN: 978-1-4577-0836-7, pp.267–274.
21. T.K. Sarker, Performance driven live migration of multiple virtual machines in datacenters, in IEEE Int. Conf. on Granular Computing (GrC-2013), Beijing, 2013, pp. 253–258.

22. A. Luntovskyy, D. Guetter, M. Klymash. Up-to-date Paradigms for Distributed Computing, Int. IEEE Conf. AICT-2017 on Advanced Information and Comm. Technologies, July 2017, Lvyv, pp. 113–119 (IEEE Xplore), ISBN: 978-1-5386-0638-4, DOI: https://doi.org/10.1109/AIACT.2017.8020078.
23. H. Jurzik, D. Ahmke. Verfuegbarkeitswerte und Verfuegbarkeitsklassen, 2007, p. 7 (in German).
24. A. Luntovskyy. Performance and Energy Efficiency in Distributed Computing, 20th Int. Multi-Conf. ACS-2016 (On Advanced Computer Systems), 12 p., in "Hard and Soft Computing for Artificial Intelligence, Multimedia and Security", Springer Int. Publishing, edited by Shinya Kobayashi, Andrzej Piegat, Jerzy Pejas, Imed El Fray, JanuszKacprzyk.
25. A. Luntovskyy, L. Globa. Performance, Reliability and Scalability for IoT, Int. IEEE Conf. IDT-2019, Zilina, Slovakia, 25–27 June 2019, 6 p. (IEEE Xplore: https://ieeexplore.ieee.org/document/8813679, DOI: https://doi.org/10.1109/DT.2019.8813679).
26. X.H. Sun, L. Ni. Another View on Parallel Speedup, in Proceedings of IEEE Supercomputing Conferences, 1990.
27. X.H. Sun, L. Ni. Scalable Problems and Memory-Bounded Speedup, Journal of Parallel and Distributed Computing, Vol. 19, p. 27–37, Sept.1993.
28. Sun-Ni Speedup Law with memory bounds and increasing workload (Online): https://www.d.umn.edu/~tkwon/course/5315/HW/MultiprocessorLaws.pdf.
29. A. Antonopoulos, G. Wood. Mastering Ethereum: Building Smart Contracts and Dapps, 2019, O'Reilly Media, 345p. ISBN: 978-1491971-949.
30. A. Luntovskyy, D. Guetter. Cryptographic Technology Blockchain and its Applications, in "Advances in Information and Communication Technologies", Springer (ISBN: 978-3-030-16770-7), LNCS "Processing and Control in Information and Communication Systems (Int. Conf. UkrMiCo-2019)" (eds.: M. Ilchenko, L. Globa et al.), 2019, pp. 14–33. (URL: https://link.springer.com/book/10.1007/978-3-030-16770-7).
31. Mining Process (Online): https://coin-hero.de/was-ist-mining-und-wie-funktioniert-es/.
32. Mining Algorithms (Online): https://coin-hero.de/mining-serie-teil-2-kryptowaehrungen-und-deren-mining-algorithmen/.
33. A. Luntovskyy, B. Shubyn.Highly-Distributed Systems Based on Micro-Services and their Construction Paradigms, IEEE Conf. TCSET 2020, Lviv-Slavske, Ukraine, Feb-2020, 8p.
34. A. Luntovskyyy, B. Shubyn, I. Scherm. Blockchaining for modern HDS, BA Magazine "Wissen im Markt", Year 3, Issue Nov. 2019, pp. 38–44 (ISSN 2512-4366).
35. Codius: Open source Hosting Platform for Smart Programs (Online): https://codius.org/.
36. Hyperledger Sawtooth (Online): https://www.hyperledger.org/projects/sawtooth/.
37. Andriy Luntovskyy, Tenshi Hara. Green IT: Energy Efficient Constructions and Applications for Data Centers and Clusters, ACS-2021 in Bydgoszcz, 28–30 June 2021 (online), in Springer Lecture Notes in Networks and Systems series, "Progress in Image Processing, Pattern Recognition and Communication Systems", Proceedings of CORES, IP&C, ACS-2021 (by M.Choraś, R.S.Choraś, M.Kurzyński, P.Trajdos, J.Pejas, T.Hyla, Eds.), due: 6 November 2021, ISBN 978-3-030-81522-6, 13p.

Part II

HDS FOR INTERNET OF THINGS AND ROBOTICS

Internet of Things: Architectures and Basic Technologies

6

IoT apps and applications are a very important part of the examined Highly-Distributed Systems. For the given chapter the following content is foreseen:

1. Motivation and Overview
2. Deployment Areas
3. Architectures and Basic Technologies at Glance
4. Development of IoT Software
5. Systems for Data Acquisition—Data Processing—Data Mining
6. IoT in Traffic Telematics
7. IoT and Smart Energy: Smart Grid Generations, Smart Home and Smart City
8. Sensor Nets and Swarm Intelligence
9. IoT and Fog Computing: Definition and Demarcation
10. Co-operation "Fog-2-Cloud"
11. IoT System Development
12. Case Study: A Freeware MQTT Solution for Fog.

The chapter examines the basics for IoT, introduces IoT devices for different use cases, as well as represents typical approaches and basic programming tools for IoT system development.

6.1 Motivation and Overview

The Internet of Things (IoT) is part of the Internet of People/Services (see Fig. 6.1). So-called IoT contains a collection of smart devices (i.e. independent, intelligent, communication-capable microsystems) and sensors/actuators. Nowadays, Wireless IoT, which is integrated via various radio interfaces, e.g. via WSN (Wireless Sensor

A. Luntovskyy and D. Gütter, *Highly-Distributed Systems*,
https://doi.org/10.1007/978-3-030-92829-2_6

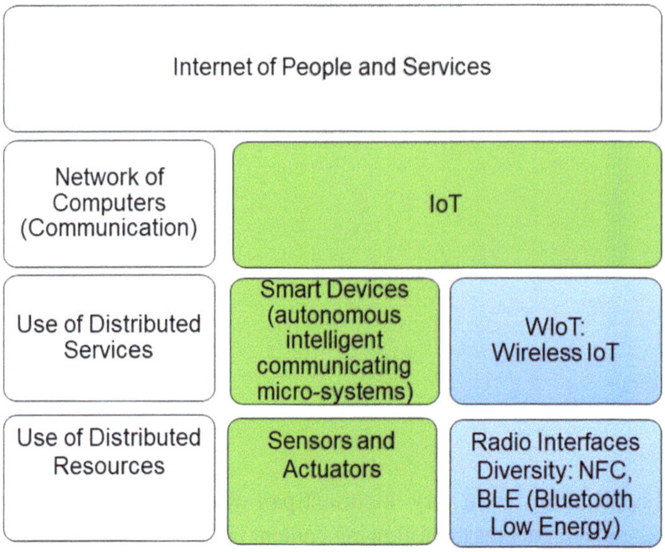

Fig. 6.1 Overall classification of IoP, IoS and IoT (Internet of People/Services/Things)

Networks), NFC (Near Field Communication), BLE (Bluetooth Low Energy), plays a growing role and are particularly up-to-date. The smart devices offer attractive smart services for [1–4]:

- Data Monitoring
- Data Acquisition
- Data Exchange
- Data Processing
- Data Mining
- Data Analysis.

The following examples of IoT deployment can be considered [5–7]:

- Smartphones
- Tablets
- Voice assistants (Alexa & Co)
- Smart watches/fitness trackers/smart glasses
- further Smart Objects and Smart Stuff.

A following layer model was established for modelling of smart service platforms (Fig. 6.2). The five layers are presented and supported for the following three strata:

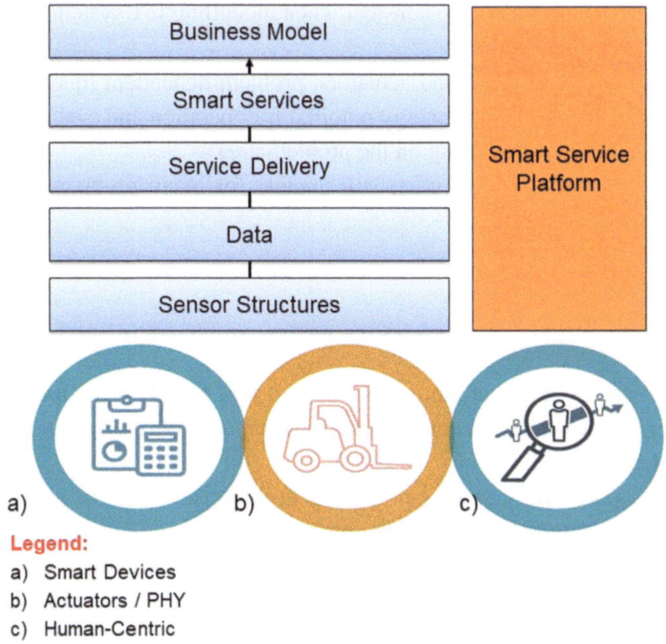

Fig. 6.2 Smart Service Platform [4]

Table 6.1 IoT project phase concept

Phase	Title	Content
Phase 1	Data Acquisition	Selection of the right data sources in terms of reliability, quality and completeness
Phase 2	Data Processing	Selection of the correct data and definition of the further processing formats
Phase 3	Data Mining	Data acquisition and pattern recognition using mathematical processes
Phase 4	Predictive Analytics	Prediction based on experiences
Phase 5	AI (Artificial Intelligence)	Process optimization [8, 9]

(a) Smart Devices
(b) Actuators/PHY
(c) Human-Centric.

The design of IoT systems includes the networking and configuration of the interactions between IoT devices. It takes place in the following rough project phases (Table 6.1), which can be divided internally into finer phases, if necessary:

There are a number of issues to consider when modelling IoT systems:

- Required number of sensors and actuators, problem-dependent (quality vs. cost), the more finely divided, the more energy required for operation and evaluation
- Real-time suitability (depending on the problem area)
- Modelling and model accuracy are still unclear for many problems (weather, landscape, concentration …).
- Operational problems
- Maintenance efforts and costs
- Disposal of the IoT devices.

In summary, it can be said that there is currently a very broad discussion of the problem with good approaches in various industries. However, a lot of research is still required on concrete modelling [1–5].

The current approaches are a lot promising in industrial/technical areas, but in some cases still very controversial in the social area.

One of the main problems for IoT systems design and deployment is often immense amount of accumulated data that is known under buzzword: so-called "Big Data". In addition, big data is usually weakly structured, short-lived and fast moving. The features of big data can be characterized using so-called "6V paradigm" [10]:

1. Volume (data are voluminous)
2. Velocity (data are generated, transferred, processed with a higher velocity)
3. Variety (multiplicity of data types and sources)
4. Veracity (trustiness of data)
5. Value (of user view point)
6. Validity (data quality).

The use of IoT systems frequently obstructed by the following frame conditions:

- thousands of sensors (huge amounts of data per unit of time)
- often no trivial parameter dependencies.

An important feature of IoT systems is the use of energy self-sufficient, energy autarky, particularly small smart devices. In particular, there are high demands on sensors and actuators for low power consumption [11–15].

Battery operation is usually only possible to a limited extent, therefore, the methods of energy harvesting, i.e. energy generating/winning from the environment are becoming increasingly important. Energy harvesting may allow years of operation without changing batteries instead of some weeks or months.

In the case of macro objects, energy is generated from alternative sources like:

- photovoltaics
- wind turbines
- waste heat removal from powerful data centres (for example, from Cloud & Heat Dresden).

So-called Micro Energy Harvesting [5–9, 16, 17] uses (maximally up to few mW), among other things, the following types of energy conversion (Table 6.2):

Industrial Internet of Things (IIoT) are special IoT and the basis for the Industry 4.0 concept [1, 2, 9, 18], which means a gigantic market for smart manufacturing as follows:

- Industrial plants
- Robotics
- Agriculture
- Trade and logistics
- Traffic
- Service industry
- Healthcare
- Weather service and hazard monitoring (tsunami, hurricane).
- Smart City.

IIoT possesses the following application benefits:

- Improving efficiency
- more flexible operation

Table 6.2 Micro Energy Harvesting and Energy Conversion Types

Energy type	Energy conversion	Convertor
Mechanical	Kinetic energy, vibration, pressure, stretching, oscillation	Electrodynamic converters such as piezo elements
Thermal	Thermal energy, heating, cooling, friction	Thermal converters such as Peltier elements, heat pipes
Light Energy	Room lighting, sunlight, other light sources	Photovoltaic converters such as solar cells, photodiodes
EM	Coils, conductors, HF radiation	Antennas, inductance coils, transformers
Natural	Wind currents, water flow, tides	Implementation depending on the type of energy
Human	Kinetic energy, thermal energy	Implementation depending on the type of energy
Other	Chemical, biological sources	Implementation depending on the type of energy

- real-time data which enable predictive production and maintenance
- better quality control
- more economical use of the machine park (short downtimes, longer unbreakable operation)
- cost reduction.

Smart grids for intelligent energy supply and remote diagnosis in telemedicine can be cited as recent IIoT examples.

6.2 Deployment Areas

IoT systems become increasingly important in multiple areas, for example:

- Smart Home
- Healthcare
- Weather service
- Traffic telematics
- Industry 4.0 and civic craft
- Military affairs
- Smart Energy.

The expenditures worldwide for IT, IoT and Blockchain (BC) are depicted in Figs. 6.3 and 6.4. As it shown, the expenditures for IT stagnate overall at a high level, but the IoT share almost doubles from 17 to 26%. The proportion of cryptographic protection by BC is interesting here (still low level of 0.32%, but rapid increase).

Let us to examine and to consider some details regarding to the following recent deployment areas 1–7 for IoT systems which are mostly known.

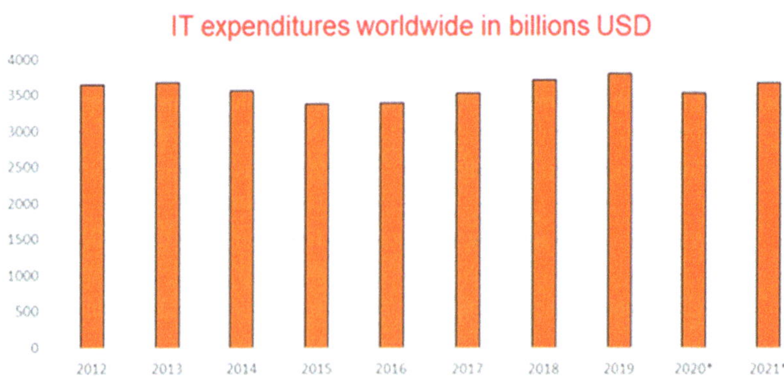

Fig. 6.3 IT Expenditures worldwide

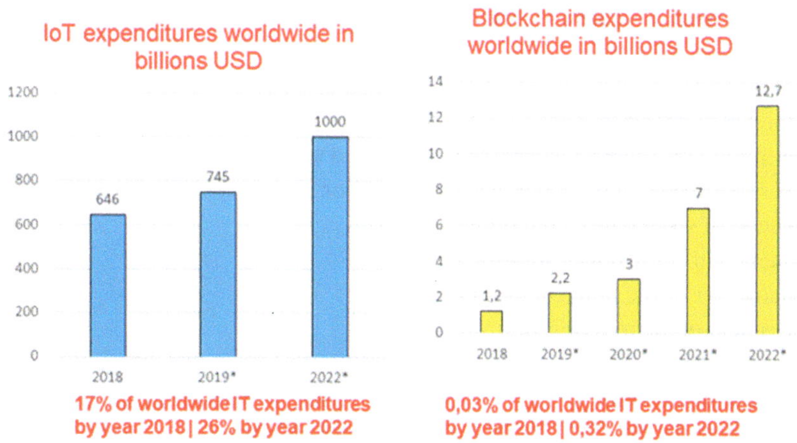

Fig. 6.4 Expenditures for IoT & Blockchain worldwide

Deployment Area 1. Civic craft and Construction

- RFID circuits for marking prefabricated components and material containers
- Control of the material transport and logistics for better overview: "what, when, where, how much"
- "Just-in-time" orders possible
- Control of the construction progress
- Occupational health and safety supporting
- Control quality for finalized structures (bridges, walls, arcs, …).

Deployment Area 2. Smart Home

- Sensors
- Presentation of the house of the future
- Consumption control and HVAC (heating, ventilation, air-conditioning, lighting, cold and hot water, …)
- Monitoring systems, alarm systems, door locking systems
- Consumer electronics, home computers
- Domestic appliances
- In-house IT infrastructure.

Deployment Area 3. Smart Metering

- Sensors on electricity meters, radiators, kitchen technology, …
- Actuators: Various switches, servomotors, Venetian blinds, louvres and jalousies
- Energy efficiency.

Deployment Area 4. Traffic Management and Autonomous Driving

- Avoidance of traffic jams
- Sensors at all road junctions
- Traffic lights, length of queues in the lanes
- Determining the intended destination of the road users (blinkers, …)
- Sensors for speed on the routes and for utilization
- Detection of special trips (fire brigade, …)
- Pollutant and contaminant sensors
- Actuators for traffic light control (priority, …)
- Assessment of the transport infrastructure
- Determination of defects (accident blackspots, traffic jam probabilities, …)
- adaptive traffic control
- Forecast calculations
- Suggestions for investments
- Autonomous driving
- Presentation of the car of the future
- Environmental compatibility (consumption, pollutants)
- Safety (driver, other road users)
- Comfort (vehicle IT, driverless, …)
- Cost reduction
- Technical requirements
- Drive concept (electric, hydrogen, eco-ethanol, …)
- Decentralized optimization of all components
- Navigation, entertainment, …
- Sensors for assessing the traffic situation
- Distance control, roadside distance, traffic signs, traffic lights
- Danger detection (children playing with a ball, traffic hooligans, …)
- On-board computer with internet connection (via 5G and beyond, …).

Deployment Area 5. Military Affairs

- Education and communication
- Sensors (enemy proximity, chemical and physical threats, radio-electronic fighting…)
- Actuators (image recording, radar control, …)
- Encrypted communication in real time
- Better friend/foe recognition in weapon systems (mines, …)
- Autonomous fighting machines
- lower risk for your own troops
- Deployment in hostile environments (radiation, …)
- Faster reactions than remotely controlled
- Economic advantages
- Possibly better acceptance of military actions (low human casualties)
- On-board computer with internet connection (via 5G, …).

Deployment Area 6. Healthcare

- Sensors for body temperature, heart pulse and blood pressure
- Control of health parameters (pulse, blood pressure, insulin deficiency…)
- Control of patient behaviour?
- Control of medical devices
- Actuators: pacemakers, active prostheses
- Applications
- Remote diagnosis
- Patient monitoring
- Monitoring the sterility of the air
- Care robots.

Deployment Area 7. Weather Service and Prediction

- Sensors:
  - Control of weather parameters (temperature, air pressure, humidity, wind direction)
  - Stationary and mobile sensors (stations, balloons, planes, drones, satellites, …)
- Actuators:
  - Cloud injections
  - Weather modification (is controversial).
  - Forecast accuracy
  - the more sensors, the better the simulation basis
  - but also, the higher the computational effort and resource consumption
  - chaotic system, long-term prediction is rather not possible.

6.3 Architectures and Basic Technologies at Glance

The following content is a brief and concise overview of common protocols, architectures and basic technologies, platforms and operating systems that are important for recent IoT systems. IoT devices and systems are supported by highly-distributed applications and apps that are, therefore, within the considered focus and of great meaning [5–7, 11–15].

For instance, the worldwide technology ZigBee, introduced since 2004, uses widespread the following proven protocol stack (Fig. 6.5):

A comparison of the widespread WSN (Wireless Sensor Networks) for IoT deployment is given in Table 6.3. Some solutions provide energy harvesting as well as use efficient IPv6. The data themselves and anonymity can be secured via AES and RSA.

Table 6.4 represents diverse basic technologies “at glance”:

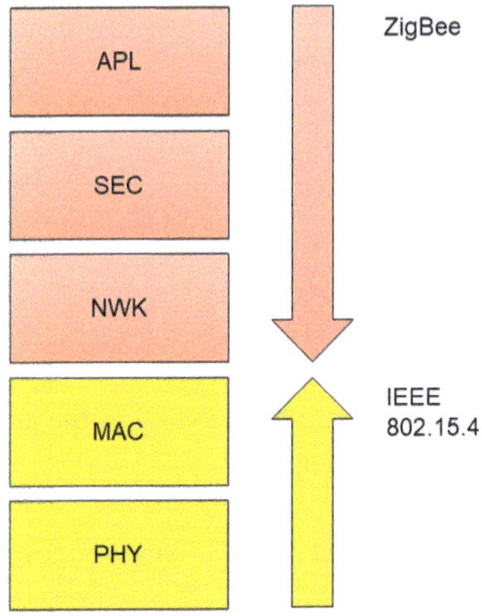

Fig. 6.5 ZigBee Protocol Stack

Table 6.3 WSN in Comparison

Feature	EnOcean	KNX-RF	Z-Wave	Zig Bee (IEEE 802.15.4)	6LoWPAN (based on ZigBee)	Nano NET
Frequency, in MHz	868	868	868	2400 int.; 868 for USA; 915 for EU	2400	2400
MAC-Layer	Beaconing	-	CSMA	Beaconing, CSMA	Beaconing, CSMA	CSMA/CA, TDMA, ALOHA
IPv6 support	-	-	-	-	available	-
Topology	Star, Mesh	Star	Star, Mesh	Star, Mesh	Star, Mesh	Mesh
Data rate, Kbit/s	125	16,4	9,6/40	250	250	2000
Node number	2**32	256	2**32	2**16 (256 in cluster)	2**16 (256 in cluster)	2**48
Security	AES	-	In mid-term	AES	-	+
Energy consumption	Very low	Low	Low	Low	Very Low	Middle
Collision probability	Very low	+	+	Low	Low	Very low
Energy Harvesting	available	No	No	No	No	No
Distance, m	30–300	10–100	20–200	10–75	10–100	40–250

Table 6.4 Basic Technologies at Glance

IEEE 802.15.4	ZigBee
• WPAN (Wireless Personal Area Networks) is a widespread radio transmission technology at PHY and MAC level • Low energy but low performance requirements with delivering of short telegrams • Long operating time without battery maintenance with implementation of low-duty cycle • Inexpensive hardware and license-free ISM bands (harmony with WLAN and Bluetooth) • Suitability for wireless sensor networks (WSN) • Technology for software support known as ZigBee	• Wireless IoT-Network in deployment since 2002 • Basis for IEEE 802.15.4 with extended layers • Sensor network (refer Fig. 6.5)
Z-Wave	**6LoWPAN**
• Wireless IoT-Network in deployment since 2001 • Frequency band between 850 and 950 MHz • ISM-Band as well as SRD-Band (Short Range Devices) • Protocol hierarchy similar to ZigBee • Home automation predominantly • Devices that are constantly active, can serve as routers • Battery-operated sensors and actuators are usually only active periodically, to perform functionality and communication	• So-called IPv6 over Low power Wireless Personal Area Network under IETF RFC 4944 • As IEEE 802.15.4 extension operated • Adaptation of the transmission format to TCP/IP • Data compression for energy efficiency • Routing for dynamic mesh networks • potentially large number of nodes due to IPv6 addresses • Mobile nodes, usually accessible via IPv6 (backwards compatible to IPv4) • The following modes are provided: newly integrated nodes, loss of nodes, orphaned nodes
IoT-Bus	**OPC—Open Platform Communication/OPC UA**
The Fraunhofer Institut IIS (Nuernberg) developed this advanced concept of the industry bus with integrated IPv6 connectivity. The overall architecture was shown in Fig. 6.6	Standard for manufacturer-independent communication in automation technology. The components are as follows: • OPC DA (Data Access)—Transfer of real time values • OPC AE (Alarms and Events) • OPC DX (Data eXchange)—direct communication between servers • OPC Command—command execution • OPC XML DA (Data Access)—coding of real-time values in XML OPC shows the overall trend towards open source. In 2009 standardization to IEC 62541 took place. Today the specification OC was summarized for OPC UA (Unified Architecture)

(continued)

Table 6.4 (continued)

	OPC UA
•	• As OPC extension owns the following components: – DA—Data Access – AC—Alarms and Conditions – HA—Historical Access – Prog—Programs • The programming architecture was depicted in Fig. 6.7. The most important areas of application are: – Industry – Agriculture – Building technology/Facility Management – Vehicle technology – Transport infrastructure – City infrastructure – Household technology – Consumer electronics

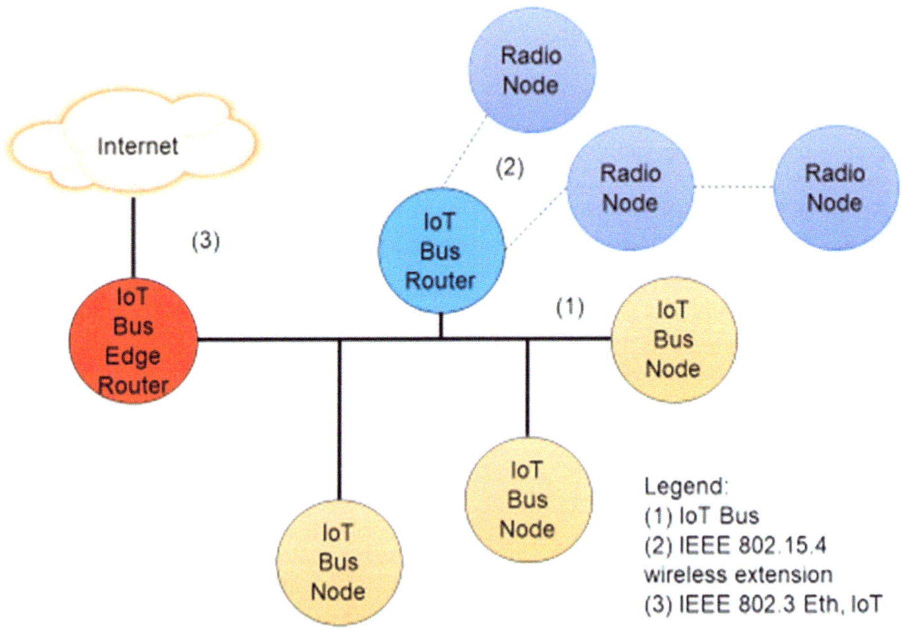

Fig. 6.6 Architecture of IoT-Bus

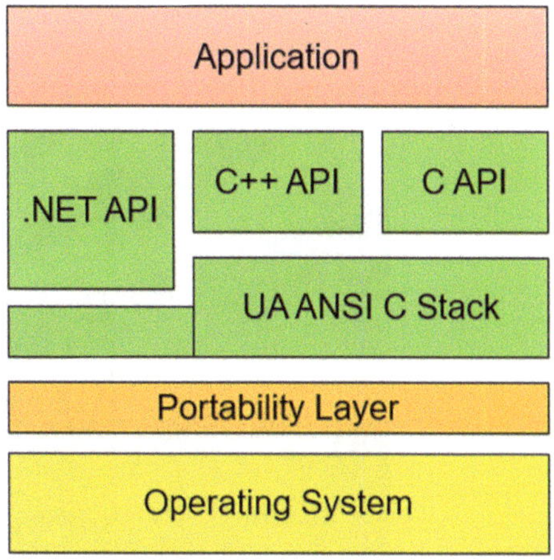

Fig. 6.7 Architecture of OPC UA

Table 6.5 Specialized IoT Operating Systems

ROS—Robot Operating System	**Further OS and derivate**
Linux-based specialized OS to supporting of robotic apps and applications. ROS is published under the open source BSD license (http://wiki.ros.org/). The features of ROS are as follows: • Libraries and tools for creating robot applications • Combined OS and IDE (Integrated Development Environment) • Hardware abstraction layer, device drivers, messaging • Visualization concepts and packet management	• RIOT OS (FU Berlin, INRIA Paris, HAW Hamburg) • QNX (Unix derivate, use e.g. in automotive engineering) • Contiki-NG (https://github.com/contiki-ng/contiki-ng/wiki) • MBedOS (https://www.mbed.com/en/platform/mbed-os/) • LiteOS (Huawei, e.g. for smart watches) • NoOS—a large number of IoT devices are used without an operating system, only with a kind of BIOS • Ubuntu for IoT (Canonical) on Intel hardware (see Fig. 6.8) for the inventory robots
Windows for IoT (Windows Embedded)	
The current versions are as follows: • Windows 10 IoT Enterprise for x86 architecture possesses improved protection against attacks compared to Win10 • Windows 10 IoT Mobile Enterprise for x86 and ARM architecture • Special development for mobile devices Windows 10 IoT Core • Special, free version for Raspberry Pi is available too (https://www.microsoft.com/en-us/software-download/windows10IoTCore)	

The following specialized operating systems (OS) are also used to support IoT-based apps and programming interfaces (Table 6.5):

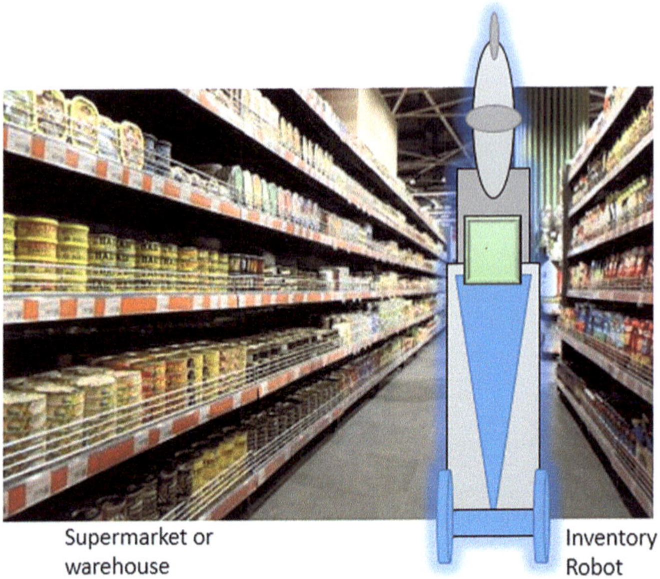

Fig. 6.8 Inventory Robot (own representation)

An inventory robot under use of OS Ubuntu for IoT (Canonical) is based on Intel hardware (refer Fig. 6.8). This robot can support IoT devices and human personal for inventory tasks in supermarkets halls and warehouse logistics.

6.4 Development of IoT Software

In principle, the development of software for IoT systems does not differ fundamentally from the development of applications for other platforms. However, there are some special features (large number of components, relatively high failure probability of individual IoTs, complex environmental conditions).

Special solutions from Eclipse are here available as the development environments, as well as various cloud-based platforms, like:

- AWS (Amazon Web Services)
- Azure (Microsoft)
- CCP (Cisco Container Platform).

Traditional languages such as C, C++, Java, JavaScript, Python etc. with special libraries for IoT devices are used as development languages.

The communication protocols that dominate in the Internet are only suitable to a limited extent for communication between IoT.

Causes are, for example, that sensors and actuators are often (for reasons of energy efficiency) in a sleep mode and that the network topology is dynamic. For this reason, technologies based on message queues are often used for communication, because they are also suitable for communication with long delays.

The protocols can usually be flexibly adapted to a variable number of recipients (load balancing).

Examples are [1, 2, 4, 8]:

- MQTT—Message Queuing Telemetry Transport
- AMQP—Advanced Message Queuing Protocol.

For communication, the XML-based data format AutomationML (IEC 62714) is used for the storage and exchange of technological data in heterogeneous environments, e.g. in modern equipment for various applications (machines, robots, smart stuff, PHY-World) and for PLC programming.

6.5 Systems for Data Acquisition—Data Processing—Data Mining

Furthermore, the systems are examined here for the IoT cycle: **Data Acquisition—Data Processing—Data Mining**. So-called **Digital Data Acquisition** System aimed to IoT cycle is represented in Fig. 6.9.

The sequential transitions between the measured values by IoT to processed date was shown.

The physical data (location, temperature, pressure …) are firstly converted into an electronically process-able form by the suitable sensors. The mapping is usually not linear and has to be calibrated specifically for the problem.

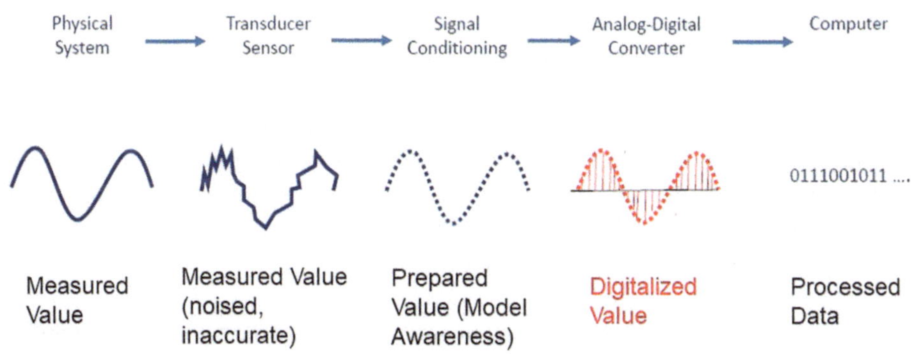

Fig. 6.9 Digital Data Acquisition System

In addition, the measured data are often subject to inaccuracies, which are corrected by model assumptions (interpolation, smoothing …). Since most of the data is of an analog nature, it must finally be digitized (collection of time and value discrete integer numbers) in order to be able to be further processed by the computer.

So-called **Data Processing** defines the type and scope of the data to be processed for specific IoT systems. A variety of problems arises here:

- Data choice
- Specification of the type, location and number of sensors
- Sensor activities for all sensors
  - Initialization
  - Sampling, or call frequency, of the data to be observed
  - Any rules for alarm messages
  - Data collection accuracy
- Further processing formats
  - Type/Accuracy
  - Integer numbers/floating point numbers
  - Text
  - Sensor-specific information (battery status, …)
  - Combined types/data structures and objects
- Priorities (real time requirements: normal—alarms, …)
- Source information
  - Sensor-ID
  - Answer to the request/message on own initiative/trap
  - timestamp
  - Transmission-specific format definitions (e.g. control data for industrial buses).

The following aspects must be considered to the conception of a system of sensors:

1. The measurement accuracy must be selected depending on the requirements of the data processing problem. The higher the accuracy, the more costs are required due to technology that is more complex.
2. The measurement duration may have to be increased and higher energy consumption accepted too.
3. The sampling frequency (number of sensors calls per time unit) must also be selected depending on the requirements of the information processing problem. The greater the call frequency, the better the time course is reflected. However, there is also more data. This increases the transmission effort and increases energy consumption. Such trade-offs have to be accurately considered.

In any case, careful modelling is required in order to optimize the resource consumption of a whole IoT system.

Data Mining can reduce the volumes of the recorded and retrieved data and improve the quality of the information (data meaningfulness and veracity). The following approaches are deployed under this aim:

- Pattern recognition using mathematical processes
  - Parameter reduction
  - Recognition of regular and periodic processes
  - Recognition of causal relationships
  - Detection of loose connections
  - Detection of exceptional situations
- Data acquisition using mathematical methods
  - Application of pattern recognition in practical operation
  - Select valid knowledge from already existing data and selectively save
  - Data reduction, e.g. with durable constant states and known periodicities as well as considering of the dependencies on complex parameters or, vice versa, even under use of non-dependence on parameters.

Correlation methods, among other things, can be used for Data Mining. Data of interest are graphically represented by scatter diagrams (see Fig. 6.10).

The correlation methods try to determine the type and strength of the relationships between the data. In most cases, one restricts oneself to normally distributed data and linear relationships (e.g. Pearson method). However, there are also more far-reaching approaches (Spearman et al.). The strength and direction of a relationship between different data are described by the correlation coefficients. Most interesting target values of a process depend on many parameters. Multiple questions have to be answered, among them:

- Which parameters are relevant for the target values of the process?
- Irrelevant parameters may even not need to be observed or, correspondently, not frequently.
- Linear relationship (simplest correlation) includes not only $y \sim x$, but also $y \sim f(x)$.
- For ideal case: strictly linear relationship between all pairs of values
- For real case: measure of correlation coefficients rxy characterizes the linear relationship between two values:
  - $r_{xy} = 0$ completely independent
  - $r_{xy} = 1/-1$ linear relationship otherwise variables influence each other, but not exclusively.

The correlation coefficient rxy of a sample is determined by minimizing the squares of deviations from a modeled dependency function. The application of correlation calculations will be discussed in the following example.

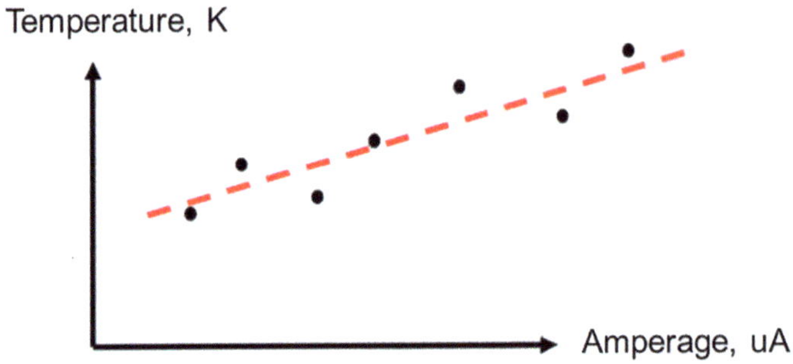

Fig. 6.10 Statistical Correlation

Example 6.1: For an ideal gas there is the following classic relationship between pressure p, volume V and the (absolute) temperature T.

We consider a measuring apparatus that permanently measures pressure and volume for a closed amount of gas, but not the temperature. Then at constant room temperature there is an ideal linear relationship between p and 1/V (correlation coefficient $r_{xy} = 1$). In this case the measurement of one parameter can be dispensed with, since after measuring one parameter the other can be calculated exactly. If, on the other hand, the room temperature fluctuates, there is no longer an ideal linear relationship (correlation coefficient<1). A project-specific decision must be made as to whether the model inaccuracy is too great; the temperature may also have to be measured.

In general, for the complex (industrial) scenarios it can be determined: There are usually diverse parameters with very different correlation coefficients.

If the correlation with the relevant result parameters is very weak, model parameters can possibly be ignored or eliminated. This means that detection may not be necessary and sensors can be saved. A very strong correlation of different parameters can mean that some parameters can be calculated from other parameters. This means that no detection is necessary either and sensors can be saved.

A process model can be modeled using the regression analysis method. The aim is to design objective functions with linear parameter dependencies and to optimally adapt the parameters of the process model ai by evaluating samples.

A process model can be modeled using the regression analysis method. The aim is to design objective functions with linear parameter dependencies and to optimally adapt the parameters of the process model a_i by evaluating samples.

$$P_{Ziel} = f(a1 * f1(P1) + \ldots + an * fn(Pn)) \tag{6.1}$$

The numerical calculation of target values can be used to reduce the effort involved in data acquisition. Furthermore, if the model quality is acceptable, it is possible to compensate the sensor errors and sensor inaccuracies via the model. The values between two measuring points can also be interpolated or future values can be predicted (extrapolated).

The regression analysis (RA) determines the dependencies between the measurement data through observation and mathematical analysis. Artificial intelligence (AI) methods grave deeper. The model behavior is evaluated over long periods of time as well as a comparison with similar systems and, as a target, a model improvement. As a result, the systems get better and better in the course of their use. In very complex cases, AI provides the only modeling dependency. Epistemologically, this can be problematic, as it is often no longer possible for people to recognize where the model limits are. That is why, human control is very important in addtion to RA.

6.6 IoT in Traffic Telematics

Modern automotive technology and cars (Fig. 6.11) use also complex IT with multiple integrated Embedded Systems and IoT devices, which are aimed to, for example, HVAC (heating, ventilation, air conditioning, windows regulators) as well as to:

- Seating mechanism
- Lighting
- Embedded electronics
- Braking systems
- Steering systems etc.

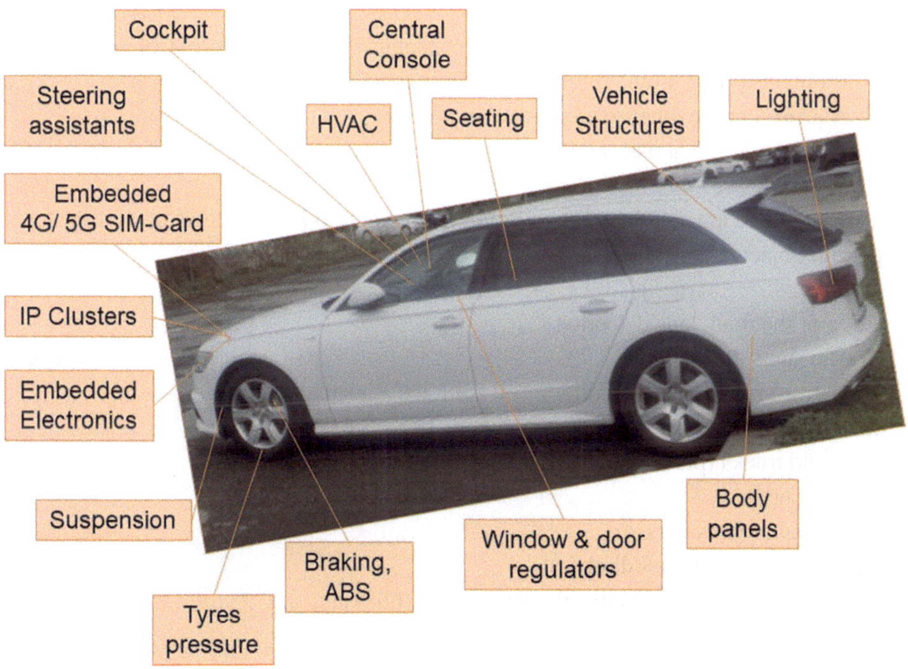

Fig. 6.11 Automotive Embedded Systems and IoT

IoT devices are also used for traffic safety control functions like:

- Monitoring of safety-relevant components (brakes, …)
- Load securing for trucks
- Condition of maintenance (tires, oil level, …)
- Optimal operation (fuel consumption, wear and tear, …)
- Driver status monitoring (fatigue, illness, alcohol, drugs, reaction time, driving errors, qualifications, driving times, braking actions)
- Security checks: OK, ALERT
  - Multi-locks
  - Manipulation attempts
  - Individual recognition of all components
  - Recognition must only be able to be manipulated if the components are destroyed, e.g. RFID circuits integrated windows, engine blocks
- Theft detection
- Are the original parts installed?

A current trend is the increasing support for autonomous driving. The classic services are integrated herewith:

- Navigation (via GPS, Google, NavGear…)
- Road condition reports, traffic jam reports via radio TC (traffic channel)
- Technical support
  - Measuring instruments for the driving situation (speed, cruise control…)
  - Problems (lack of fuel, oil pressure, cooling water…)
  - Alarms (engine oil pressure, brake oil pressure, …)
  - Active aids and assistance (ABS, …)

In addition, there are solutions for environment recognition too, such as position/value of important objects).

- Edge of the road, lane detection
- Traffic signs, traffic lights
- Dangerous road surface condition (oil spill, ice, aquaplaning…)
- Information about other traffic participants like
- Cars and trucks (predecessor, successor, neighbors)
- Pedestrians
- Bikes and moto-rollers
- Relative speed (amount/direction)
- Expected intentions of the traffic participants (blinking, …)
- Special situations
- Fire brigades, police, First Aid…

- Children, disabled persons.
- Modern systems also support active driver actions.
- Driver information (visual/acoustic)
- automatic speed limit
- automatic braking (traffic signs, dangers, ...)
- Choice of traffic route
- Choice of traffic speed
- Decision to change lanes (navigation, overtaking).

In addition to the technical problems, there are also problems of social acceptance, e.g. attribution of blame in the event of accidents. Who is to blame? Who is responsible for the offense? There are some possible cases:

- To driver?
- Then it is not autonomous driving!
  To manufacturer?
- High financial risk by multiple precedents and law cases on prominent car vendors are already known.
- To car insurance?
  High cost of policies as a sequence!
- To the state?
  Makes only sense for a test period atmost.

In general, extreme real-time requirements and very high traffic safety requirements are applied to autonomous driving. Use of the navigation algorithms for route planning is already state-of-the-art. The timing is not critical, as the routes are planned in advance and, in the worst case, driving errors lead to detours.

On the other hand, the driving itself is time-critical. At a vehicle, velocity of approx. 100 km/h is equal to ~28 m/s. It means that 1 m is covered in 36 ms. Such comprehension leads to the fact that the detection of the traffic situation must be steady updated to a cm-accuracy in time intervals of at least 1 ms and all this with absolute reliability (higher radio network QoS requirements).

Let us consider some sub-problems, e.g. the problem of recognizing traffic signs. This must be done with extreme security, even if the traffic signs arc not always in optimal condition (dirt, corrosion, damage, overridden by hand-regulating police officers).

The traffic signs can be covered by trees or trucks in side lanes etc.

A related problem is the detection of traffic lanes. The borders must be reliably detected, even in problematic situations (no color borders, yellow temporary repair tracks, locking cones, delimitation by leaves, snow as well as not easily recognizable, damages).

Both problems require complex, intelligent sensors that reliably guarantee measurement and transmission operations in the ms time range.

The intentions of other traffic participants must also be recognized in real time, for instance:

- A car in the back in the same lane starts to change its lane, possibly without blinking.
- Playing children
- Unsafe traffic participants and drivers (beginners, seniors, disabled persons)
- Vehicles with technical defects
- Extra-long vehicles with swinging out
- Extremely powerful vehicles and sport cars (with macho drivers at the steering wheel).

Due to the risk of rear-end collisions, traffic jam detection is important. In urban areas, limited control is possible through communicating induction loops in the road surface or light barriers on the road.

These are relatively sluggish because only counts are made and queue lengths are not measured directly.

Much more powerful is wireless communication between control systems and traffic participants/road users who know the position of each participant and, if possible, their intended destination and, in principle, enable optimal control.

One of the important research areas is decentralized support for autonomous driving, which can be implemented, for example, under use of 5G radio technology and beyond. On this basis, it is possible to realize a comfortable detection of traffic jams for suburb highways and country roads. Traffic jams in curved areas are particularly dangerous (see Fig. 6.12) in such cases.

In the given example, vehicle A wants to take a left turn without having a view of the near end of the traffic jam. The example makes use of the fact that vehicle B saw the traffic jam in the opposite direction while driving past. Vehicle B informs vehicle A that there is a risk of congestion in its direction. Vehicle A can then reduce its speed punctual.

At present, motorways are heavily loaded by truck traffic. There is practically a permanent truck lane. Despite the large number of drivers, there is a high risk of rear-end collisions and jams.

With the help of the so-called electronic drawbar, columns of trucks can be formed and shaped under distance approx. 2 m (so-called Platooning). A driver drives in the first truck of the convoy (platoon), the following trucks are controlled by the lead vehicle. You drive at a short distance without a driver (see Fig. 6.13). The advantage of such solution is that there are no rear-end collisions and jams, less traffic space is required, only one driver drives and the others can relax and rest.

The prerequisite is that all participants in the column are equipped with automatic driving systems that communicate via 5G and beyond radio technology under extreme real-time conditions (latency <1 ms).

With individually assembled trains, variable train lengths would be possible. However, this would require a fundamental change in the safety concept of the railways.

The greatest benefits can be achieved through integrated transport systems.

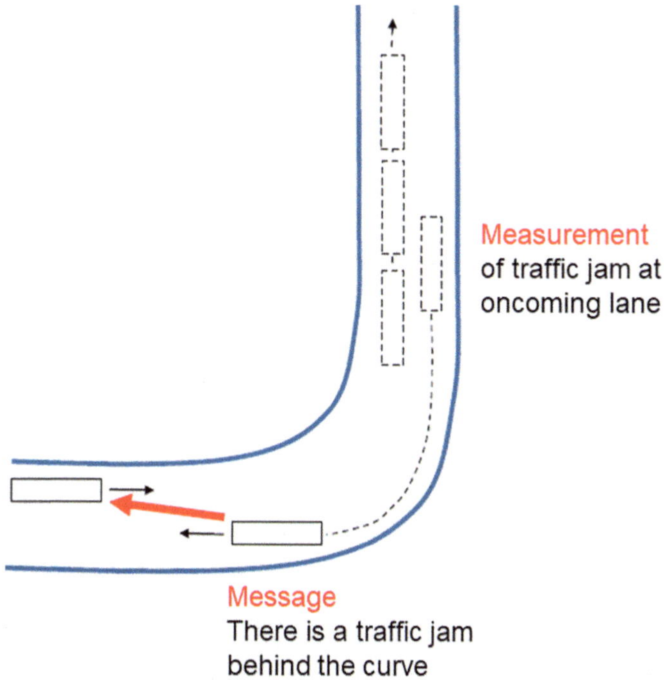

Fig. 6.12 Sensors for Traffic: Warning Messaging

Fig. 6.13 Sensors for Traffic: Platooning

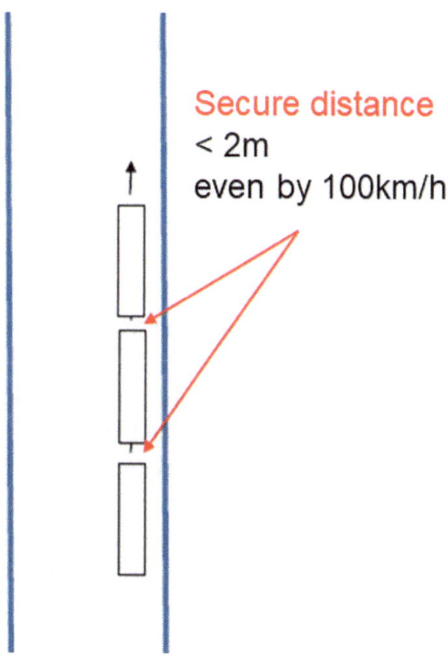

6.7 IoT and Smart Energy: Smart Grid Generations, Smart Home and Smart City

Smart Energy is a collective term for intelligent technologies within the energy industry (generation, storage, conversion, transmission, consumption control). At any point in time, the energy generation must correspond to the energy consumption. Deviations lead to voltage fluctuations or even to fluctuations in the current network frequency (power grid frequency). Guaranteeing this balance is a complex problem, due to absence of the storage facilities for electrical energy in the power grids.

In the convenient power grids (in mid-term, Smart Grids) there are base load generators and regular load generators. The base load generators run for a long time with maximum performance and optimal operating parameters. The regulating load generators cover the fluctuations in the energy demand and therefore work with utilization below 100% and thus with lower degrees of efficiency.

The classic 1G power grids have a relatively easily predictable course of electricity demand. The proportion of (inefficient) balancing energy can thus be kept lower. The generation of electricity from renewable sources, on the other hand, cannot be controlled according to demand; it depends on environmental factors (weather, daily rhythm…). This leads to the absurd situation that a reserve energy generation with classical power plants has to be kept, which is very uneconomical and polluting nowadays.

The 2nd generation Smart Grids (so-called 2G Smart Grids) have to implement a complex control system that still guarantees a constant balance between production and consumption in the situation of electricity production and fluctuating electricity consumption what is difficult to predict. This situation requires a more flexible control of the consumption according to the current network load. This is possible, for example, through situation-adjusted energy prices and consumption that is controlled according to economic interests.

So-called Smart Metering is required for these aims, i.e. intelligent meters and sensors are accounting and measuring not only the integrated consumption of the users, but also the time history of consumption and any feed-in by the users (solar power generation). This approach enables flexible billing using an up-to-date electricity price. Technically and economically, this approach offers enormous advantages; any privacy and data security problems must be resolved through growing social recognition and consensus.

In the middle term, 3rd generation Smart Grids (so-called 3G Smart Grids) will be required. The strongly fluctuating electricity production from renewable energy sources must be resolved. This production can be partially achieved by regionalizing of electricity production. In this way, a temporary regional wind calm can be compensated by increased production in other regions. However, this approach only works if the transmission capacities of the power grid are significantly expanded. Furthermore, it is necessary to integrate high-performance electrical energy storage devices in order to cushion short-term load peaks. However, the capacity of such storages is not sufficient to compensate for larger fluctuations. To do this, an integrated system of different processes must be developed.

The technologies for converting various forms of energy are becoming increasingly important, such as: power to gas, power to liquid as well as several inverse techniques. In contrast to electricity, energy can be stored very well in the form of gas and liquid fuels. With this process, superfluous wind power is stored in the gas network, for example, and gas-fired power plants can be dynamically deployed or step in as peak load power plants when there is peak demand situation in the electricity network. Unfortunately, the losses resulting from the energy conversion have to be also accepted.

The decentralized consumers can also improve the economy of the network.

With Smart Metering, larger devices can be switched on at a low electricity price and electricity can even be returned at a very high price. The intelligent charging of the batteries of electric vehicles will be of great importance for smoothing out peaks in demand in the power grid.

Smart Home is other name for Home Automation technology that has long been established in large office buildings and is currently also becoming attractive for residential buildings. The goals of the Smart Home are to improve living comfort, to provide more effective methods for heating, electricity management, lighting and domestic safety. Well-known manufacturers of Smart Home technology in Germany are, for example, Innogy SE, Technisat, Conrad. There is a wide range of controls, sensors and actuators, controlled e.g. via special remote controls and televisions, smartphone apps, laptop software. Figure 6.14 shows a selection of IoT equipment for the Smart Home concept.

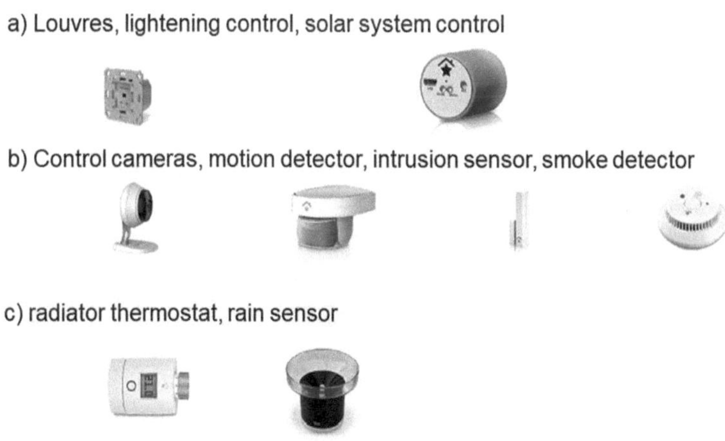

Fig. 6.14 Sensors and actuators by Smart Home

A consistent further development leads to the concept of the Smart City with the following fixed core aims:

- Efficiency of services
- High level of municipal services
- Optimizing the use of public resources
- Sustainability (population development, environment …)
- Citizens' mobility
- Safety and protection in everyday life and in the event of a disaster
- The attractiveness of the municipality (image for tourists and investors).

The use cases include i. a. a smart traffic concept (parking, directions, traffic light controls), a smart waste management (optimization of routes, collection frequency, recycling centre organization), a smart security concept (ensuring short reaction times for First Aid, police, rescue troops or German THW) and smart management for urban property (building use, street lighting).

The Smart City concept in Germany is politically supported by the federal government, the state governments and local authorities. Many scientific and social organizations are also involved. There are also diverse projects at the level of the European Union (e.g. Madrid).

6.8 Sensor Nets and Swarm Intelligence

The most important use cases for the IoT devices arise within the Industry 4.0 projects. If there are a particularly large number of sensors in a system, one speaks of so-called sensor swarms. The sensors are used for flexible, dynamic production.

A permanent status recording of the goods to be produced and the machines used is necessary. The prerequisite for this production method is that the production systems and the products are intelligent and capable for communication.

In this context, one speaks of swarming sensors (and actuators) with sensory and communication skills.

Due to extreme miniaturisation, the swarms are also designated as Smart Dust.

The sensors often act completely autonomously and are equipped with their own energy supply. The requirements for the sensors and actuators are extremely low costs, low energy requirements, and small size and weight. It may be possible that performance is reduced too. Distributed algorithms are widely used and a high level of redundancy is planned inclusively. An important planning criterion is also the definition of the time points of activities and their scope. This essentially determines the life cycle of the sensors, i.e. the operating time until the next charging process and the overall service life, i.e. system aging. Since battery technology has not made any significant progress in the last decade, sensor node energy management is extremely important. The sensor node must always have sufficient energy! The following scenarios exist:

- Energy is always enough
- Dynamic energy management
- Self-control of energy availability.

Normally, when modelling a system, a forecast of the future must be provided about the energy consumption of future measurement tasks and the next recharging process. The sensor works until the energy is exhausted and then pauses for recharging. Battery-oriented power supply means a restriction in terms of working time and sensor volume or weight. This is why Energy Harvesting processes in which the energy is obtained from environmental sources are gaining in importance (see Fig. 6.15).

As a rule, a sensor possesses earlier a small buffer storage and a permanent energy generation module. The energy demand must then be optimized so that the sensor can fulfil its tasks. In the low energy mode, parameters of the register memory and the fastest CPU near SRAM are often lost. Therefore, saving into a non-volatile memory area is necessary. In a similar way, the energy demand must be optimized at all levels (energy architecture model, Fig. 6.16).

We can find the most important applications of IoT technology in the Industry 4.0 concept.

For instance, a central control station can monitor a system of machines, which each have several sensors and actuators that can be combined to a swarm system of S sensors (sensor swarming). The sensors can interact within their swarm. Swarms can be hierarchically combined to larger swarms. The control center interacts with the swarm as

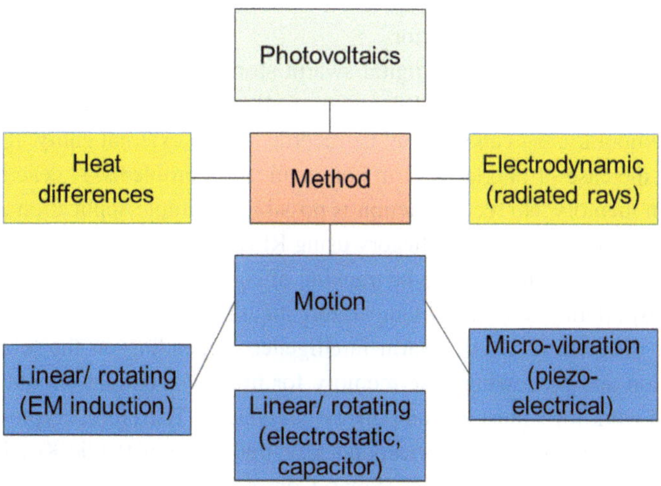

Fig. 6.15 The mostly-used methods for Energy Harvesting

Process Model for IoT Programming		
Task for developers	Problem	
	Algorithm	
	Programming Style	
Task for compiler	High-Level Instruction Set	*Controller-specific*
	Assembler Command Set	
	Controller for Internal HW	
Communication method	Processor Communication	
	Operating Hardware	*Operational Hardware Environment*
	Functional Hardware	

Fig. 6.16 Energy Architecture Model

a holistic structure! The swarm carries out instructions collectively (with independent organization) as it was shown in Fig. 6.17. Use cases of such systems are as follows:

- Driverless transport systems
- Intelligent postage stamps
- Container logistics and forwarding.

Each of swarm participants performs a fixed, simple algorithm and their own perception is characterized with only limited overview. Overall, the swarm is an intelligent system with a not clearly predictable behavior.

The sensor swarm possesses a digital swarm storage of the knowledge and skills of each individual swarm element and the sum of all information in the swarm. A swarm memory is temporary, but can possibly be recorded by an external entity. It may be contradictory in the case of internal swarm communication problems. A deactivated swarm has no swarm memory. No reconstruction is possible. A simple application example consists in the realization of a product history using RFID circuits.

A more complex example would be tracking of container shipments with providing of contents, transportation routes, and temperature measurements as well as alarming.

There are different levels of swarm intelligence, depending on the level of internal communication and processing. The memory for product identities should be managed securely. The workflows must be compulsory guaranteed. Therefore, crypto-technologies for swarm sensors are required, correspondently, under use of Public Key Infrastructure (PKI) and Blockchain (BC) processes (Smart Contracting).

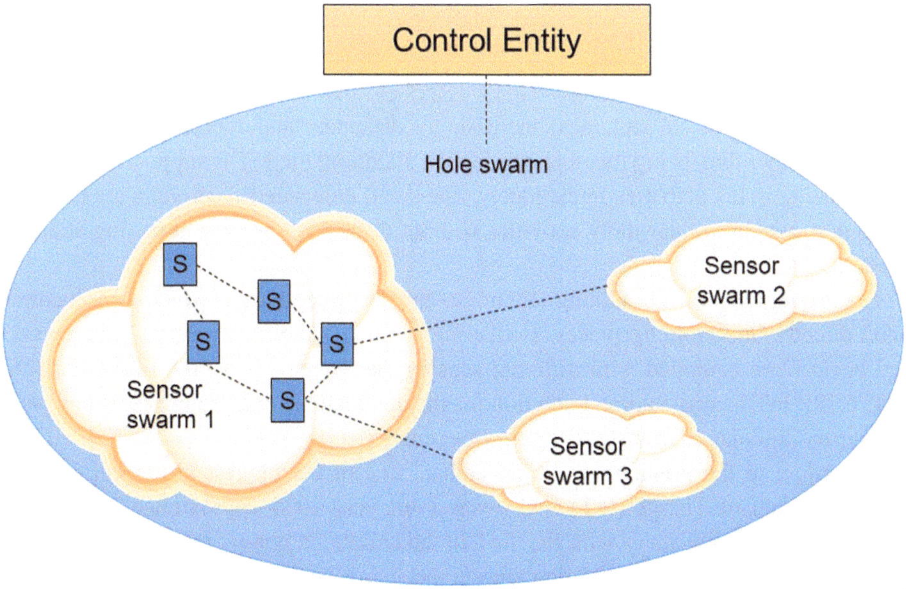

Fig. 6.17 Sensor Swarms

Indoor location can be organized using sensor swarms too. The basis is so-called lateration, in which several stationary beacons periodically send their ID, position and other parameters.

The distances to the beacons of other sensors (approximate) can be determined by measuring the signal strength (receipt power in mW) or transmitting time (in ms).

The own sensor position can be determined by intersecting the circumference of the distance. At least three beacons are required to determine the position in the plane. In practice, the determination is relatively imprecise due to measurement errors, so that more than three beacons are used and an optimal value is determined from several calculated positions using the method of minimizing the squares of deviation.

This is computationally and energy-intensive. For this reason, the mathematical process is simplified in the case of low-energy chips while accepting a lower accuracy of the location. The position values can be improved through communication between the participants because the measurement errors partially cancel each other out. Further advantages of swarm communication:

Creation of swarm maps, the participants know the positions of the other participants.

Additional information on the spatial geometry and the propagation conditions of the radio waves.

6.9 IoT and Fog Computing: Definition and Demarcation

The concept „Internet of Things" (IoT) describes, that the per IP networking gadgets (laptops, tablets, smartphones) increasingly disappear and replaced by "intelligent objects". Rather than being the object of human attention, the IoT is supposed to help the human being in his activities immediately. The small embedded controllers are designed to support people in their daily activities without distracting or even attracting attention at all.

For example, we are talking about so-called "wearables", which are incorporated directly into clothing pieces. Furthermore, in an "Intelligent Home", the embedded devices are equipped with different sensors like Bluetooth, RFID, 6LoWPAN etc. (Fig. 6.18). IoT is also a correct solution for the applications that require real-time, such as industry automation, transport, video streaming etc.

According to the prediction of Gartner Inc., one of the biggest IT-consulting companies, Internet of Things (IoT) will connect till year 2025 approximately 25 billion devices. But even nowadays with the "IoT in the childhood phase" numerous networked devices compete around the data pipelines in the clouds because of the large data volume (cp. "big data" problematics).

In other words, myriads of small, energy-efficient nodes are interlinked to form a "foggy" infrastructure. Multiple devices are almost permanently connected to the cloud and more and more are being moved to the cloud according to the existing strategy of services' construction. This migration creates the next problem of lack of data rates and higher processing delays. At this point fog computing comes into play. The term "fog computing" was originally introduced by Cisco. Fog operates mainly like a bridge between IoT with connected gadgets and intelligent nodes and remote data centers (clouds). As a solution for these huge data volumes, which have to be retrieved and processed in the big data centers and clouds, the small intelligent controllers and gateways acts in their immediate proximity. The data are efficiently preprocessed via these active low-energy devices [1–3].

6.10 Co-operation "Fog-2-Cloud"

The term "Fog Computing" extends the paradigm of the Cloud Computing towards to an intelligent network node (microcontroller based on Radio Network Edges) and enables therefore a diversity of new applications, apps and services. The fog computing solutions are characterized via the following characteristics [1–3, 8, 18]:

- low-latency, location-awareness (fast reactivation of nodes)
- wide geographical distribution
- very large number of nodes and mobility, IPv6 recommended
- leading role of "Wireless Access"

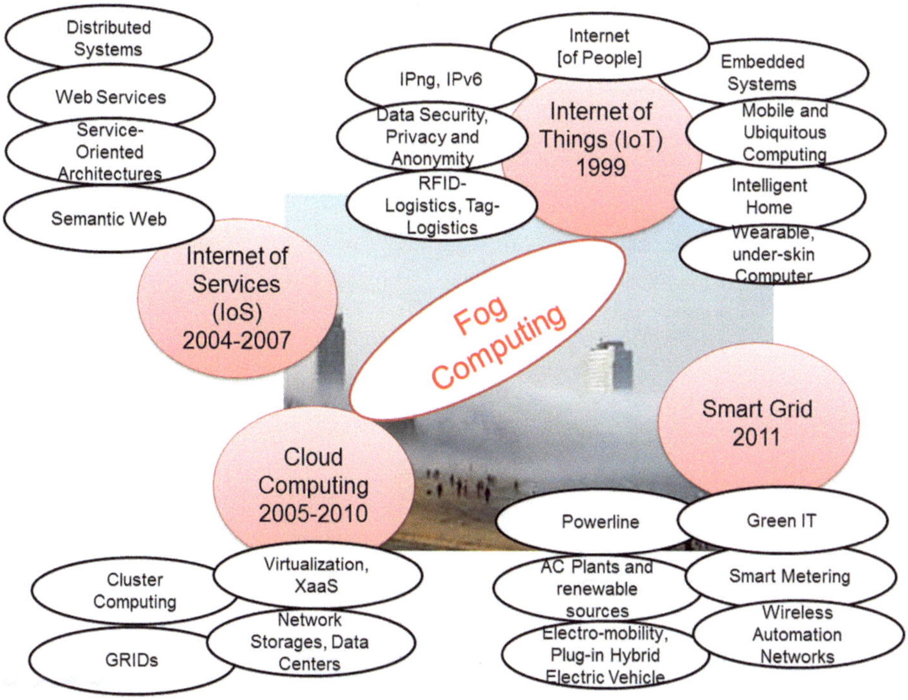

Fig. 6.18 Internet of Things in the context of clouds, web services and energy efficiency [1–3]

- streaming and real time applications
- node heterogeneity.

The most important functionalities of fog computing are (everything on the client site):

1. Data acquisition on the site
2. Data buffering
3. Runtime for compact applications (apps)
4. Rapid pre-processing on the site.

Fog Computing provides a suitable platform for further development of IoT services on the basis of the following known and new network technologies [2 , 21]:

1. Wireless Sensors and Actuators Networks (WSNs): ZigBee, Bluetooth, EnOcean
2. RFIDs
3. Wi-Fi/WLAN (IEEE 802.11 ac, ad)
4. 6LoWPAN
5. 5G-Mobile (IMT2020) and beyond.

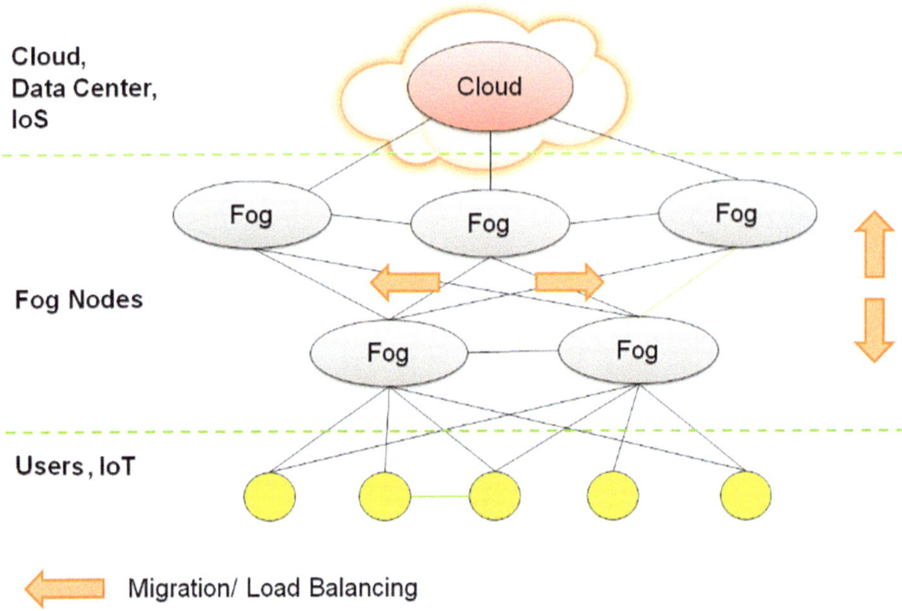

Fig. 6.19 Architecture of fog computing in the coexistence to the clouds (users + fog + cloud)

With the fog computing the services and computing processes are shifted on the edge of the networks (closely to the users). A compromise must be found between partial or complete displacement between the clouds and fog. Fog computing can be also called as "edge computing". The company IBM is also trying with a similar initiative to move the traditional cloud-based Internet to the "edge".

When is talking about "edge computing," it literally means the edge of a network, the periphery where the Internet ends and the real world begins.

Data centers remains in the focus of the networks. The small controllers and typical gadgets like laptops, smartphones, tablets, multimedia players, watching cameras are positioned "on the edge". How will be fog computing developed in the medium term? The fog will not push the clouds. The question is not so: "fog computing vs. cloud computing", but rather about their co-existence (Fig. 6.19).

The architecture includes the following three hierarchical planes [2, 4, 16, 17, 19]:

- Plane 1: The clouds and data centers, which build an IoS with typical access via web service protocols.
- Plane 2: The fog nodes which are involved to the virtual environments for data preprocessing, functionality migration and load balancing with the clouds (refer plane 1).
- Plane 3: The users with end devices which build an IoT and are placed on the edge of the fog infrastructure.

Table 6.6 Comparison of Fog and Cloud Techniques

Cloud or network storage challenges	How can fog help herewith?
Cloud or network storage challenges	+ less hops
Latency as a big problem	+ data locality and local caches
Limited mobility by data acquisition	+ on-the-site-processing
Data rate limitations	+ fast failover
Reliability and robustness acceptable but face-to-face "big data" questionable!	+ location awareness
Plain data storage without meta-information for search	+ less hops

Such kind of the distributed architectures for combined use of cloud and fog computing offers several clear advantages have been discussed in [2, 4]. The comparison of fog to fog + clouds architecture is summarized in Table 6.6:

There are a lot of architecture examples for a fog computing platform and applications [1–3, 16]. Fog computing offers a lot of advantages:

- supports "Big Data" and real time analysis
- energy efficiency
- performance based on physical proximity
- smaller data streams over Internet than by cloud computing
- Reduced costs for data processing (Big Data)
- Computing performance in relation to bandwidth is always less expensive (thanks to Raspberry Pi, Arduino Uno, Banana Pi etc.)
- Replicability in case of the failures
- Lower transport costs and lower latencies
- improved QoS,
  but security and privacy is questionable.

The encryption and use of CIDN (Collaborative Intrusion Detection Networks) [2] is strongly recommended. This makes fog computing the driving force for IoT. Medium-term development is the IPv6 deployment for more efficient addressing and increasing of data security. The security challenges are growing. The authentication of connected devices in the combined structures (user + fog + clouds) must be guaranteed. The encryption and digital signature are achieved by a robust combination of AES + RSA + PKI or, correspondingly, Blockchain (BC) and the use of new generation firewalls and CIDN [6, 7, 20].

6.11 IoT System Development

Proven methods, models, platforms and tools with specific extensions for the IoT are used for the development of IoT systems [1–7]. In general, the procedure can be as follows.

The conventional ER models (Entity-Relationship) are used for communication between users and developers.

The de facto standard for data modelling is called SysML (Systems Modeling Language). The language can also be used successfully for IoT system development. The structure of the Systems Modeling Language was shown in Fig. 6.20. It is based on version 2 of UML with the associated structure and behavior diagrams and with additions to other diagram types such as “Requirements” and “Parametric”. SysML offers numerous applications for systems engineering for modelling of complex systems. Complete solutions are available for IoT development engineers, e.g. from DoubleSlash (https://www.doubleslash.de/).

Furthermore, not only systems for data transfer, data collection and processing are required but also the systems for analysis, evaluation and visualization are also very useful. The techniques of AI and ML are becoming more and more important.

A typical workflow for IoT development is presented in Fig. 6.21. Via multiple stages with feedbacks, an optimal workflow for IoT devices integration can be established.

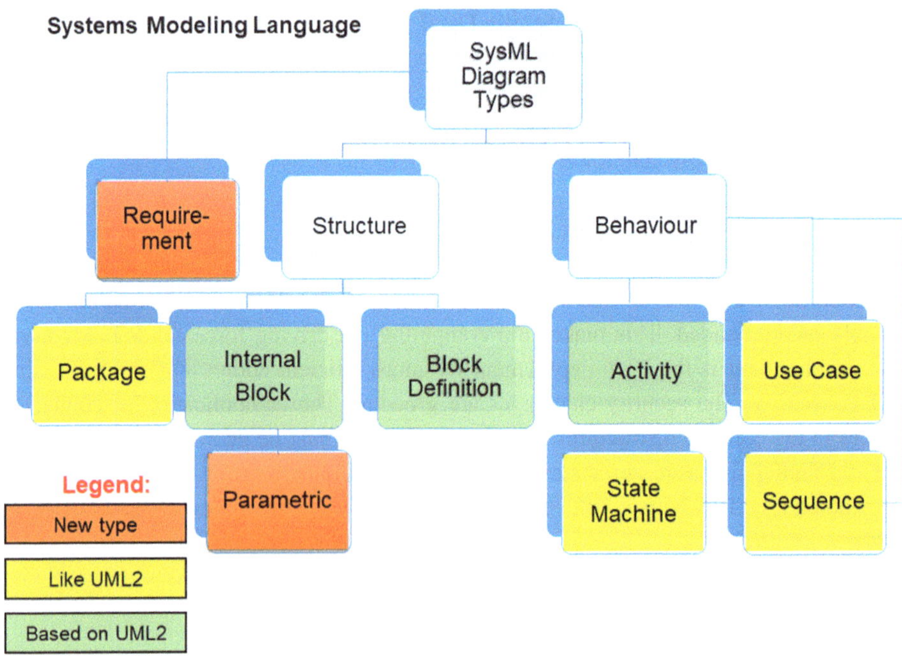

Fig. 6.20 Architecture of Systems Modeling Language

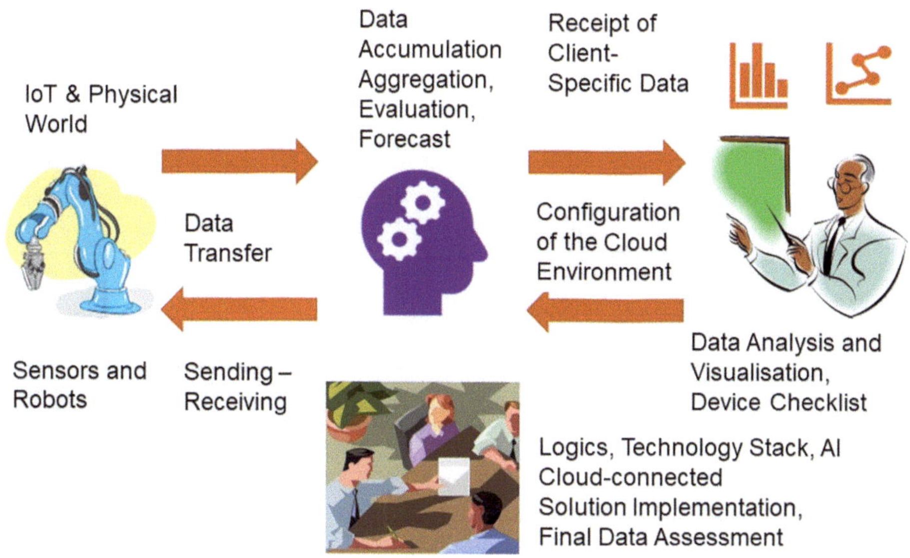

Fig. 6.21 Typical Workflow for IoT Development

A further selection of certain development platforms is listed below:

1. IBM Watson: Universal support, i.e. for IoT-Problems too
2. Rockwell Automation: Manufacturing support software
3. LabviewNXG: software-centric platform for automated measure and test applications.
4. The platform includes setting up and configuring systems, developing measurement and test codes as well as creating web-enabled applications.
5. Pfannenberg Signal Analytics (PSA): This system represents a Plug & Play solution for analysis tool for real-time monitoring of signal devices in heterogeneous machinery/machine parks.

IBM Watson IoT Platform and IBM IoT for Automotive are represented correspondingly in Figs. 6.22 and 6.23.

Rockwell Automation Tool is represented in Fig. 6.24 (source: https://literature.rockwellautomation.com/idc/groups/literature/documents/rm/2080-rm001-de-e.pdf).

Well known Java-based IDE Eclipse IoT provides multiplicity of tools (refer Fig. 6.25) to programming of IoT-Apps in Java (https://iot.eclipse.org/). These platforms offer API access to IoT hardware interfaces via integrated libraries. This above-mentioned cluster of IoT Eclipse tools includes Eclipse IoT, Eclipse Kura and Eclipse fog05. Let us to discuss it in detail.

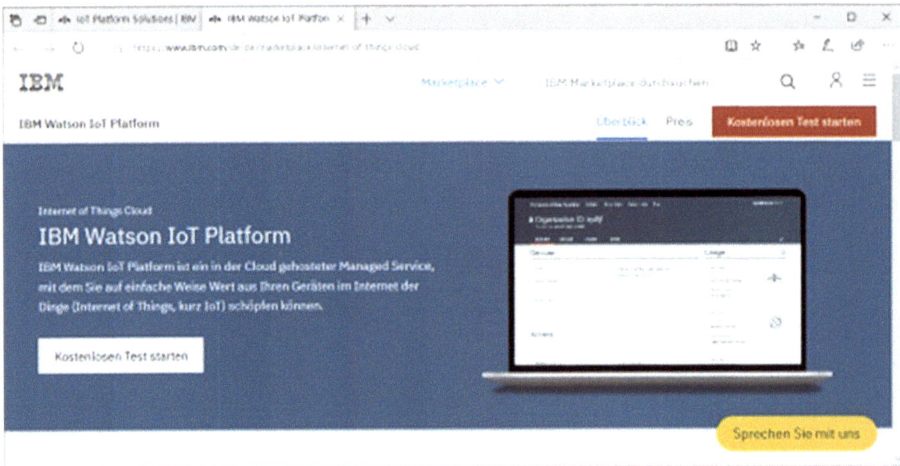

Fig. 6.22 IBM Watson IoT Platform (Screenshot)

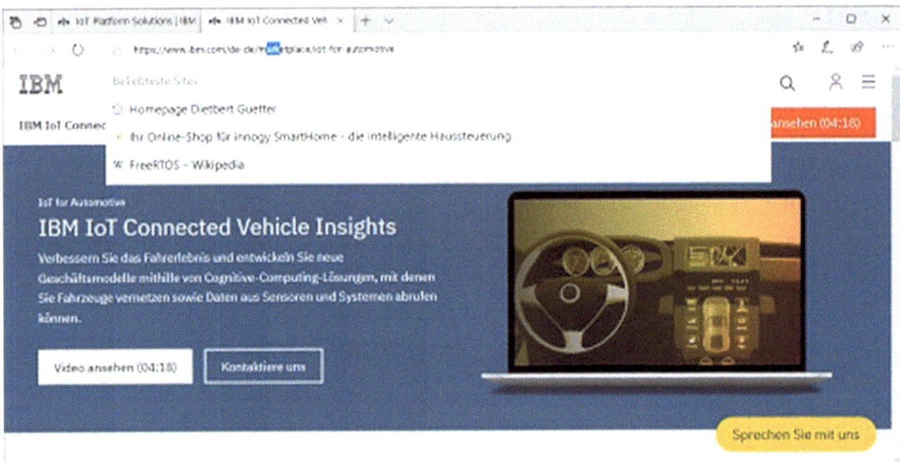

Fig. 6.23 IBM Watson for Automotive (Screenshot)

1. Eclipse Kura is an advanced IoT framework, based on Java/OSGi specification, which offers API access to various hardware interfaces (USB, RS-485, BT, BLE) Via IoT gateways. Protocol support for the field level of production systems is also integrated, as well as support for MQTT communication and communication between the cloud and Thing (see Fig. 6.26). A central component of this efficient development system is also the support of dynamic resource management. For application-oriented

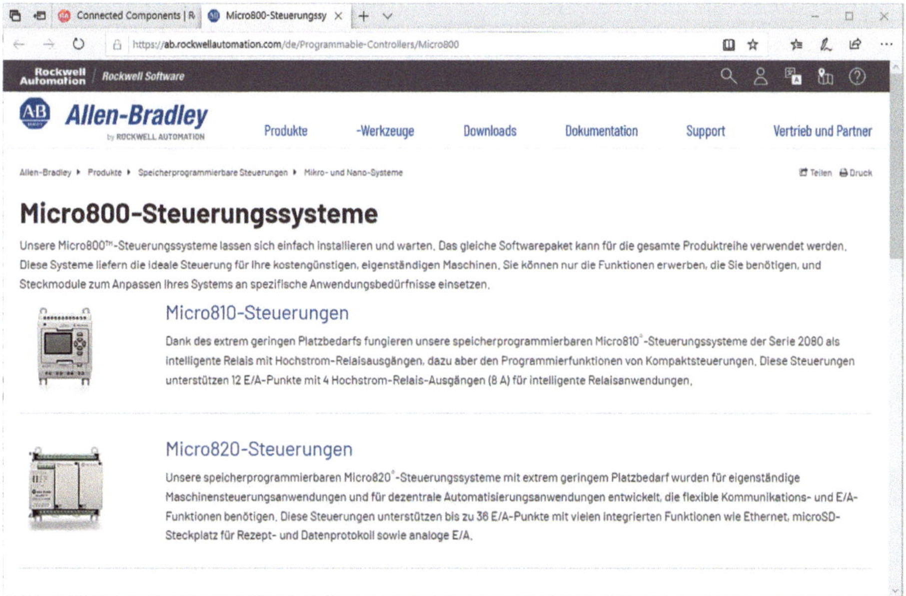

Fig. 6.24 Rockwell Automation (Screenshot)

programming, there is a trend towards web-based visual data flow programming. Eclipse Kura is compatible with practical IoT cloud platforms thanks to OPC UA support.

2. Eclipse fog05 (refer Fig. 6.27) offers service-oriented support for cloud-fog solutions (such as SaaS, IaaS, PaaS) as well as end-to-end communication from cloud to thing (IoT device). The platform is largely scalable with visualized monitoring and dynamic resource management. The following service-oriented components belong to the architecture:

- T—Thing, or IoT device in a cell
- E—Edge, Fog
- B—Binary
- C—Container
- UK—Uni Kernel
- uS—Micro Services
- VM—Virtual Machines.

Numerous VMs in the cloud can call up further containers, image files, autonomous kernels as well as the responsible services and micro-services. They all are flexible building materials for efficient HDS, i.e. IoT devices and highly distributed apps.

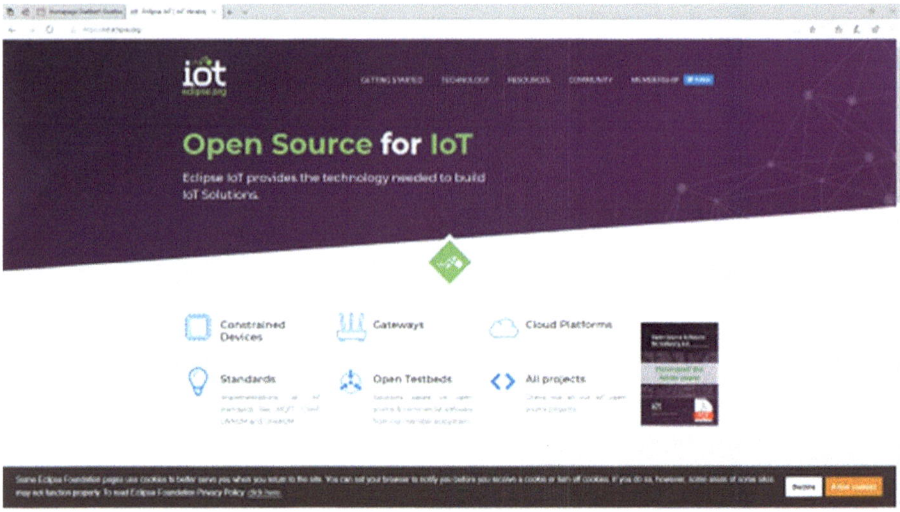


```
import paho.mqtt.client as mqtt

# The callback for when the client receives a CONNACK.
def on_connect(client, userdata, rc):
    print("Connected with result code "+str(rc))
    client.subscribe("$TEMPERATURE/#")
    # subscribing to all topics startings with "TEMPERATURE"

# The callback for when a PUBLISH message is received.
def on_message(client, userdata, msg):
    print(msg.topic+" "+str(msg.payload))

# MAIN
client = mqtt.Client()
client.on_connect = on_connect
client.on_message = on_message

client.connect("iot.eclipse.org", 1883, 60)
client.loop_forever()
```

Fig. 6.25 Eclipse IoT Platform (Screenshot with Source Code Excerpt)

6.12 Case Study: A Freeware MQTT Solution for Fog

A typical deployment scenario [8] within computing centers is considered [8] and co-operation "fog—private cloud" is examined. The usage scenario is targeted to middle-range cloud or computing center parameter polling (air temperature, humidity, voltage analysis). Furthermore, the scenario can be extended via physical entry control and access detection within the middle-range cloud centers [8]. The measured data from the cloud (computing, networking, storage, environmental) are acquired and analyzed. Let us to discuss how small sensors detect overheat in provider racks. The provider hosts

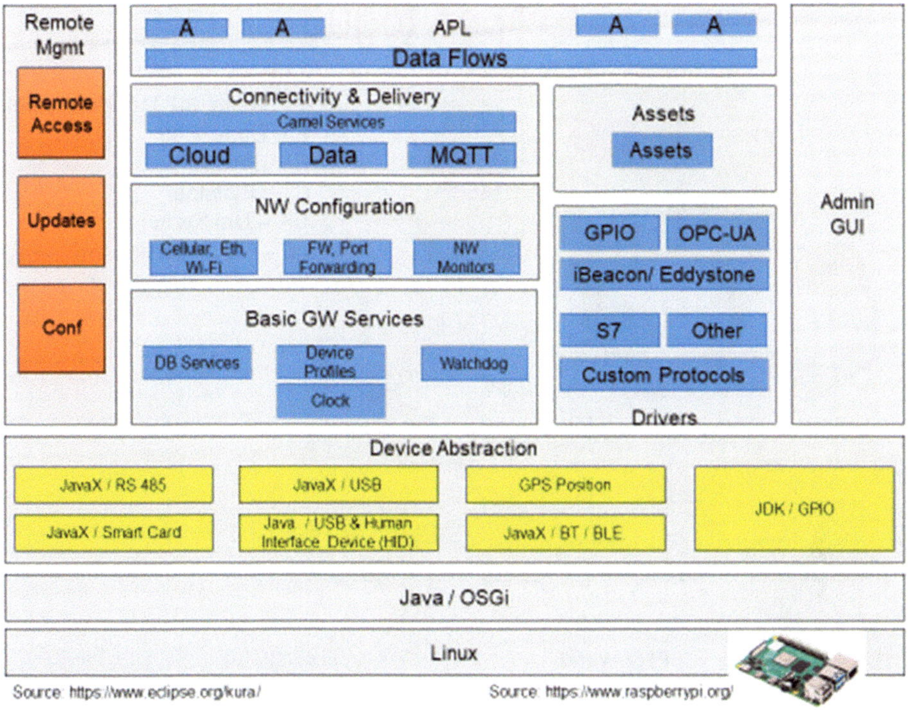

Fig. 6.26 Architecture of Eclipse Kura

are managed via SNMP and Nagios-tools. The sensors are managed via MQTT, AMQP under use of queue-based „Publisher-Broker-Subscriber-Model" (PBSM). Multiple API are available, i.e. Web Services over OASIS, Eclipse, JMS, Apache, RabbitMQ, Windows 10 IoT Core etc. So-called PBSM allows to the MQTT clients to communicate in P2P as well as broadcast modes (Fig. 6.28).

Example 6.2: The typical workflow consists of the following steps:

- All three clients (so-called subscribers) open TCP connections with the Broker.
- Clients B and C subscribe to the topic "Temperature T" and wait for the messages.
- At a later stage Client A publishes a value to the topic T = 22.5 °C.
- The Broker redirects the message to the topic to all Subscribers.

The corresponding sequence diagram for the publishing and subscripting is depicted. All the subscribers are with the actual temperature delivered. A freeware MQTT-based solution for IoT and Fog is shown below (cp. Fig. 6.28). The important components of the examined architecture are as follows: Mosquitto, Nagios (SNMP-based tool),

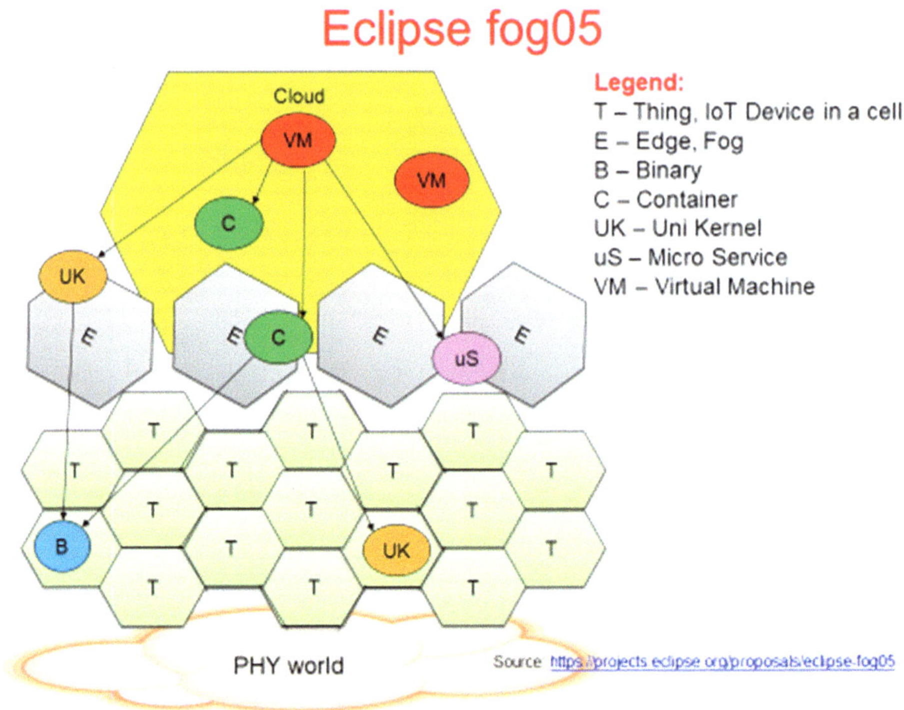

Fig. 6.27 Architecture of Eclipse fog05

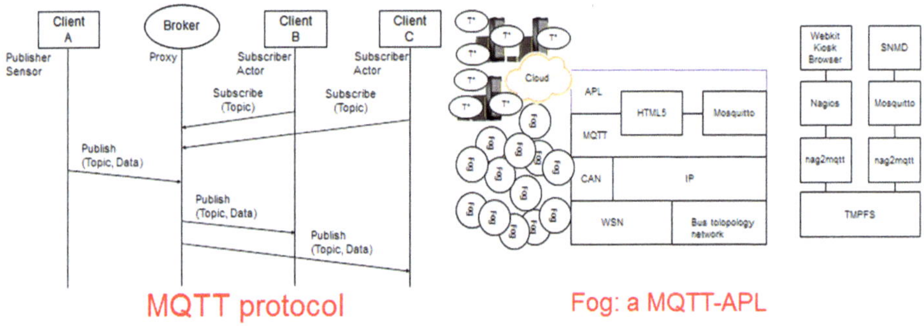

Fig. 6.28 Case Study: MQTT-solution for IoT and Fog (own representation based on [7, 11])

TMPFS as well as interfaces to Web and Nagios. The Mosquitto tool is an IoT-project (iot.eclipse.org) and provides a broker for MQTT and other protocols and tools. SNMD (Sensor Node Management Device) is SNMP-based tool as well as an efficient interface to Nagios and Moscuitto. The TMPFS (temporary filesystem) is a file system which is used in many Unix-like operating systems to create a RAM disk. With TMPFS some parts of the real working memory can be integrated and described as a hard disk [8, 18]. The examined scenario shows the energy-efficient co-operation "fog—private cloud provider" as well as provides PW-factor increasing.

References

1. Handbuch IoT (Online): https://www.handbuch-iot.de/.
2. Mahmood Zaigham (Ed.): Fog Computing: Concepts, Frameworks and Technologies, Springer 2017, London, ISBN 978-3-319-94890-4.
3. Jamil Y. Khan, Mehmet R. Yuce (Eds.). Internet of Things (IoT): Systems and Applications, 2019, New York, Jenny Stanford Publishing, ISBN 9780429399084, 366 p.
4. Lueth, Knud Lasse. The 25 best IoT Platforms 2019 - Based on Customer Review (Online): https://iot-analytics.com/the-25-best-iot-platforms-2019/.
5. A. Luntovskyy, J. Spillner. Architectural Transformations in Network Services and Distributed Systems: Service Vision. Case Studies, Springer Nature, 2017, 344p. (Monograph, ISBN: 9-783-6581-484-09).
6. Luntovskyy, A., Gütter, D. Moderne Rechnernetze - Lehrbuch, Modern Computer Networks - Handbook, Springer (2020), 481 p. + 265 pict., ISBN: 978-3-658-25616-6 (in German, 11 July 2020), vol.1.
7. Luntovskyy, A., Gütter, D. Moderne Rechnernetze - Übungsbuch, Modern Computer Networks - Exercises, Springer (2020), 145 p. + 44 pict., ISBN: 978-3-658-25618-0 (in German, 11 July 2020), vol. 2.
8. Workshop der Fa. IBH IT-Service GmbH Dresden am 16.3.2017.
9. Machine Learning (Online): https://computingeducation.de/proj-ml-uebersicht/.
10. A. Luntovskyy, L. Globa, B. Shubyn. From Big Data to Smart Data: The Most Effective Approaches for Data Analytics, in Advances in Information and Communication Technology and Systems, Springer Nature, Part of the "Lecture Notes in Networks and Systems Series", MCT 2019, Vol. LNNS 152, Chapter 2, pp. 23–40, Sept. 2020 (ISBN: 978-3-030-58359-0, DOI: https://doi.org/10.1007/978-3-030-58359-0_2).
11. Andrew S. Tanenbaum, David J. Wetherall (2012): Computernetzwerke, 5. aktualisierte Auflage; Pearson Studium, 2012; ISBN: 978-3-8689-4137-1, 1040 Seiten.
12. Andrew S. Tanenbaum and Maarten Van Steen (2013): Distributed Systems: Principles and Paradigms. Pearson, 2013, 633 p.
13. Andrew S. Tanenbaum, Todd Austin (2014): Structured Computer Organization, Prentice Hall, revised edition, 2014, 802 p. (ISBN 978- 0132916523).
14. Andrew S. Tanenbaum, Herbert Bos (2016): Modern Operating Systems, 4th Edition, Pearson, 2016, 1313 p. (ISBN 978-0133591620).
15. Alexander Schill, Thomas Springer (2012): Verteilte Systeme - Grundlagen und Basistechnologien; 2. Auflage, Springer-Verlag, 2012, 433 S., ISBN: 978-3-642-25795-7.
16. Luntovskyy, Andriy. Energy-Efficient Network Services as Smart Grid Issue, Soft Computing in Computer and Information Science, DOI 978-3-319-15147-2_25, 2015.
17. Luntovskyy, Andriy, Dyadenko, Alexander (et al.). Multimedia Services Differentiation in 4G Mobile Networks under Use of Situational Priorities, Energy-Aware Communications, DOI 978-3-642-23541-2_21, 2011.

18. Jörn Lehmann. Wirtschaftsverbände arbeiten bei Industrie 4.0 eng zusammen, 2017, VDMA, ID 979115 (Online): http://www.vdma.org/article//articleview/979115/.
19. Shanhe Yi, Zijiang Hao, Zhengrui Qin, Qun Li (Online): Fog Computing Platform and Applications, Dept. of Computer Science, College of William and Mary, 2017: http://www.cs.wm.edu/~zhengrui/.
20. C.Fung, R. Boutaba. Intrusion Detection Networks: A Key to Collaborative Security (ISBN: 978-1466564121), 2013, 261p.
21. Andriy Luntovskyy, Dietbert Guetter, Igor Melnyk (2011): Planung und Optimierung von Rechnernetzen: Methoden, Modelle, Tools für Entwurf, Diagnose und Management im Lebenszyklus von drahtgebundenen und drahtlosen Rechnernetzen. V+T/ Springer Fachmedien Wiesbaden GmbH, 2011, 435 Seiten (ISBN 978-3-8348-1458-6).

Intelligent Networking and Bio-inspired Engineering

7

The given chapter discusses how active deployment of modern networks in Industry 4.0 stimulates the further development of the novel software technologies for IoT, Fog Computing and Robotics [1–7]. Such novel applications and mobile apps operate nowadays on advanced hardware and software platforms. The approaches to their development are often aimed on productive and energy efficient Cloud and Fog cooperation scenarios. The important paradigms and relevant approaches for development of so-called novel SLMA (Server-Less Mobile Apps) have been discussed. An overview of the up-to-date robotics applications and platforms was given. The existing platforms and frameworks for their development are considered as well as analysis of the properties of the future robot platforms is offered. The specials kind of the advanced web apps, so-called PWA (Progressive Web Apps) are suitable for the purpose of spared data use, autonomous access and participation on clouds, increasing of performance, fault tolerance, as well as of online activity of the users [1–9].

The energy-efficient intelligent networking becomes step-by-step one of the most important components of the 4th-industrial revolution called "Industry 4.0". The chapter discusses how active deployment of modern networks stimulates the further development of IoT and fog computing. The processes of the involving of IoT and fog will be curry out in mid-term, within a few next years. Some case studies are provided. Bio inspired networking and engineering find their spread deployment in "Industry 4.0" too.

7.1 Backgrounds: "Industry 4.0" and Intelligent Networking

The algorithmically and energy-efficient intelligent networking becomes step-by-step one of the most important components of the 4th-industrial revolution called "Industry 4.0". The chapter discusses how active deployment of modern networks stimulates the

A. Luntovskyy and D. Gütter, *Highly-Distributed Systems*,
https://doi.org/10.1007/978-3-030-92829-2_7

further development of mobile apps, IoT and fog computing (Fig. 7.1). In the future "green" technologies will be wide-spread. The services will be often shifted from the clouds on „the network edges". The energy autarky can be realized via self-sufficient and low-energy nodes. Such interoperable software solutions and platforms help to the providers with energy saving and enhancing of resource efficiency. Together with the client–server and cloud-centric applications and apps the novel applications are permanently developed. These apps must also provide autarky in sense to be only loose coupled to the servers. One of the further goals of the chapter was to introduce the readers the basic knowledge about the modern, self-organizing networks and robots. These contents will be an important part of the teaching at the BA Dresden as well as TU Dresden in the future. Modern intelligent and energy-efficient networks, IoT and fog stimulate the further development of "Industry 4.0". The further works are focused on programming of new scenarios on bio-inspired engineering, deployment of the robots to pedagogical aims as well as behavioral human modeling. The "Industry 4.0" (2011) is one of the future important aims in the German Federal Government's high-tech strategy. The driving force [10] for the new generation of the industry (new industrial revolution) is the further automation and computerization of production processes via IoT, fog and cloud computing based on intelligent networking. The goal is as follows: smart factories have to be characterized by

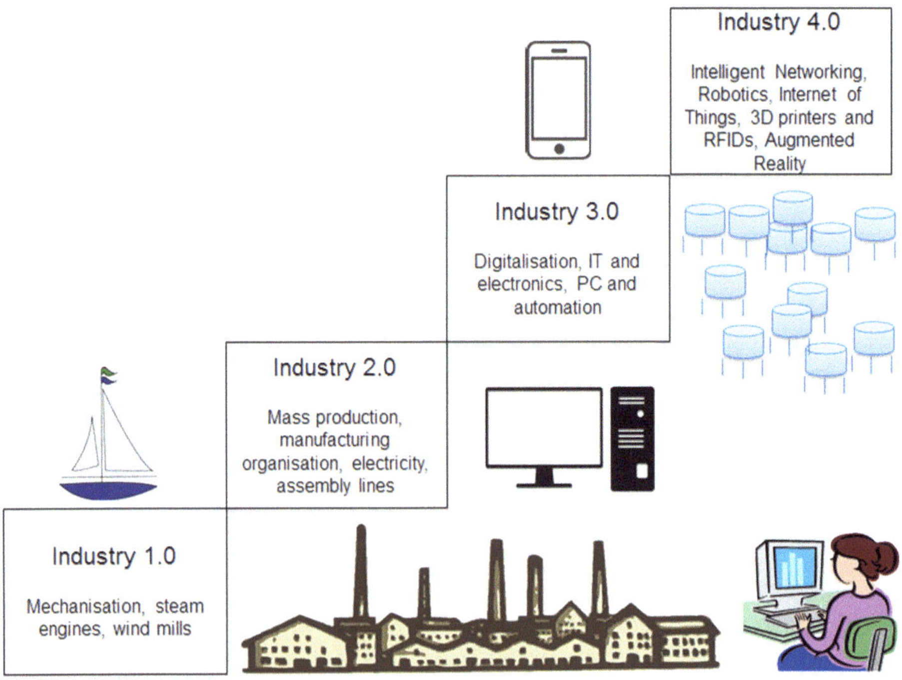

Fig. 7.1 Context "Industry 4.0" or the four industrial revolutions (own presentation)

adaptiveness, resource efficiency and ergonomic working conditions as well as the integration of customers and business partners into the business value chain [10].

Within Industry 4.0, information and communication technologies as well as automation and production technologies become increasingly and more than ever interfaced to each other. The political ambition is to defend and extend the traditional core of the German industry with its internationally outstanding positions. The manufacturing processes are not only automated (like in Industry 3.0) but also the processed components obtain their additional intelligence via the equipment with not expensive chips (processors, sensors, wireless mini-senders and RFIDs). As further important aim the integration of customers and business partners in an optimized value chain has to be reached through their extensive networking and under taking account of aspects of data security, privacy and anonymity [10–13] in these chains. Therefore, we discuss the novel software solutions in this scope: mostly mobile and interfaced to the PHY-world; compact, energy- and algorithmic-efficient; fog and cloud based; progressive, loosely-coupled to the server part, often server-less.

The next sections of the chapter are organized as follows:

1. A periodization for the novel software.
2. Techniques and trends to PWA, mobile OS and SLMA.
3. Robotics SLMAs and platforms.
4. Case studies for fog and robotics.

7.2 Trend to the Server-Less Mobile Apps

7.2.1 A Periodization for the Novel Software

Within the history of the development of the server based and serverless applications there were permanently the fluctuations between fat and thin client pattern. The both possess some cons and pros in dependence from the requirements, network QoS and use context. E.g. in 1994 Sun Java terminal as thin clients were favorited, but the lack of network performance forced to change to Web Services in 2003. Then P2P systems as serverless structures gained on popularity, then the Cloud based solutions became a trend (2011) under predominant use of the load balanced thin clients with functionality delegation to the clouds. A periodization of development of the server based and serverless apps is depicted in Fig. 7.2.

7.2.2 Distribution Techniques and a Trend to the Serverless Mobile Apps

So-called Serverless Mobile Apps (SLMA) are examined below. The algorithmic efficiency and the performance are the mostly important conditions for so-called serverless

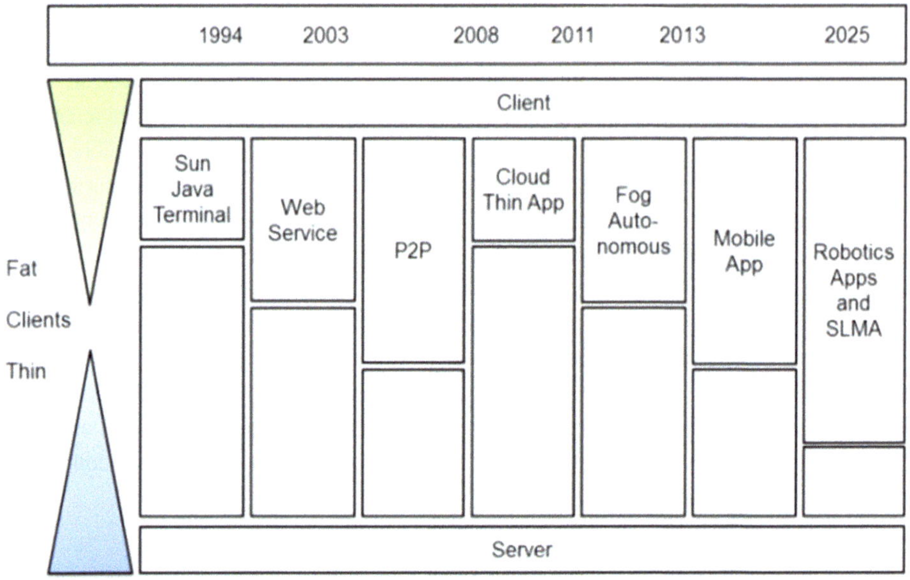

Fig. 7.2 Periodization of the server based and serverless applications (own presentation)

mobile apps. The illustration to the performance optimization for client–server systems is given in Fig. 7.3. The following main approaches are generally used to provide the increasing of the QoS in the novel software systems, i.a. for Serverless Mobile Apps, i.e. SLMA [14–17]:

- Server replication (S) via the available gateways (GW) for more availability and reaction time minimization.
- Multithreading on the client (C) for fine parallelization with the goal of the speedup of the run time.
- Caching of URLs, objects, components, data on the client often used aimed for shorten of the network access.
- P2P systems without fixed ordering of servers (P—peer) and with so-called Autonomous Participation in Cloud Services.

SLMAs can be divided as follows:

1) conventional or traditional CPS, i.e. the applications embedded in machines;
2) cloud centric IoT;
3) serverless Fog computing as well as
4) the robots.

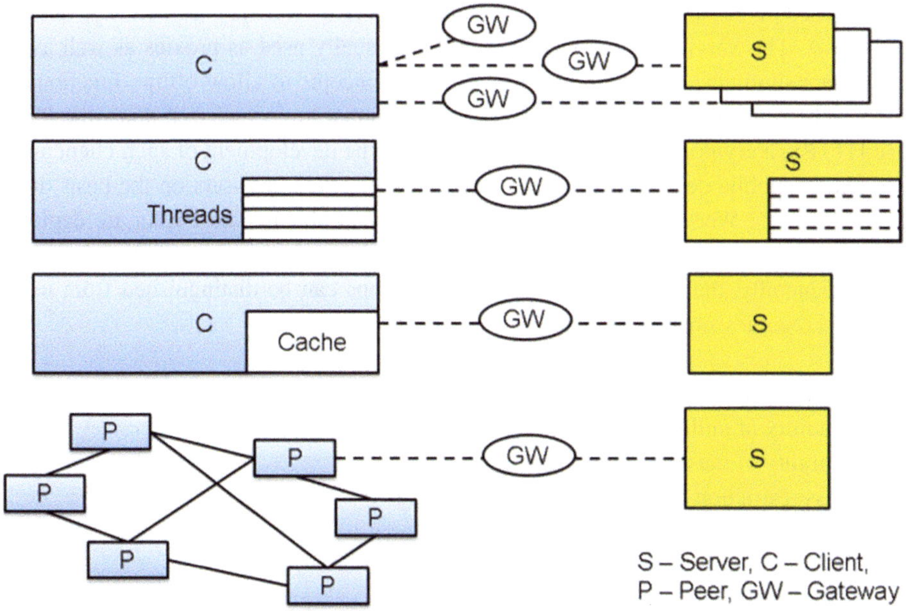

Fig. 7.3 Performance optimization methods for the server-based and server-less applications

The mostly used client–server processing paradigms are as follows: fat and thin clients. The last mentioned are nowadays frequently in use in context of the cloud technology. The modern cloud services provide high performance for the distributed computing and support multimedia data transfer from hidden clusters and storage media to the outside. By leveraging of virtualized computing and storage resources as well as of Web technologies (Web clients, Web services, file system WebDAV etc.), the clouds permit scalable, network centric abstracted IT-infrastructures, platforms and applications "on-demand". The billing of these services is userdependent, while the access is realized by the thin clients. As a disadvantage for cloud technology, the following must be noted: non-uniformity of the data security and safety aspects [1, 12, 14, 18].

7.2.3 Progressive Web Apps

Use of so-called "Progressive Web Apps" (PWA) can now become an efficient alternative to native mobile apps. Instead of the complex development and installation of native apps, the web browser can be used together with optimized web apps. PWAs are also more efficient than plain web apps [7]. PWA is a website with multiple characteristic functions of a native application. Such an app is normally implemented using the standard techniques for RWD (Responsive Web Design) like HTML5, CSS3, JavaScript and can be used across all platforms. The so-called “Service Workers” are herewith the core of a PWA. The Service Workers [1, 4, 7, 12] are usually used as proxies as well as OS containers simultaneously with different caching concepts to allow offline functionality for a web app or web application (Fig. 7.4). In contrast to normal web apps, the PWAs use HTTPS, a cryptographically secured protocol. The development of such client applications for mobile devices for existing versions of the web browsers on the basis of so-called “Service Workers” is an up-to-date technical problem. The PWAs are deployed with the purpose of the increasing of performance, fault tolerance, online activity by the users. Generally, the so-called Progressive Web Apps can be distinguished from native apps and regular web apps by the following quality criteria:

- Reduced conversion by installation and update Process
- Findability of online content
- Linkability of the available information
- Access restriction and optimized memory usage.

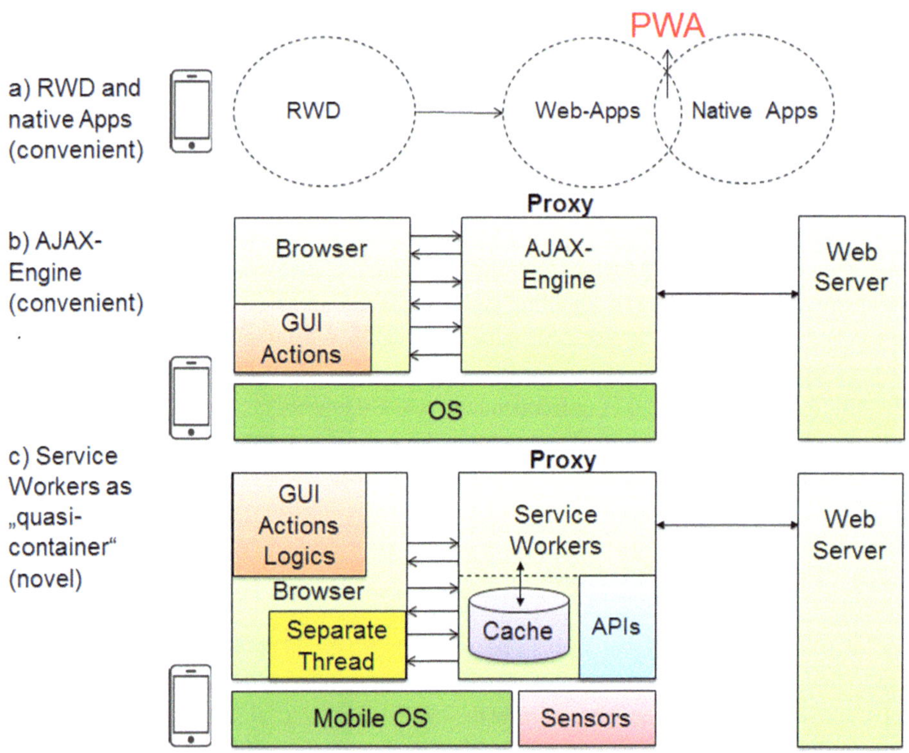

Fig. 7.4 Comparison and demarcation of PWA to the conventional technologies [1, 13, 14]

Figure 7.4 depicts a comparison and demarcation of PWA from regular RWD, AJAX concepts as well as from container technology (sandboxes). The complexity of the installation process for novel PWA is significantly reduced for a pair clicks in comparison to the native apps. The advantages of PWAs under use of Service Workers are as follows:

- Web App in form PWA behaves like a native app and can be stored on the smartphone's home screen.
- Service Workers enable the offline use of a PWA even by partial network decoupling (3G, 4G, Wi-Fi).
- Data volume and traffic consumption can be reduced.

Additionally, it will profit by the commercial and marketing effects. The users are better connected to the advertised product through efficient push notifications which are sent via PWA. The positive user experiences are improving the user loyalty to the advertised products by increasing of the conversion rate via spared data rates and volumes. The development tools for PWAs with Service Workers can be supported by Ionic Lab framework as well as by the specialized Google Lighthouse and Chrome Canary. The mentioned frameworks offer many interfaces (APIs), which are necessary for the use of PWA.

7.2.4 Co-operation Architectures and Technical Platforms "Fog-Cloud"

The last times the new paradigm, called "Fog Computing" (also "Edge Computing") obtains more and more meaning. Fog Computing is also a distributed computing environment that supports the IoT approach and brings processing closer to periphery with the augmentation of the cloud functionality [1–6, 12]. The fog energy autarky can be realized via the wireless self-sufficient low energy nodes on "the edge" e.g. RFID, NFC, ZigBee, EnOcean, 6LoWPAN which are interoperable with 5G nets. The autarky fog nodes provide the monitoring of technological processes, of hospital areas, of environmental contamination, and facility management as well as can cooperate with the clouds and deliver the acquired data. The services and preprocessing are shifted on „the network edge", i.e. closer to the users (partially or completely). The communication between a cloud and the fog nodes is performed via the three planes (refer the figure) aimed to flexible data preprocessing, VM and functionality migration as well as load balancing. The first Fog Cloud cooperation platforms exist already [1, 4].

The demarcation of the Fog to other systems like traditional CPS (cyber-PHY-systems) and cloud centric IoT with their interoperability aspects are shown in Fig. 7.5. Traditional CPS (cyber-PHY-systems) are compared with cloud centric IoT and the loosely coupled Fog solutions based on light weighting protocols like REST, MQTT etc. The distinguishing between the PHY-world, heterogenic interfaces, analytic blocks, coordination agents, adaptive role-based interfaces as well as a flexible data flows to the fog and cloud demarcation are used within the presented architectural solutions for the novel SLMA. The used "Fog-Cloud" platforms (refer Figs. 7.5 and 7.6) provide cloud and fog computing interoperability together with efficient desktop and mobile access via smartphones and tablets. The small intelligent "stuff" cooperates with "big clouds" and provides optimal load balancing as well as costs expenditures [1–6]. For deployment of such kind of cooperation platforms is supported via the industrial IoT gateways from Siemens, Intel, Bosch etc. Fog applications enable optimize service latencies, network bandwidth and available QoS. The security, privacy and anonymity in fog solutions are an open common research topic nowadays [1–6], and, therefore, require more detailed discourse. Concrete technical platforms and software solutions for Fog Computing are rare. They remain mostly a vague technical concept to be fully realized within the next years. Still, a few preliminary architectures exist. One such implementation platform to cloud and fog computing interoperability is offered in [1–6]. Such platform (Fig. 7.6) possesses the following structure: five layers (0–4), inclusive SDN and hypervisors, as well as multi-layered stack (3–4 layers) for digital appliances and sensor nodes. The platform is oriented to heterogeneous applications. The cloud contains primarily all necessary elements: CPUs, RAM, HDD, SSD, DAS as well as interfaces and combines the

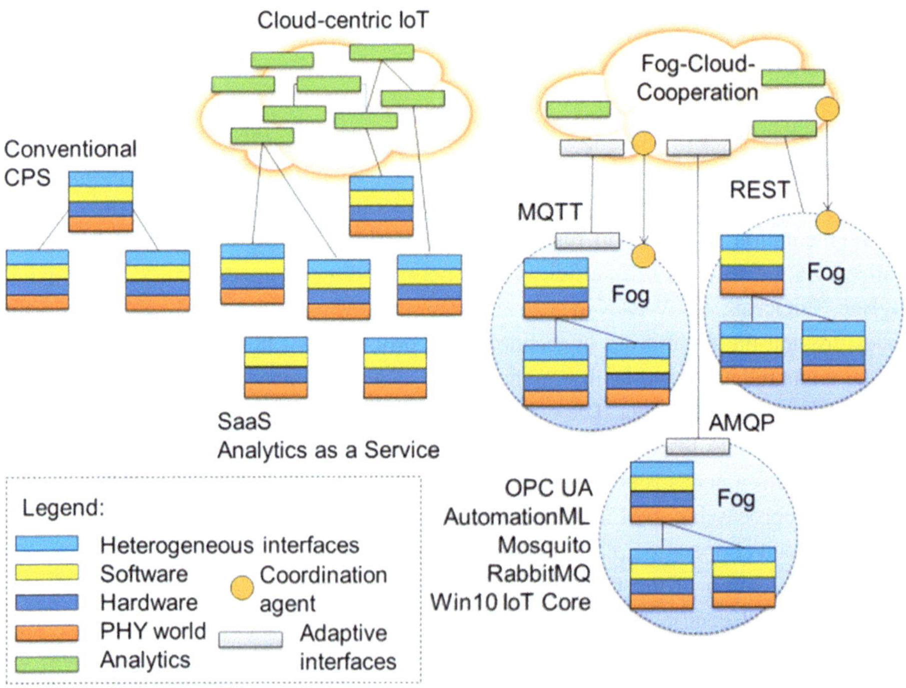

Fig. 7.5 Fog demarcation and interoperability aspects [1, 13, 14]

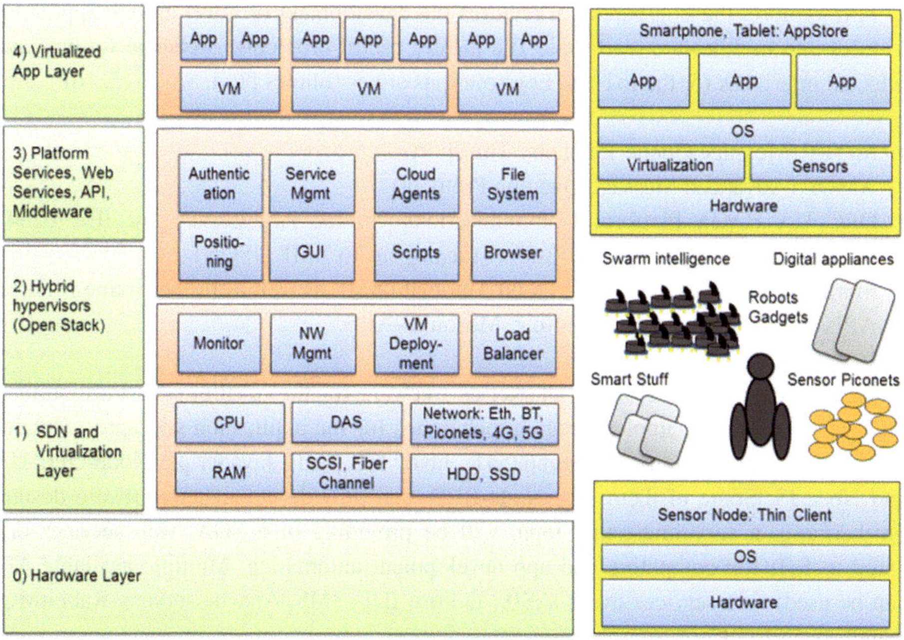

Fig. 7.6 Cooperation Architecture for Fog (Smart Stuff, IoT) and Clouds [13, 14]

advantages of image-streaming with template-based management. The configuration has to be built on the fly, based on the configuration profiles, just, like e.g. by HPE [1, 4, 5, 13, 15, 16]. The usable server profile templates are as follows:

- shared hypervisors, VM and physical OS and containers
- computing nodes, storages as well as DAS (Direct Attached Storage)
- coupling interfaces like Ethernet, SCSI, FC, iSCSI, FCoE etc.
- The fog part can contain as a rule:
- thin clients on the sensor nodes under tiny OS (ARM).
- thin clients for small robots (swarm intelligence).
- the mobile apps offered via digital appliances and so-called Smart Stuff.

By the co-operation between Clouds and Fog systems, the influence of transfer times and delays as well as service failure probabilities should be taken into account. The transfer of a computing task to a cloud can cause real time problems because the execution can be delayed. One of the important construction principles for Fog Computing is namely real time capability. In the face of the latencies within mobile networks which were significantly reduced down to several microseconds: e.g. 5 ms for 4G mobile radio as well as 1 ms for the 5G, which will be deployed by year 2020.

Furthermore, the Fog paradigm requires the development of novel mobile app architectures and middleware platforms. The mostly useful protocols, exchange formats and selected interfaces for the IoT and Fog solutions are as follows [1, 4, 5, 13, 15, 16]:

- MQTT (Message Queue Telemetry Transport)
- AMQP (Advanced Message Queuing Protocol)
- OPC UA (Open Platform Communications Unified Architecture) or IEC 62541 Specification Google Protocol Buffers (analogous to XML, JSON)
- Markup Languages, e.g. XML-based AutomationML as well as further frameworks
- APIs and IoT-oriented services (e.g. Mosquito-API).

They are today available on the market as freeware for the practitioners. Furthermore, there is a diversity of implementation techniques for the application and apps: containers in C++, applications in Open62541, Java, MS.NET, JavaScript, Node-OPC-UA and Node.JS. Some of them, e.g. Node-RED, can provide automated software design. Furthermore, a very interesting topic will be providing of a SOA/ web services oriented to IoT, fog computing and app development automation. Multiple available API can be used: web services over OASIS, Eclipse IDE, JMS, Apache servers, RabbitMQ, Windows 10 IoT Core etc.

7.2.5 New Paradigms with Google Fuchsia

Nowadays, along with the existing OS from Google called Google Android and Google Chrome OS, a completely new, next generation real time OS is created to supporting apps from such domains like desktop, mobile, Smart Stuff, CPS and IoT. Google Fuchsia OS [19] promises a significant contribution to the development paradigms for mobile apps in midterm [17]. One of the prompting factors for the development of the Fuchsia project was the clear desire of Google to get rid of the dependence on Java as soon as possible. Fuchsia OS is built as a real time OS on an efficient microkernel Zircon but not a Linux kernel (cp. Google Android and Google Chrome OS). A hybrid architectural model (refer Fig. 7.7) based on microkernel and quasiconsistent layers enable efficient memory management, communication between the expanding server and application processes, as well as network communication.

Fuchsia OS provides more data security based on the concept of microkernel with surrounding server processes and extending quasiconsistent layers. The application processes (A) order to the server processes (R) the requests (Req) for the app execution. The execution follows only through the secured microkernel. The microkernel delivers the responses (Rsp) with the resulting data and visualizations itself. In addition, the HAL (HW Abstract Layer) was aimed to supporting of hardware and software drivers for sensors and OS components. The apps development is based on BSD, MIT, Apache licenses and mainly oriented to Dart (proprietary client sided language for native apps within

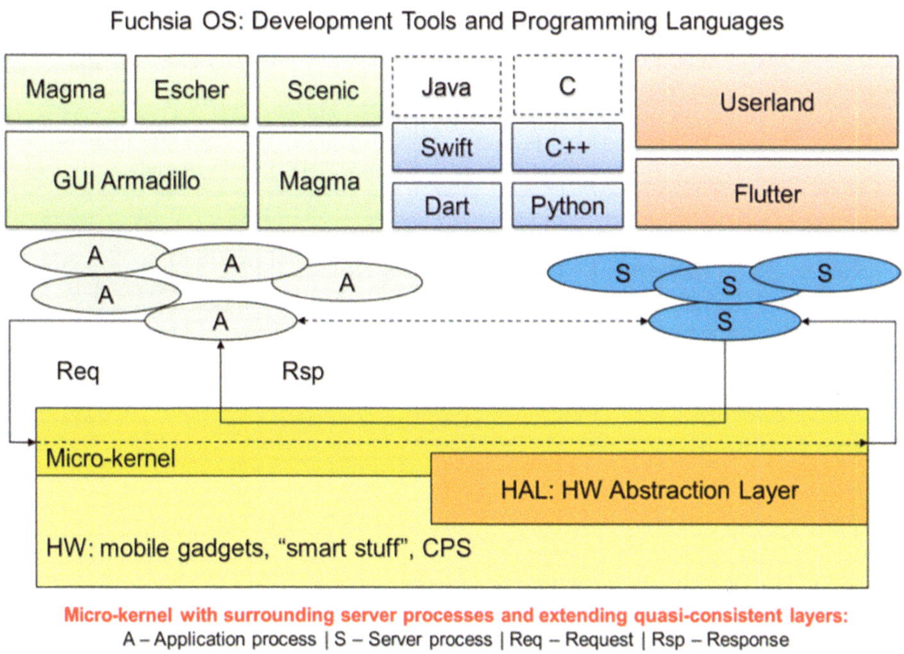

Fig. 7.7 Fuchsia architecture based on microkernel [14]

Fuchsia), Swift (Apple's for compatibility with iOS), Python and some further programming languages. GPL Java will be used no more within Fuchsia as well as some further PL like e.g. "plain old C". This offers more freedom in the use of the development tools and frameworks as well as affects positively the endprice of apps to be developed. The current state of the Fuchsia OS development is as follows: voluminous documentation and an existing working prototype are already available.

7.3 Server-Less Apps for Robotics

So-called SLMA (Server-Less Mobile Apps) is per definition a composite application executed in parallel across physical and virtual spaces [5–14]. The SLMA are nowadays only partially cloud based and based on the CPS (cyber-PHY-systems). SLMAs are divided as follows: conventional or traditional CPS, i.e. the applications embedded in machines, cloud centric IoT, serverless Fog computing, as well as the robots. The following Fig. 7.8 depicts the SLMA-related terms and architecture components for robotics [5–9, 14].

On this place, I would like to cite the following opinion of [3, 7, 9, 16]: "I had a strong feeling that a meaningful robotics scientific and ethic laws could be developed,

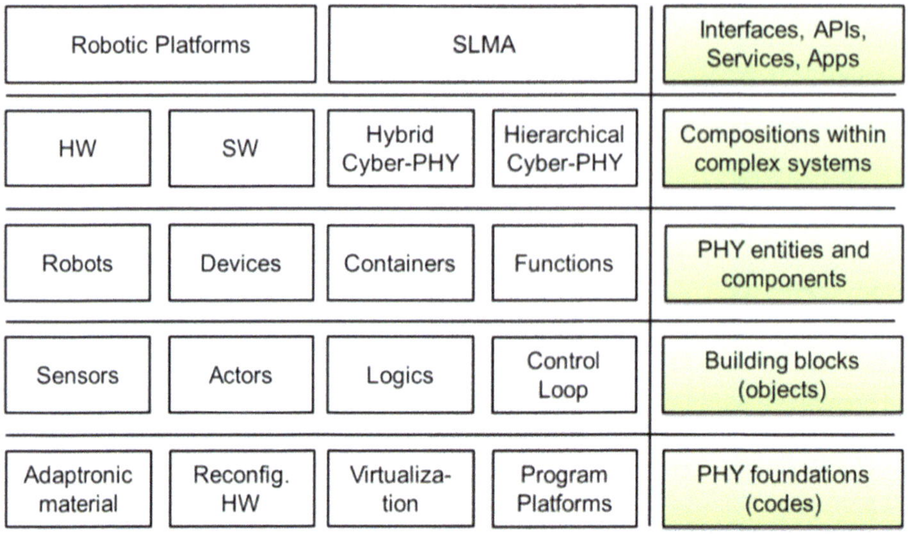

Fig. 7.8 SLMA and robotics: composite applications across physical and virtual spaces [1, 13, 14]

and that it would be best to think in terms of general concepts rather than concentrate exclusively on particular devices" (refer Table 7.1).

Based on the available statistics [5–9, 14] a fastest growth of professional service robot sales till 2020–2025 can be recognized, i. a. mobile service robots with SLMAs. There are already ca. 5 million personal/domestic robots and ca. 50,000 service robots and drones in professional domains have to appear till 2020. Due to their diversity let's to give a shorten classification. The following professional domains are recognized with the approximated robotics quota: defense (45%), agriculture (21%), logistics (7%), medical (5%), as well as industrial (22%). The robot architectures and platforms possess [3–14] the following specifics:

- Client centric, loosely coupled to the server
- Dominant "Sense-Plan-Act"-paradigm with mobility
- Standing in interaction, often asynchronous, and in real time under uncertainty.

Furthermore, the mostly used robots' taxonomies are as follows (cp. Fig. 7.8):

- programmable, multifunction devices
- domain of use: industrial, service, entertainment, medicine, household
- appearance: hard vs. soft hardware
- movement: walking, driving, flying, stationary (refer Fig. 7.8).

Table 7.1 The codex of the robot ethic laws

The "Three Laws of Robotics" (1942) by Isaac Asimov from his story "Runaround" as appropriate inspiration can be cited as follows:	In 1974–1986 the additional laws 4–6 of robotics were introduced by L. Dilov, N. Kesarovski, V. Ivanov, H. Harrison:
1) A robot may not injure a human being or, through inaction, allow a human being to come to harm	4) A robot must establish its identity as a robot in all cases
2) A robot must obey the orders given it by human beings except where such orders would conflict with the First Law	5) A robot must know it is a robot
3) A robot must protect its own existence as long as such protection does not conflict with the First or Second Laws	6) A robot must reproduce. As long as such reproduction does not interfere with the First or Second or Third Law

Due to the shifted accent on the performance, energy and algorithmic efficiency for the SLMAs the following important algorithms have to be optimized [3–14]:

- Robot movement and trilateration in 3D-space (unrestricted movement: communication—computation; stationary and mobile robots)
- Robot body Collision Detection with Deadlock Prevention (sensing and routing; autonomous relative positioning)
- Path planning autonomous driving.

The specific methods there are too:

(1)no planning, only reaction or
(2)mapping and localization + path planning.

The responses are waiting in varying temporal scopes: ms for feedback (deployment by 5G mobile is good possible), hours for complex tasks (often to clouds delegated). Commonly, the used nowadays robotics platforms can by classified [3–14] to the four following categories (refer Fig. 7.8):

- Hardware platforms (for walking, driving, flying, stationary programmable and multifunction robots)
- Coordination platforms (e.g. for swarm robots)
- Development platforms (language specific, oriented to SLMA for all types: industrial as well as service robots)
- Runtime platforms (SLMA, all types).

SLMAs are supported via multiple robotic platforms, which are so much important for the further development. Some examples of robot kinds and maintaining platforms are as follows [11–15]:

- Harvard Swarm Robots supported via Amtel WinAVR and Amtel AvrStudio tools
- Robotis BIOLOID robots with RoboPlus Motion platform
- Aldebaran Robotics with NAO Emotion robots
- Webots development platform
- RobotOS runtime architecture (ROS with so-called Autonomous Participation in Cloud Services)
- extended Texas Instruments Architecture for robotics.

as well as the recent development standards for robotics (s. below).

To the recent official and industrial development standards for robotics, belong as follows: Texas Instruments Architecture, ISO 8373/TC 184. ISO 8373 contains a definition for industrial robots: automatically controlled and reprogrammable multipurpose manipulators in three or more axes. ISO TC 184 is a movement-focused standard under development. The future requirements are as follows: robot swarms; code offloading from robots; virtual robots and machine learning [8, 16].

7.4 Case Study: Kilobots and "Bio-inspired Engineering"

"Bio-Inspired Engineering" investigates how to use the biological methods and models in industrial processes generally as well in "industry 4.0". Inter alia the swarm intelligence of socialized animals has been taken into account. The scientists from Harvard University (Cambridge, Massachusetts) discovered how to increase the efficiency of a swarm of 1024 = 1 K small robots (so called "kilobots"). The kilobots [8] are the small collective operating robots which can model a kind of "swarm intelligence" that is the property of socialized animals like fishes, ants, cranes, bees etc. Under use of specialized programming tools, "the machine armada" can arrange itself different 2D-forms (Fig. 7.9). The kilobots can be programmed to imitate the forms of fish and bird swarms as well as other socialized animals and biological forms. In their construction, the kilobots are not expensive microprocessor boards with three twiggy legs, two vibrating motors which help them go forward or rotate. The both motors can be configured using the 255 available power values and give the robots a differential drive. Each kilobot possesses few advanced sensors to be coordinated, e.g. a RGB LEDs and IR-sensors, as well as a small CPU ATmega-328 (8-Bit/8 MHz clock) [20–25]. The kilobots negotiate and communicate with their neighbors on distances till to 7 cm via IR-reflection on the ground. Via receiving a message per sensors, the distance to the kilobot that it sent can be determined by their signal power. The ambient light can also be detected by kilobot. As IDE in C and open source GUI the software Amtel WinAVR and Amtel AvrStudio

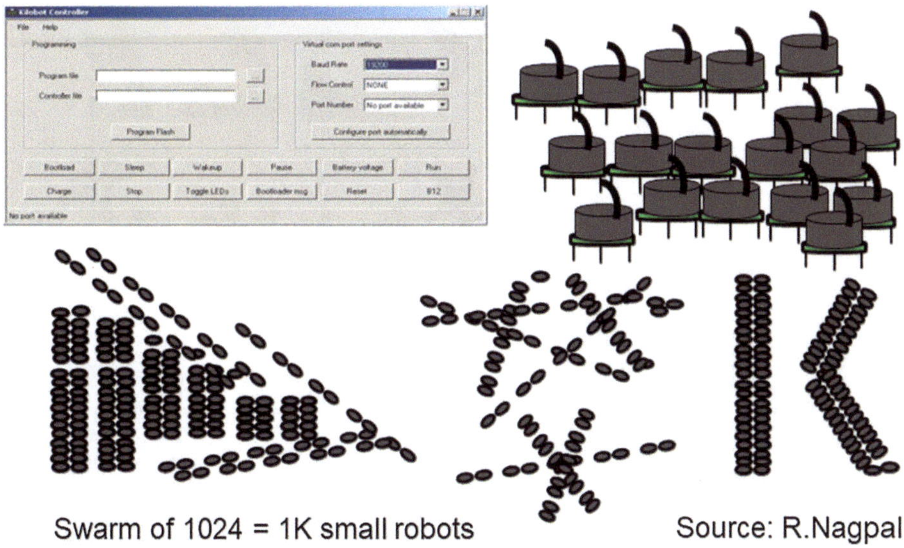

Fig. 7.9 Case Study: Swarm robots 2D-configurations and programming tools [9–14]

is available (cp. Fig. 7.10). A swarm of kilobots is programmed together as a wireless cluster. The kilobots can interact, avoiding each other, playing follow-the-leader, or even performing simulations of nutritional processes. The deployment aims for kilobots are quite different:

- The modeled 2D-structures allow comprehending the behavior of human masses in order to avoid the emergences or in the mass panic situations as well as help to handle easier such situations.
- Furthermore, the behavioral aspects of the colonies of the microorganisms or animals can be modeled.
- As further deployment examples certain urban as well as suburb construction and development scenarios can be considered as well as some historical scenarios of the propagation of the human populations too.
- The collaboration in the clean-up actions for environmental protection or in the periods after disasters and after environment contaminations can be reconstructed under use of the kilobots.
- And last but not least, in midterm the modeling of the millions of selfdriving cars (Google's Waymo or Tesla autopilots) on multiple roads will be possible via the discussed techniques.

The main goal of such experiments with small und inexpensive robot consists of modeling of technical processes and engineering objects ("Biologically Inspired Engineering"

Fig. 7.10 The kilobot swarms and their possible 2D-configurations [8, 20–25]

by M. Rubenstein) [11–15]. The robots help to obtain an inexpensive and adequate perception about real of technical processes and engineering objects. The modeling of socialized behavior of some animals or even humans in a crisis or catastrophe situation can be reached via providing of large groups of robots working together to complete a formulated task that they can't do on their own (cp. Fig. 7.10).

"Bio-Inspired Engineering" [14, 20–25] investigates how to use the biological methods and models in industrial processes generally as well in "industry 4.0". Inter alia, the swarm intelligence of socialized animals has been taken into account. The scientists from Harvard University (Cambridge, Massachusetts) discovered how to increase the efficiency of a swarm of 1024 = 1 K small robots (so called "kilobots"). The kilobots [8] are the small collective operating robots which can model a kind of "swarm intelligence" that is the property of socialized animals like fishes, ants, cranes, bees etc. Under use of specialized programming tools "the machine armada" can arrange itself different 2D-forms (refer Figs. 7.9 and 7.10). The kilobots can be programmed to imitate the forms of fish and bird swarms as well as other socialized animals and biological forms [20–25].

In their construction, the kilobots are not expensive microprocessor boards with three twiggy legs, two vibrating motors which help them go forward or rotate. The both motors can be configured using the 255 available power values and give the robots a differential drive. Each kilobot possesses few advanced sensors to be oriented and coordinated e.g. a RGB LEDs and IR-sensors. The technical characteristics of the kilobots are [20–25]:

- Kilobot-CPU: ATmega 328 (8-Bit-processor with 8 MHz tact frequency).
- Kilobot-RAM: 32 KB flash for applications and the bootloader, 1 KB EEPROM for retrieving of the calibration values and other non-volatile data as well as 2 KB SRAM.

The kilobots negotiate and communicate with their neighbors on distances till to 7 cm via IR-reflection on the ground. Via receiving a message per sensors, the distance to the kilobot that it sent can be determined by their signal power. The ambient light can also be detected by kilobot.

How to program the kilobots?

As IDE in C and open source-GUI for kilobots the software Amtel WinAVR and Amtel AvrStudio is available (Fig. 7.11). A group of kilobots—a swarm—is programmed together as a wireless cluster.

The kilobots can interact, avoiding each other, playing follow-the-leader, or even performing simulations of nutritional processes. The deployment aims for kilobots can be quite different [8, 20–23]:

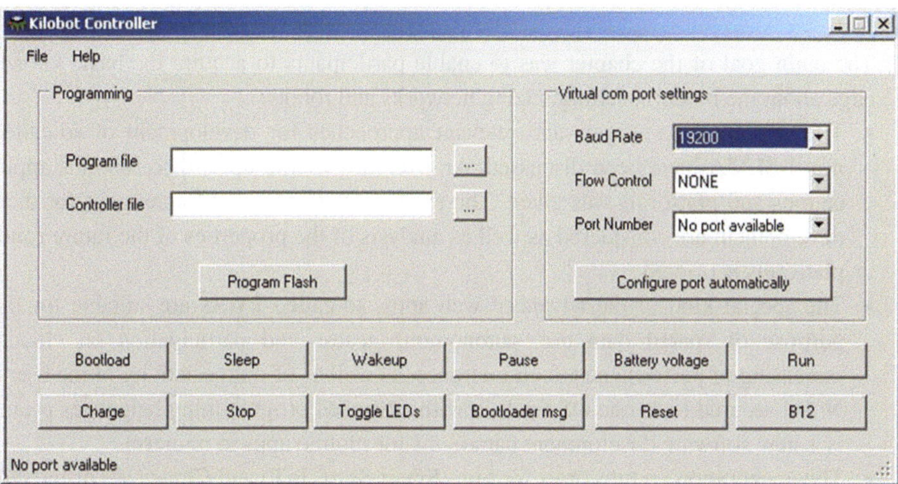

Fig. 7.11. The kilobots can be programmed

1. The modeled 2D-structures allow to comprehend the behavior of human masses in order to avoid the emergences or in the mass panic situations as well as help to handle easier such situations.
2. As further deployment examples certain urban as well as suburb construction and development scenarios can be considered as well as some historical scenarios of the propagation of the human populations too.
3. Furthermore, the behavioral aspects of the colonies of the microorganisms or animals can be modeled.
4. The collaboration in the clean-up actions for environmental protection or in the periods after disasters and after environment contaminations can be reconstructed under use of the kilobots.
5. Moreover, finally and yet importantly, in midterm the modeling of the millions of selfdriving cars (e.g. the Google's project titled Waymo [24] or Tesla autopilots) on multiple roads will be possible via the discussed techniques.

The main goal of such experiments with small und inexpensive robot consists of modeling of technical processes and engineering objects ("Biologically Inspired Engineering" by M. Rubenstein) [25]. The robots help to obtain an inexpensive and adequate perception about real of technical processes and engineering objects. The modeling of socialized behavior of some animals or even humans in a crisis or catastrophe situation can be reached via providing of large groups of robots working together to complete a formulated task that they can't do on their own (cp. Figs. 7.9 and 7.10).

7.5 Conclusions

1) The main goal of the chapter was to enable participants to acquire the basic knowledge about the modern, selforganizing networks and robots.
   - The important paradigms and relevant approaches for development of so-called novel SLMA have been discussed. An overview of the up-to-date robotics applications and platforms was given. The existing platforms and frameworks for their development are considered as well as analysis of the properties of the future robot platforms is offered.
   - The special kind of the advanced web apps, so-called PWAs are suitable for the purpose of spared data use, autonomous access and participation on clouds, increasing of performance, fault tolerance, as well as of online activity of the users. Novel secured real time OS Fuchsia with advanced programming languages promises new software development paradigms for mobile apps in midterm.
   - The cooperation architectures for Fog (Smart Stuff, IoT) and Clouds are discussed. The efficient access to the platform must be provided via mobile apps. Then it was concentrated on the case studies.

- In the future “green” technologies will be widespread. The services will be often shifted from the clouds on „the network edges”. The energy autarky can be realized via selfsufficient and low energy nodes. Such interoperable software solutions and platforms help to the providers with energy saving and enhancing of resource efficiency. Hence, such platforms are often heavy weighing.

2) The following case study regarding to the kilobots and “bio-inspired engineering” was examined. The further works are focused on programming of new scenarios on bio-inspired engineering, deployment of the robots to pedagogical aims as well as behavioral human modeling.
3) Our further goal is to provide the readers with a brief introduction to the latest state of robotics.
   - Within “Industry 4.0” more and more intelligent robots are used as well as the information, communication, automation and production technologies are increasingly interlinked.
   - These contents will be an important part of the teaching at the BA Dresden as well as TU Dresden in the future.
   - Modern intelligent and energy efficient networks, IoT and fog stimulate the further development of “Industry 4.0“.
4) The further works will be focused on programming of new scenarios on bio-inspired engineering, deployment of the robots to pedagogical aims as well as behavioral human modeling.

References

1. A. Luntovskyy, D. Guetter, M. Klymash. Up-to-date paradigms for distributed computing, 2nd Int. IEEE Conf. AICT-2017, July 2017, pp. 113–119, IEEE Xplore (DOI: https://doi.org/10.1109/AIACT.2017.8020078).
2. A. Luntovskyy, L. Globa, O. Stepurin. Performance-energy tradeoff models for distributed computing, 13th Int. IEEE Conf. TCSET-2016, Febr. 2016, pp. 613–617. IEEE Xplore (DOI: https://doi.org/10.1109/BlackSeaCom.2016.7901554).
3. J. Spillner. Autonomous Participation in Cloud Services, Dresden, 2012.
4. PWA (Online): http://t3n.de/news/wahre-konversionskiller-lange-594340/.
5. B. Siciliano, O. Khatib: Handbook of Robotics, Springer, 2008, 1624 p. (ISBN 978-3-540-30301-5).
6. O. Khatib, Stanford Univ. CA, USA, International Foundation of Robotics Research (Online): http://www.ifrr.org/.
7. J. Spillner. Current and Future Platforms for Robotics, Manuscript to a probe talk for Habilitation, Dresden, 2015, 33 p.
8. A. Cornejo, R. Nagpal. Kilobots: Distributed Range-Based Relative Localization of Robot Swarms. Algorithmic Foundations of Robotics XI, April 2015.
9. J. Spillner. ROS and Cloud Robotics, Teaching Lecture, Zurich, 2017.
10. Jörn Lehmann. Wirtschaftsverbände arbeiten bei Industrie 4.0 eng zusammen, 2017, VDMA, ID 979115 (Online): http://www.vdma.org/article//articleview/979115/.

11. NoDNA.de (Online): https://nodna.de/Bioloid-Premium-Robot-Kit-V2.
12. A. Luntovskyy et al. Intelligent Networking and Bio-Inspired Engineering, 2nd Int. IEEE Conference UkrMiCo'2017, Sept. 2017, Odessa, Ukraine, 4 p. (IEEE Xplore: http://ieeexplore.ieee.org/document/8095421/).
13. IBH IT-Service GmbH Dresden Workshop on 16.3.2017.
14. A. Luntovskyy. SLMA and Novel Software Technologies for Industry 4.0, on Int. Conf. ACS - 2018 on Advanced Comp. Systems, Szczecin - Międzyzdroje, 24–26 Sept. 2018, In: Pejaś J., El Fray I., Hyla T., Kacprzyk J. (eds) Advances in Soft and Hard Computing. ACS 2018, "Advances in Intelligent Systems and Computing", vol 889. Springer, Cham., Springer Int. Publishing, 12p. (Part of the Advances in Intelligent Systems and Computing book series; AISC, vol. 889, DOI: https://doi.org/10.1007/978-3-030-03314-9_16, ISBN: 978-3-030-03313-2).
15. Alexander Schill, Thomas Springer (2012): Verteilte Systeme – Grundlagen und Basistechnologien; 2. Auflage, Springer-Verlag, 2012, 433 S., ISBN: 978-3-642-25795-7 (in German).
16. Andriy Luntovskyy, Josef Spillner (2016): Architectural Transformations in Network Services and Distributed Systems: Service Vision. Case Studies, 350 p., April 2017, Springer Nature Verlag, ISBN: 978-3-658-14840-9.
17. Andriy Luntovskyy, Dietbert Guetter, Igor Melnyk (2011): Planung und Optimierung von Rechnernetzen: Methoden, Modelle, Tools für Entwurf, Diagnose und Management im Lebenszyklus von drahtgebundenen und drahtlosen Rechnernetzen. V+T/ Springer Fachmedien Wiesbaden GmbH, 2011, 435 Seiten, ISBN 978-3-8348-1458-6 (in German).
18. Shanhe Yi, Zijiang Hao, Zhengrui Qin, Qun Li (Online): Fog Computing Platform and Applications, Dept. of Computer Science, College of William and Mary, 2017: http://www.cs.wm.edu/~zhengrui/.
19. Google Fuchsia OS GitHub (Online): https://github.com/fuchsia-mirror/.
20. Radhika Nagpal. Kilobots, Harvard Univ., Cambridge, Masschusets (Online): http://www.eecs.harvard.edu/ssr/projects/progSA/kilobot.html.
21. Robotics for 10 Dollars (Online): https://sites.google.com/site/afron10dollarrobot/.
22. Swarm Lab (Online): http://www.theswarmlab.com/.
23. K-Team Corporation (Online): http://www.k-team.com/.
24. Google selfdriving cars (Online): https://waymo.com/.
25. Michael Rubenstein. A selforganizing thousand-robot swarm, Biologically Inspired Engineering (Online): https://wyss.harvard.edu/a-self-organizing-thousand-robot-swarm/.

8 Robotic Apps and Platforms: Mobility, Localization, Management and Security Aspects

The given chapter was dedicated to the analysis of the most important up-to-date trends in robotics. The following four aspects are discussed: Mobility, Localization, Management and Security. The aspect of security can be specially divided into Data Security and Safety, as well as Privacy. The above mentioned aspects are very important for the conception and development of the performant robotic platforms. They have to be implemented under consideration of the efficient distribution models and analytics placement options within the cloud and fog environments. The problematics of the unmanaged and unstructured "Big Data" must be overcome by the actual robotic platforms too. Additionally, they are oriented more and more to support of so-called virtual robots.

8.1 Motivation and State-of-the-Art

Nowadays, the number of robots that cooperate with the humans, machines and with each other in various fields of manufacturing and economics, everyday life, entertainment industry, social pedagogy is growing rapidly. The chapter is dedicated to the important scientific practical problem of the use of modern mobile serverless apps, which are capable of autonomous operation, and robotics supporting platforms, in particular, to their classification and standardization. The scenarios for the use of robots that are focussed not only on manufacturing and on modelling processes in the industry but also for behavioural modelling, among them on the use of robots in pedagogy and sociology, are considered [1, 2].

Cyber-PHY, IoT, sensor networks, Robotics (thick and serverless mobile applications), real time network applications (thin clouds clients) can generate large arrays of unmanaged, weakly structured, and nonconfigured data of various types, known as Big Data. With the acceleration of industrial development "Industry 4.0" processing of such

A. Luntovskyy and D. Gütter, *Highly-Distributed Systems*,
https://doi.org/10.1007/978-3-030-92829-2_8

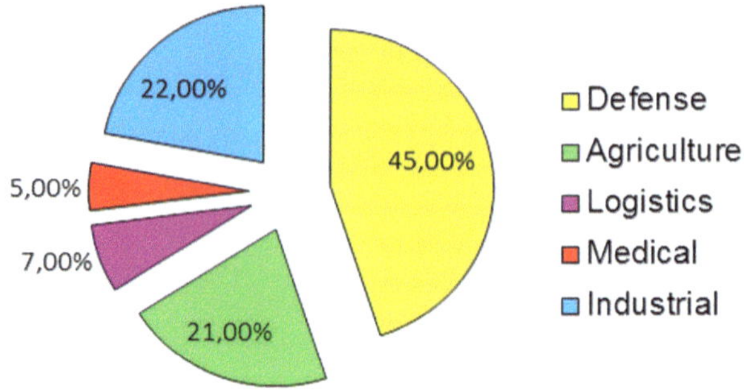

Fig. 8.1 Approximated distribution of the deployment fields for the robots

data became considerably more complicated. However, the so-called problem Big Data is hard to solve or resist nowadays! The chapter discusses the Best Practises and Case Studies aimed to overcoming of the Big Data problematics too.

Additionally, the robots, IoT, so-called Wearables are according to Gartner Consulting Corp. "potentially dangerous technologies" regarding the security and, especially, privacy [1, 2].

Based on statistics [3, 2] from IFR (International Federation of Robotics), rapid growth of sales of professional work robots is expected till 2020–2025. In the midterm, up to 5 million personal or home robots are needed [1–3]. Until 2020, it is possible to put into operation up to 50,000 service robots and drones in various fields of application for the following approximated distribution (Fig. 8.1): agriculture (21%), defense (45%), healthcare (5%), industry (22%), and logistics (7%).

8.2 Modern Robots Diversity and the Taxonomies

Table 8.1 represents the criteria for demarcation between the machines, computers, robots and, namely, humans by not only optical look and appearance. Due to the robots heterogeneity, the following classification of modern robots was offered (Robots Taxonomies) [1–5]:

1. By use area: domestic, service, industrial, medical, social and pedagogical.
2. By appearance: hardware based, software based, programmable, multi function devices.
3. By movement type: walking, motorized, driving, flying, mobile, and stationary.
4. By optical look: engines, gadgets, humanoids.
5. By control type (refer Fig. 8.2).

Table 8.1 Comparison "Machine–Computer–Robot–Human"

Criteria	Machine	Computer	Robot	Human
Mobility and motoric	0	-	+	+
Mechatronics	+	-	+	-
Freedom grades by movement	0	-	0	+
Programmed control	0	+	+	+ (instincts, reflexes)
Logic Output, Artificial Intelligence and ML	-	0 (software)	0	+
Micro-programming, flexible changeable instruction set	-	+ (micro-programming) assembler	+	+
Sensors availability	-	0	+	+ (senses)
Mechanical components	+	-	0	0
Notice:	0—partially compliant		+—fully compliant	

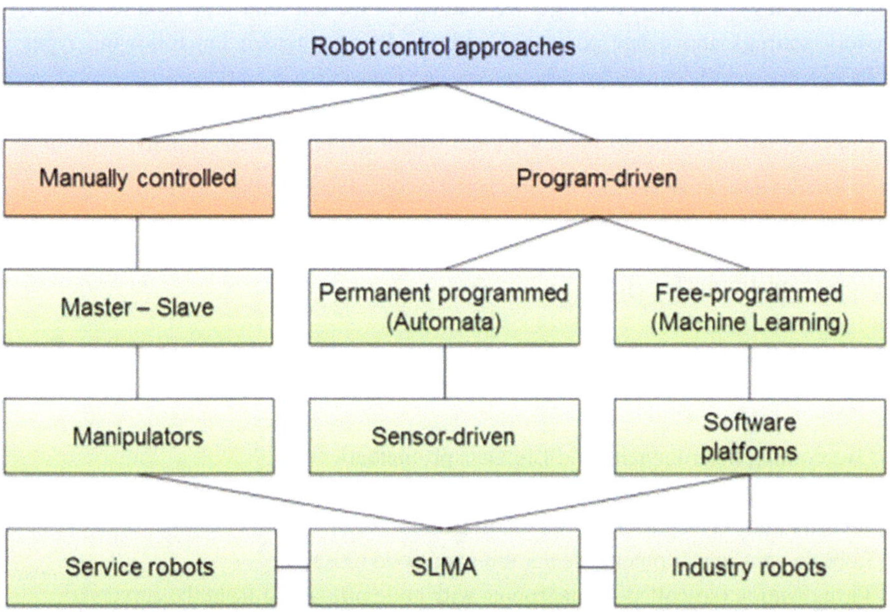

Fig. 8.2 Robots Taxonomies

Some examples of robot kinds and maintaining platforms are as follows [1–5]:

1. Harvard Swarm Robots supported via Amtel WinAVR and Amtel AvrStudio tools
2. Robotis BIOLOID robots with RoboPlus Motion platform
3. Aldebaran Robotics with NAO Emotion robots
4. Webots development platform
5. ROS runtime architecture with so-called Autonomous Participation in Cloud Services
6. extended Texas Instruments Architecture for robotics as well as the recent development standards for robotics (s. the examples in the next chapters).

8.3 Robotic Platforms and Server-Less Mobile Applications

Nowadays the current tendency of creating of so-called serverless mobile applications (SLMA) should be noted. Such mobile apps are widely used in the robotics. Within the history of the development of server and non-server applications, there were constantly fluctuations between the templates "fat client" (rich client) and "thin client". Both have some advantages and disadvantages depending on the app and API requirements, QoS network settings and usage context. Conditionally distinguish the following types of SLMAs:

1. Conventional CPS (cyber-PHY systems) which are integrated explicitly into the IoT, smart stuff und gadgets, as well as sensors and actuators (servomotors, engines)
2. Cloud centric IoT and robot apps
3. Serverless apps for Fog Computing
4. Robotic platforms [6, 7].

Why is the deployment of robotic platforms so important today? First, the conception and use of robotic platforms is led under considering of the following factors [1, 2, 8, 9]:

1. Overcoming heterogeneity and Big Data problematics.
2. Standardization of basic functions for Mobility, Localization, Management.
3. Security and privacy routines standardization.
4. Robotic apps development acceleration and code reusability.
5. Virtualization (so-called virtual robots with code offloading from the servers).
6. Autarky and energy efficiency
7. Machine Learning (ML).

Commonly, the used nowadays robotics platforms can by classified to the four following categories [1, 2]:

1. Hardware platforms (for walking, driving, flying, stationary programmable and multi function robots)
2. Coordination platforms (e.g. for swarm robots)
3. Development platforms (language-specific, oriented to SLMA for all types: industrial as well as service robots)
4. Runtime-platforms (SLMA, all types), the SLMAs are supported via multiple robotic platforms, which are so much important for the further development.

8.4 ROS—Robot Operating System

The ROS (Robot Operating System, 2007) is aimed to so-called personal robots and social humanoids by Stanford Artificial Intelligence Laboratory in 2007 [1, 2]. ROS serves simultaneously as a development framework (FW) for the apps under BSD license and uses open source. The OS is wide spread in industry, by academicians as well as is supported by OSRF (Open Source Robotics Foundation).

The system includes the following mostly important architecture components (Fig. 8.3): HAL (so-called HW Abstraction Layer), OS kernel, drivers, package management, APIs, as well as the frequently used components with communication and messaging functionality between apps or their parts (so-called Middleware and web services). In addition, ROS pursues a SOA approach to optimize the communication between individual robotic components. The ROS for personal robots and social humanoids can be used in real time scenarios too. The framework is oriented to the following robot types and platforms: PR1, PR2, TurtleBot, STAIR I, STAIR II (Stanford), Nao (Aldebaran, SoftBank), Husky A20, Lego Mindstorms NXT, Z-Robotics etc. The ROS architecture

Robot Operating System (ROS)	A FW with OS properties
SLMA: C++, Python, GUI	Autonomous client parts and GUI mostly, the server parts are mostly relatively small
Message Queuing Layer	Message exchange between the programmes and components
SOA Layer	SOAP and REST Web Services
Components and Middleware	Frequently used functions and communication components
Package Layer	APIs and Package Management
Programme Libraries	Use and robot SW deployment
OS Kernel Layer	Linux, Mac OS (experimental), MS Windows (rudimentary)
Device Layer	PHY device drivers
HAL	HW Abstraction Layer
Hardware	PHY-environment

Fig. 8.3 ROS Architecture

(refer the Fig. 8.3) combines HAL and the further layers with GUI and more than 2000 libraries for C++ and Python. The network with robotic nodes with the installed ROS transmits the executable files. The advantages of this OS are as the follows: fault-tolerance, reduced development complexity, availability of the examples on functionality and usage, documents and DB, video streams with tutorials, navigation via way-point navigation and location tracking [2]. The system is also configured to face recognition and sample calibration.

ROS uses the main catalogue (Master Registry), parameter servers as well as the "Publish-Subscribe-Request-Reply" application protocol. For Messaging the both conventional transport protocols UDP and TCP are used for the exchange of the unstructured, raw and plain data, data arrays and structures. The system services are generally divided into:

(a) synchronous, persistent by scene planning;
(b) asynchronous, monitorable, interruptable by certain drivers and aimed to task planning.

8.5 Swarm Robots, Bioloids and Further Platforms

So-called "Bio-Inspired Engineering" investigates how to use the biological methods and models in social and industrial processes generally as well in the "Industry 4.0". So-called swarm robots are depicted in Fig. 8.4a, b. The swarm intelligence of socialized animals has been taken into account. The scientists from Harvard University discovered how to increase the efficiency of a swarms of 1024 = 1 K small robots, i.e. "kilobots".

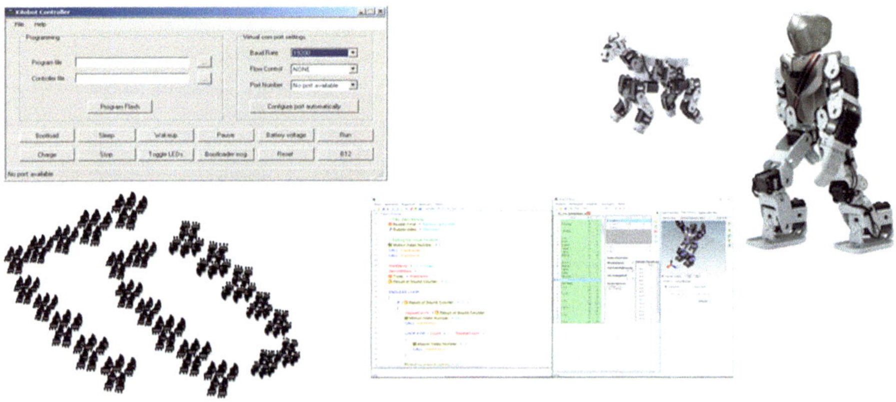

(a) (b)

Fig. 8.4 Programming tools: **a** Amtel AvrStudio for swarm robots 2D-configurations; **b** RoboPlus for bioloids

The kilobots [2, 1, 5] are the small collective operating robots which can simulate a kind of "swarm intelligence", i.e. the property of socialized animals like fishes, ants, cranes, bees etc. Under use of specialized programming tools "the machine armada" can arrange itself different 2D-forms. A swarm of kilobots is programmed together as a wireless cluster. The kilobots can interact, avoiding each other, playing follow-the-leader, or even performing simulations of nutritional processes.

The main goal of such experiments [2, 5] with small und inexpensive robot consists of modelling of technical processes and engineering objects ("Biologically Inspired Engineering" by M. Rubenstein and R.Nagpal, the kilobots founder).

The Bioloid robots [4] belong to the Robotis Corporation from South Korea (1999). The following motivation for the developers was established: "a beautiful future in the happy world of merged robots and humans,"—so Robotis CEO B. Soo Kim said. The robots are used at world's leading universities for the student trainings such as U.S. Naval Academy, Leipzig HTWK, Tokyo University, Indianapolis Institute of Technology, Kanpur Institute of Technology etc. They are widely deployed for the studies in mechanical engineering, mechatronics and robotics. Especially, they are also available at Saxon Academy of Studies BA Dresden [2, 4, 5]. The "creatures" with different complexity levels can be delivered. Some further examples on advanced robotic platforms are provided in Table 8.2 alphabetically.

Future robotic platforms offer virtualization technologies and can include so-called virtual robots over the network. The distinguishing features of such robot platforms are presented in an overview in the next chapters of this work.

8.6 Big Data Problematics and Machine Learning

Huge and heterogeneous data volumes are nowadays acquired by robotics and IoT. The number of robots that co-operate with a person, the machines and with each other in various fields of industry and economy, everyday life and the entertainment industry, social pedagogy is growing rapidly. The wide deployment of robots and robotics data create the right conditions nowadays for Big Data too.

Often the data research cannot be completely studied or their trends recognized due to the complexity and heterogeneity. Robotics Big Data became so immobile that their access and management cannot be processed without use of special techniques for data structuring [1, 2, 8–10]:

1. Taxonomies and predictive classification.
2. Preliminary clustering and compression.
3. Predictive deviation detection (based on regular statistics).
4. Descriptive association rule discovery (fuzzy based).
5. Ontologies and knowledge data bases (KDB).
6. Regression Models.
7. Sequential Pattern Discovery (Machine Learning).

Table 8.2 Overview of the Systems and Platforms [2]

Systems and platforms	Programming languages	Availability and licence
Amtel AvrStudio/WinAVR (Kilobots)	Eclipse, C++	Service
Apache Open Whisk	Node.js, Switch, Docker, Python	OSS
AWS Lambda	Node.js, Java, Python, C#	Service
Azure Functions	Node.js, Python, C#, F#, PHP	Service
Effe	Go	OSS
Fission	Node.js, Python	OSS
Google Cloud Functions	Node.js	Service
Hook.io	Node.js, ECMAScript, CoffeeScript	OSS + Service
IBM Cloud Functions	Node.js, Swicht, Docker, Python	Service
IronFunctions	Node.js, Java, Python, Go	OSS
Kubeless	Python	OSS
LambCI Docker-Lambda	Node.js	OSS
Lever OS	Node.js, Go	OSS
Naoqi, RoboCup (Aldebaran, SoftBank Nao)	Python, C++, Java, JavaScript	Service, Framework, OS
OpenLambda	Python	Academic + OSS
OVH Functions	Node.js, Python, Perl, Go, Bash	Service
RoboPlus (Robotis BIOLOID)	icon-driven C++	OSS + Service
ROS	C++, Python	BSD license, OS + framework
Webtask.io	Node.js	OSS + Service
Notices: OSS—open source system	Go, Golang—programming language of Google	

The analytic blocks can be placed on cloud and fog, which disassemble the complexity and heterogeneity (s. next chapter). Depending on the event frequency (event driven, periodically, permanent) and the behaviour of the sensors (push, pull), a sensor survey and data accumulation are performed. For example, let a WSN to consist of 15.000 sensors. Each sensor can transfer a short telegram up to 100 Bits. Thereby:

- The survey for each sensor is conducted 20 times per hour: 2000 Bits/h
- x 24 h = 48000Bits/daily = 6000 Bytes/daily for each sensor
- In general, an average sensor wireless network accumulates experimental data for 3 years × 365 days:

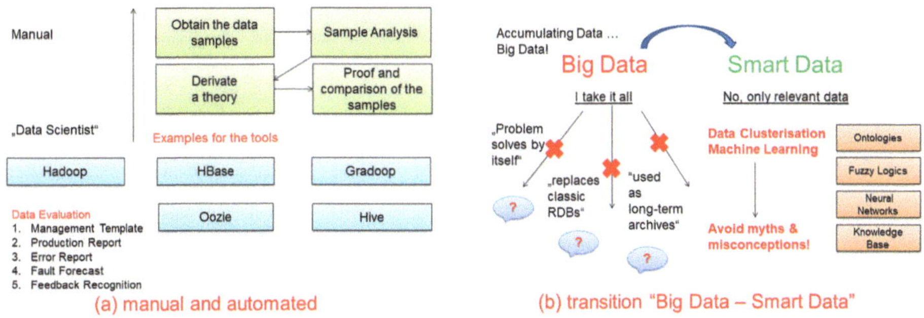

Fig. 8.5 Data Analytics: **a** manual and automated; **b** transition "Big Data—Smart Data"

- $\times 365 \times 3 = 6.57$ MB for each of the sensors
- The overall-data for the mentioned network:
- 6.57 MBytes $\times$ 15,000 sensors = 100 GB of raw data!

The research of the accumulated data must be carried out automated methods (Fig. 8.5a, b). As a result, of such research data evaluation the following common documents can be issued [8, 9]: management templates, production and error reports, fault forecasts and feedbacks.

The overcoming of myths and misconceptions by data processing is possible due to use of Machine Learning (ML). The (cloud-based) ML system enables artificial creation of knowledge from the obtained voluminous experimental data in background mode. An artificial system is "learned" from samples and examples and can summarize them after the completion of the study and evaluation (training) phase [10]. The ML system recognizes templates and trends in research data.

8.7 Analytics Placement Options

Up-to-date robotic platforms and IoT apps (Fig. 8.6) are able to use virtualization technologies and virtual robots over the network [1, 2, 6]. These applications can be classified into three groups (Fig. 8.7).

For performance increasing and Big Data overcoming there are the following approaches to intelligent distribution of the data analytics [2, 6, 7]:

1. Conventional Robots.
2. Cloud-Centric Robots.
3. Distributed (Fog-Cloud-cooperating) Robots.

Only the cloud centric solution (2) and further distributed, fog-cloud-cooperating solution (3) both are able to overcome the discussed problem nowadays and in the future.

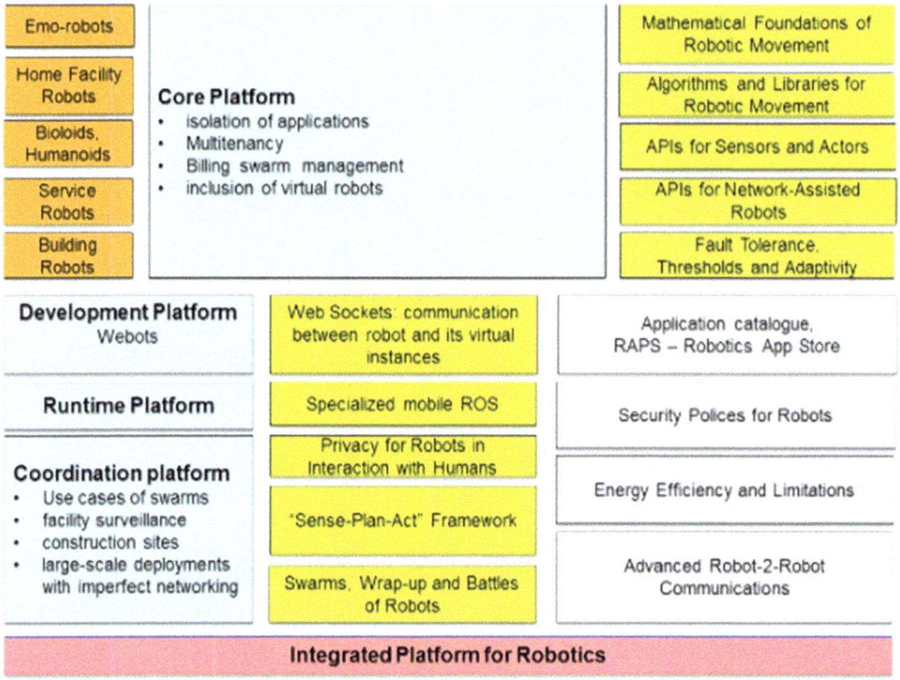

Fig. 8.6 The properties of the future robotics platforms [2, 3]

The analytic blocks, migration agents as well as further adaptive interfaces are delegated to the clouds and, possibly, after pre-processing and clustering backwards to so-called "fog" (refer Fig. 8.7) under use of the mentioned solutions and protocols (AutomationML, Mosquito, OPC UA, RabbitMQ etc.). The virtual analytical components (middleware, web services, and mobile software agents) are placed in the cloud and fog environment. Virtual cloud and fog solutions contain software components for the robots that implement rebootable (virtual) business processes [2, 7]. Thus, we are talking about replaceable and customizable robotic algorithms in various fields of application (industry, medicine, communication and telecommunication, entertainment), which can collect process and retrieve voluminous heterogenic Big Data. The following requirements to the future robotic platform can be taken into account additionally:

- Functional requirements (what does software do?) based on the processed tests (usually as a "black box") like its dimensions, scalability and amount of users, tuples, records, queries, transmitted video volumes.
- Non-functional requirements and corresponding "white box tests" are as follows: on performance, aimed to load proving and so-called "stress tests", on usability, maintainability, reliability and source code portability between the robotic nodes.

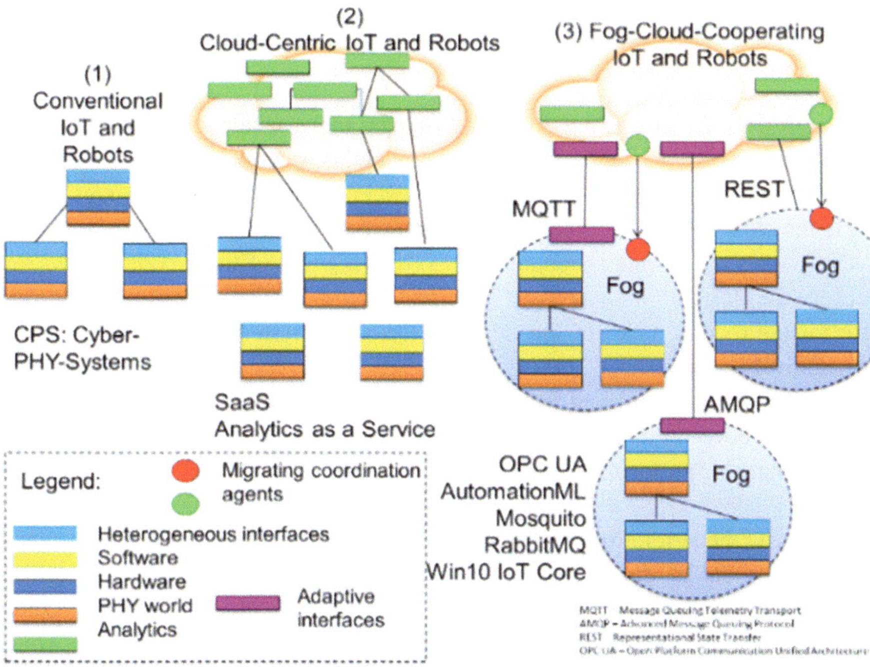

Fig. 8.7 Analytics Placement Options: three communication ways [3]

8.8 Security in Robotics

The aspect of security can be specially divided into Data Security and Safety, as well as Privacy. According to Gartner Corp. and some further experts [1, 2, 11, 12] the robotics and IoT became together more and more "a dangerous technology" because of the standing messaging between the intelligent computing nodes, e.g. usual and ordinary everyday "things" like private cars, walls of apartments and offices, electrical goods, commodity packaging, furniture, valuable papers etc. Any confidential data about the contactor of these things can be immediately transferred in plain text through the available channels: LAN, 3-4G mobile, Wi-Fi, BT and NFC.

Such technologies can concern the privacy of citizens and, even, harm the state security interests. Therefore, nowadays the above mentioned technologies are thoroughly studied by the leading political and security structures worldwide, in particular, by the EU commissions and in USA. The solution is possible through the use of the corresponding cryptographic protocols, firewalls, intrusion detection systems, that provide reliable protection, but … not only of them.

Therefore, let us discuss the security aspect for up-to-date robotic apps and platforms more in details. The robotic platforms and apps have to consider the security aspect too and provide the necessary functionality. The following approaches for can be used in this way [1, 2, 11, 12]:

1. The partition of the interacting robots and "things" to the smaller clusters (piconets).
2. LEACH method for voting for the wireless or mobile cluster heads.
3. Choice of the secured gateways (application level firewalls with antimalware) to linking to the secured clouds for the secured analytic processing.
4. Security risks monitoring (virus, intrusions) on the edges of the robotic clusters and between the clusters.
5. Use for the robotic nodes of the Intrusion Detection Systems (IDS) which operate cooperatively each other.
6. CIDN technology deployment for the interacting robotic nodes.
7. The traffic between the interacting robotic nodes can be encrypted and authenticated via cryptographic protocols like VPN, IPsec, TLS/SSL and/or HTTPS.

The widespread modern Intrusion Detection Systems evaluate and prohibit the potential hackers' attacks that are directed against a computer systems or a network. IDS increase data security significantly in opportune to the classical firewalls which lonely deployment is not satisfying. Intrusion Prevention Systems (IPS) are the enhanced IDS which provide the additional functionality aimed to discovering and avoiding of the potential attacks [1, 2, 11, 12]. Nevertheless, as a rule the classical IDS/IPS are operated autonomously. They are not able to detect temporary unknown hackers' threats which became year by year more sophisticated and complex. Those dangerous threats can serve to disorder the operation of data centres and robotic clusters round-the-clock in 24/7-mode. Therefore, the cooperation and collaboration of the IDS within a network is of the great meaning [10]. A CIDN is a further concept for a collaborative IDS/IPS network intended to bridge over the disadvantage of the standalone defence against the unknown dangerous attacks (Fig. 8.8). The CIDNs allow to the participating IDS as the network peers to share the detected knowledges, experiences and best practices oriented against the hackers' threats [1, 2, 11, 12].

The main requirements to the construction of a CIDN and the support of such functionality are as follows: efficient communication at short up to middle distance, robustness of the peers (IDS) and links, scalability and mutual compatibility of individual participating peers (single IDS). The typical interoperable networks are as follows: LAN, 3-4G mobile, Wi-Fi, BT and NFC. Collaborative intrusion detection networks (CIDN) consist from multiple IDS-solutions under use of multiple "things", robots, gadgets, PC, end radio devices and installed firewalls as well of the groups of users which are divided into the clusters peers (titled as users Alice, Bob, Charlie, Dave etc.). The coupling between the groups is loosely or tightly. However, the reputation of the users is quietly different (cp. Fig. 8.8): good, compromised (refer Buddy), malicious (refer Trudy and Mallory). Additionally, the insider attacks to CIDNs are possible (e.g. by Emmy with temporary "good" reputation). The CIDN can efficiently prevent such multiple attacks

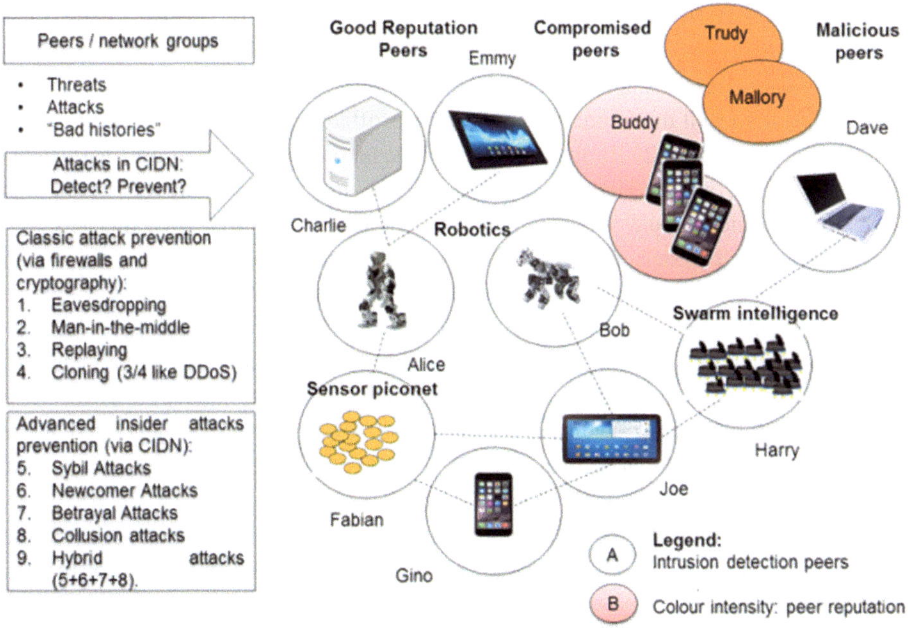

Fig. 8.8 Example of peer cooperation within a CIDN

A1-9 (cp. Fig. 8.8) by peer-to-peer cooperation. This type of networking improves the overall accuracy on the threats assessment. The cooperation among the participating single peers (IDS-collaborators) became more efficient within of a CIDN. Unfortunately, the CIDN can become itself a target of the attacks and malicious software. Some malicious insiders within the CIDN may compromise the interoperability and efficiency of the intrusion detection networks internally (Table 8.3). Therefore, the following tasks must be solved: selection of the peers (collaborators), resource and trust management, collaborative decision making [1, 2, 11, 12].

Table 8.3 CIDN Functionality

Certain CIDN examples	Detection and prevention of the attacks A1-A9	Topology type	Focus	Specialization on the further threats
Indra	+	Distributed	Local	SPAM
Domino	+	Decentralized	Global	Worms
Abdias	+	Centralized	Hybrid	Trojans
Crim	+	Centralized	Hybrid	Social Engineering, web appl. firewalls

8.9 Conclusions and Outlook

1. The existing platforms and frameworks for robotics are considered as well as analysis of the properties of the future robot platforms is offered within the given work.
2. Modern robots diversity for the important areas, taxonomies and best practices (industry, sociology, and pedagogics) were discussed under considering of the efficient distribution models and analytics placement options within the cloud and fog environments. Such models are aimed to overcoming heterogeneity and acquired unmanaged and unstructured Big Data. One of the most important paradigms of the software development for robotics are SLMAs, which cooperate with cloud and fog scenarios.
3. This work can be qualified as a "work-in-progress". The authors try to find new efficient properties for the future robotic components distribution as well as for virtualization of their functionality (virtual robots). They have to be integrated to the components, APIs, libraries, examples of sources which are oriented to localization, coordination, mobility support and management.
4. The above-mentioned aspects (Mobility, Localization and Management) are implemented in the most important software for the robotics under considering of the next aspect of Security.
5. The aspect of security that can be specially divided into Data Security and Safety, as well as Privacy is analysed. Some approaches and solutions are offered. Very important the observance of the data security by communication between the robots and "things", the protection of enterprise and personal data of citizens became by interacting with this "potentially dangerous" autonomously working equipment, as well as the anonymity and protection of the private sphere.
6. And last but not least: the robot engineering practices, trainings and student projects address and support today the specific MINT disciplines (Math–IT–Natural sciences–Technology) immensely. Since 2016, some modules are deployed at the Saxon Academy of Studies (BA Univ. of Coop. Education in Dresden) under use the cost and complexity moderate robots, which were included to the bachelor curricula.

References

1. Int. Federation of Robotics: http://www.ifr.org/.
2. J. Spillner. Serverless Cyber-Physical Applications, Keynote Talk at Workshop "Science meets Industry", 2018, Fraunhofer-Institute Centre Dresden, 20 p.
3. A. Luntovskyy. Modern Apps and Platforms for the Robots, Int. Conf. on Theoretical and Applied Aspects of Modern Infocommunications, Vyshkiv-Carpathians-2019, Jan. 2019, 4p.
4. A. Cornejo, R. Nagpal. Kilobots: Distributed Range-Based Relative Localization of Robot Swarms. Algorithmic Foundations of Robotics XI, April 2015.
5. NoDNA.de: https://nodna.de/Bioloid-Premium-Robot-Kit-V2/.

6. A. Luntovskyy et al. Intelligent Networking and Bio-Inspired Engineering, Int. IEEE Conf. UkrMiCo-2017, Odessa, Ukraine, 4p. (IEEE Xplore: http://ieeexplore.ieee.org/document/8095421/).
7. A. Luntovskyy. Advanced Software-Technological Approaches for Mobile Apps Development, 14th Int. IEEE TCSET-2018 Conf., Lviv-Slavske, 2018, 6 p. (IEEE Xplore: https://ieeexplore.ieee.org/document/8336168/), DOI: https://doi.org/10.1109/TCSET.2018.8336168.
8. A. Luntovskyy. SLMA and Novel Software Technologies for Industry 4.0, 21-st Int. Conf. ACS-2018, Szczecin- Międzyzdroje, 2018, in: J. Pejaś, I. El Fray, T. Hyla, J. Kacprzyk (eds.). Advances in Soft and Hard Computing, Springer Int., 12p. (Part of the AISC book series, vol. 889, DOI: https://doi.org/10.1007/978-3-030-03314-9-16, ISBN: 978-3-030-03313-2).
9. D.Blankenberg. Big Data in der Industrie 4.0, TIQ Solutions Leipzig, Workshop an der IBH Dresden, 13.3.2018 (in German).
10. L.Globa, I. Svetsynska, A. Luntovskyy. Case Studies on Big Data, Journal of Theoretical and Applied Computer Science, JTACS, Polish Academy of Science, Gdansk, No. 2, 2016, ISSN 2299-2634, 10p.
11. C.Fung, R. Boutaba. Intrusion Detection Networks: A Key to Collaborative Security (ISBN: 978-1466564121), 2013, 261p.
12. A. Luntovskyy, M. Klymash. Examination of Modern Concepts for Firewalls and Collaborative Intrusion Detection, International Research Journal Telecommunication Sciences, Kiev (Registration Certificate KB No. 15064-3636P, ISSN 2219-9454, UDC 621.39), 2016, Issue 14, No.2, pp. 44–49.

Energy Efficient IoT

9

The given chapter is aimed to investigation of the energy efficiency issues for IoT solutions, based on combined sensor and contactless systems. The combined energy efficient solutions are namely the hierarchical organised infrastructure WSNs and selforganised ad-hoc WSNs, which are widely interoperable with the 4G/5G base stations and microcells, the backboned Wi-Fi access points as well as with inexpensive and energy efficient RFID/NFC reader farms. In regard to provide the energy efficient WSN protocols, a holistic multilayered approach is used, which is based on Low-Duty-Cycle- Principle, energy harvesting, LEACH Clustering and topology optimisation, efficient OS and software frameworks, enabling data reduction.

9.1 Motivation

Since 2003, the P2P systems (Internet of Things, Fog Computing) in combination with convenient C-S communication model based on IoP (so-called Internet of People) as well as serverless structures (SLMA, robotics) have gained on popularity. Then the Cloud based solutions became a trend (2010) under predominant use of the load balanced "thin clients" with functionality delegation to the clouds. Under use of fog computing the IoT solutions are constructed. The workload is shifted on the edge (minor QoS requirements) to the energy autarky and resource economizing small nodes (higher energy efficiency requirements) [1–3].

9.1.1 Demarcation Between IoP, IoS and IoT

Firstly, let us to give a demarcation between so-called IoT, IoS and IoP concepts (Fig. 9.1). Internet [of People] consists of up to ~1–5 billion of devices nowadays. Since

A. Luntovskyy and D. Gütter, *Highly-Distributed Systems*,
https://doi.org/10.1007/978-3-030-92829-2_9

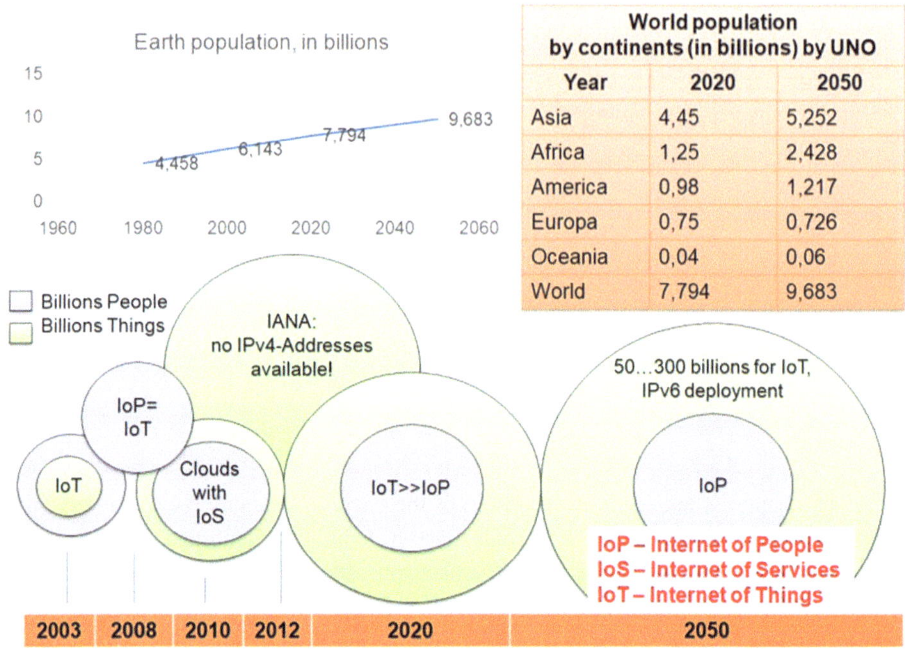

World population by continents (in billions) by UNO		
Year	2020	2050
Asia	4,45	5,252
Africa	1,25	2,428
America	0,98	1,217
Europa	0,75	0,726
Oceania	0,04	0,06
World	7,794	9,683

Fig. 9.1 Demarcation and Development Phases for Mobile Internet and IoT

2003 in so-called Internet of Services (IoS, Internet of Services) the innovative technological developments have been driven with the aim of the creation of new delivery channels for services and new business models. The creation of such services based on cloud computing in 2010 became possible under use of the open platforms and interface architectures like SOA (Service Oriented Architectures), respectively Enterprise SOA, which provide IoS for companies.

What does the term “'IoT” actually mean? The origins of the concept can be founded at Auto-ID-Lab@MIT (RFID research and design). The concept clarification belongs to Kevin Ashton (SAP, Telefonica) in 1999. In midterm (approx. in 2025), the complete transition to the IoT is possible. Based on the prediction, IoT will possess more than N>50–300 billion devices. Under use of IPv6 with 2^{128} available addresses, the number of devices can grow up to N aprox. 340 * 10^{36}! It means approx. N per_user = 35 * 10^{27} “things” per user in 2050 theoretically. Table 9.1 intents to clarify differences between the above-mentioned concepts IoP, IoS, and IoT. IoT (Internet of Things) is a collective term for technologies of a global information society infrastructure that enables physical and virtual objects to be networked with one another and allows them to work together through IoS and enabling network technologies (refer next sections). The fog computing is a necessary basis for IoT development.

Demarcation and Development Phases for Mobile Internet and IoT is given in Table 9.1.

Table 9.1 To the Terminological Clarification

Period (past and in mid-term)		2003	2008	2020–2023	2050
Population on the Earth		6e+9	7e+9	7794 e+9	9683 e+9
Techno-logy and quantity	Intelligent Things (Robotics, Wearable, Intelligent Stuff, Embedded)	–	7e+9	10e+9	50–300 e+9
IPv4 and IPv6 addresses		4294e+9 (Version 4)		3402e+38 (Version 6)	
Format 2**32	4.294.967.296= ~4.3 billions	IoP (Internet of People)	IoS (Cloud Computing)	IoT (Internet of Things)	IoT2.0 (What does it mean?)
Format 2**128	3,40282366920938463463374607431177e+38= ~340 sextillions				

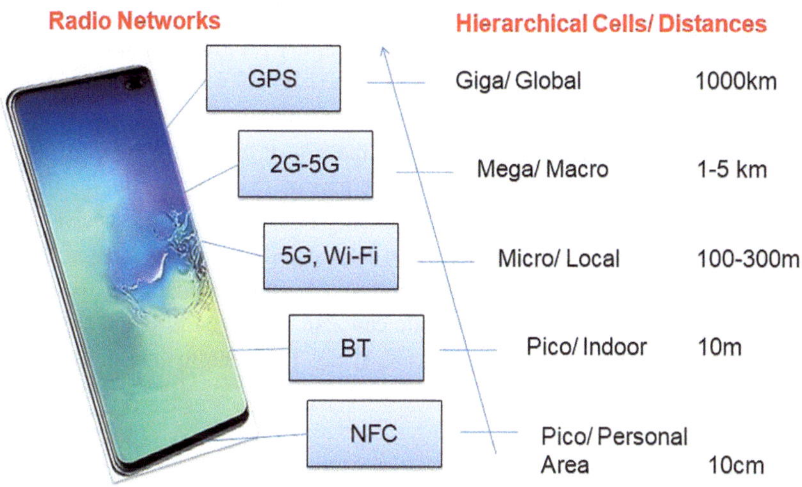

Fig. 9.2 Smartphone as "All-in-One" [1–3]

9.1.2 IoT EnablingNetwork Technologies

Which IoT relevant network technologies do we know? As an example, a smart mobile device can be considered as "All-in-One" (Fig. 9.2).The following IoT relevant network technologies are working together and can be considered as IoT Enabling Network Technologies [4–17]:

1. Mobile Networks (LTE, 5G)
2. WSN (Wireless Sensor Networks, ZigBee/IEEE 802.15.4, EnOcean, Z-Wave)

3. 6LoWPAN (IPv6 over Low power Wireless Personal Area Networks, based on ZigBee)
4. Wi-Fi (Wireless Fidelity IEEE 802.11ac, ax, ah)
5. Bluetooth v5 (IEEE 802.15.1), IrDA (Infrared Data Association)
6. RFID (Radio Frequency ID) and NFC (Near Field Communication)
7. GPS (Global Positioning System)
8. Powerline, Homeplug, PoE (Power over Ethernet)
9. KNX (Konnex), LON (Local Operating Network)
10. Infrequently, QR (Quick Response) and Watermarks (as steganografic applications) can be considered.

Modern IoT devices are interoperable with 2G-5G mobile networks under use of the convenient hierarchical cell concept (pico-cells, micro-cells, mega-cells and giga-cells, refer Fig. 9.2), e.g. NB-IoT or LTE-CatM sensor networks. These networks are completely interoperable (Table 9.2).

Some examples of actual sensors are depicted in Fig. 9.3. These can support cooperation with RFID, position awareness, are efficiently programmed with Java (a–d) as well as provide energy harvesting (e–g). The Preon32 sensor node from Virtenio is based on an energy efficient ARM processor with higher computing power which enables to calculate even complex algorithms directly on the sensor and to use a comfortable programming language (PL) such as Java. A large part of the energy reserves of a sensor node is required to keep the radio chip periodically ready to send and receive or to wait for incoming requests. In another approach, a sensor node was implemented that can be woken up with a semi passive RFID and therefore uses almost no energy as long as it is not addressed from outside. The sensor node can also measure the signal strength of the wakeup signal along the three spatial axes {X, Y, Z}. This feature is used to estimate the position of the sensor node.

Table 9.2 Overview of hierarchical cells

Type	Distance	Mobility (km/h)	Deployment by …
Giga-cell	~1000 km	~4700 (~1,3 km/s)	International providers, satellite radio
Macro-cell	~1–5 km	up to 500	National and regional providers
Micro-cell	~100–300 m	up to 120	Metropolitan areas, city districts, campus, office area
Pico-cell (Indoor)	~10 m	~10	Hotspots at railway stations, airports, hotels, restaurants, clubs, home area
Pico-Cell (Personal area)	~10 cm	Stationary	PAN, Wearable, Smart Stuff

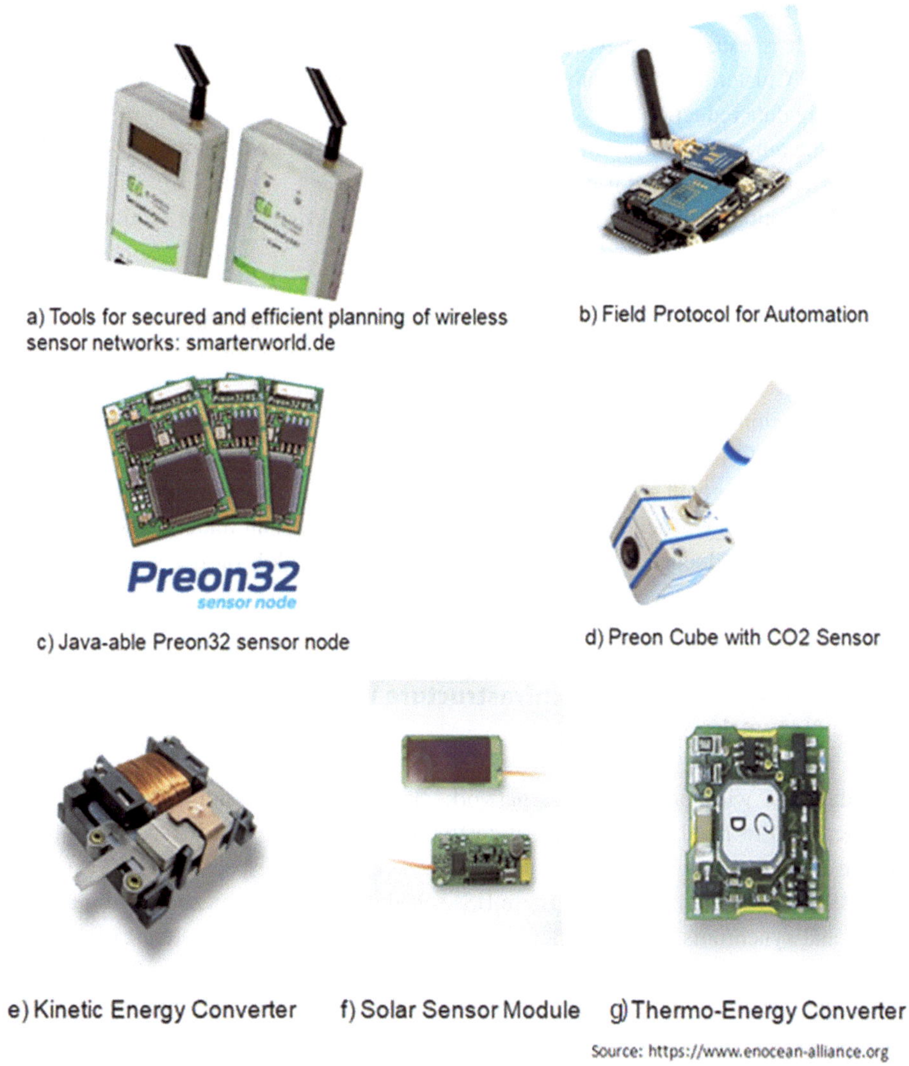

Fig. 9.3 Sensor Examples (**a**)–(**g**) [8, 9]

9.1.3 To the Structure of this Chapter

The remainder of this chapter's content follows as:

- State-of-the-art for energy-efficient approaches and solutions and approaches for hierarchically organised infrastructure WSNs and selforganised ad hoc WSNs. They are widely interoperable with the 4G/5G base stations and micro cells, the backboned Wi-Fi access points.

- In regard to provide the energy efficient WSN protocols, an holistic multilayered approach is used, which is based on Low-Duty-Cycle-Principle, energy harvesting, LEACH Clustering and topology optimisation, efficient OS and software frameworks, enabling data reduction.
- With the aim of further optimisation, the inexpensive and energy efficient RFID/NFC reader farms can be used in combination to WSN.
- The evaluation of the solutions and the case studies is offered in next sections.

9.2 State-of-the-Art for Energy-Efficient Approaches and Solutions

The demarcation for hierarchically organised infrastructure WSNs and selforganised ad hoc WSNs as well as state-of-the-art for energy efficient approaches and solutions is offered within the given section. The section includes a brief overview of common methods for WSN energy efficiency respectively. The most important tradeoffs between various factors that influence the WSN energy efficiency and QoS on the various network layers are discussed.

9.2.1 Energy Efficiency for Infrastructure Wireless Sensor Networks

Nowadays the widespread wireless networks are based on hierarchical structures and are integrated to fixed (hierarchical) infrastructures, based on Wi-Fi-Access Points and Mobile Base Stations. Mobile end devices are assigned to a specific 4G/5G base station (BS) and supplied with data. These BSs schedule the communication: radio channel access and synchronisation. A variety of WSN technologies and types exist (Table 9.3). The most important problem in all of these technologies is the energy efficient operation of the sensor networks. Energy efficient sensor nodes are characterised by durability, interoperability and guarantee of quality of service (QoS) requirements in the constructed WSN. In addition, they provide long life; they have high reliability and inexpensive options for customisation. Some examples of the most commonly used wireless sensor networks are as follows:

- Z-Wave
- EnOcean
- ZigBee
- 6LoWPAN (refer Fig. 9.4).

Table 9.3 .WSN in Comparison [6–11]

Feature	EnOcean	KNX-RF	Z-Wave	ZigBee (IEEE 802.15.4)	6LoWPAN (based on ZigBee)	Nano NET
Frequency, in MHz	868	868	868	2400 world-wide;	2400	2400
MAC-Layer	Beaconing	–	CSMA	Beaconing, CSMA	Beaconing, CSMA	CSMA/CA, TDMA, ALOHA
IPv6 support	–	–	–	–	available	–
Topology	Star, Mesh	Star	Star, Mesh	Star, Mesh	Star, Mesh	Mesh
Data rate, Kbit/s	125	16,4	9,6/40	250	250	2000
Node number	2**32	256	2**32	2**16 (256 in cluster)	2**16 (256 in cluster)	2**48
Security	AES	–	In mid-term	AES	–	+
Energy consumption	Very low	Low	Low	Low	Very Low	Middle
Collision probability	Very low	+	+	Low	Low	Very low
Energy Harvesting	available	No	No	No	No	No
Distance, m	30–300	10–100	20–200	10–75	10–100	40–250

9.2.2 Energy Efficiency for Self-Organizing Wireless Sensor Networks

In contrast, selforganizing ad hoc networks do not require any predefined assignments of the communication paths andno fixed infrastructure too. All network components have equal rights and communicate directly with one another like the peers. In this case, the coordination of the channel access and synchronisation of the sensors, APs and BSs with one another must take place in a selforganizing form. This results in special requirements for protocol development. The energy efficiency for selforganizing wireless sensor ad hoc networks will be of great importance in the future in regard to support of IoT solutions. Such sensor networks are mainly used where long term, selfsufficient operation with low data rates and energy consumption is important. In addition, the selforganisation properties allow the use of sensors in large numbers and in otherwise inaccessible places. This usage enables diverse applications: medicine patient monitoring, high performance sports and wearable, environmental observations, contamination and forest fires prediction, improvement of weather and earthquake predictions, animal

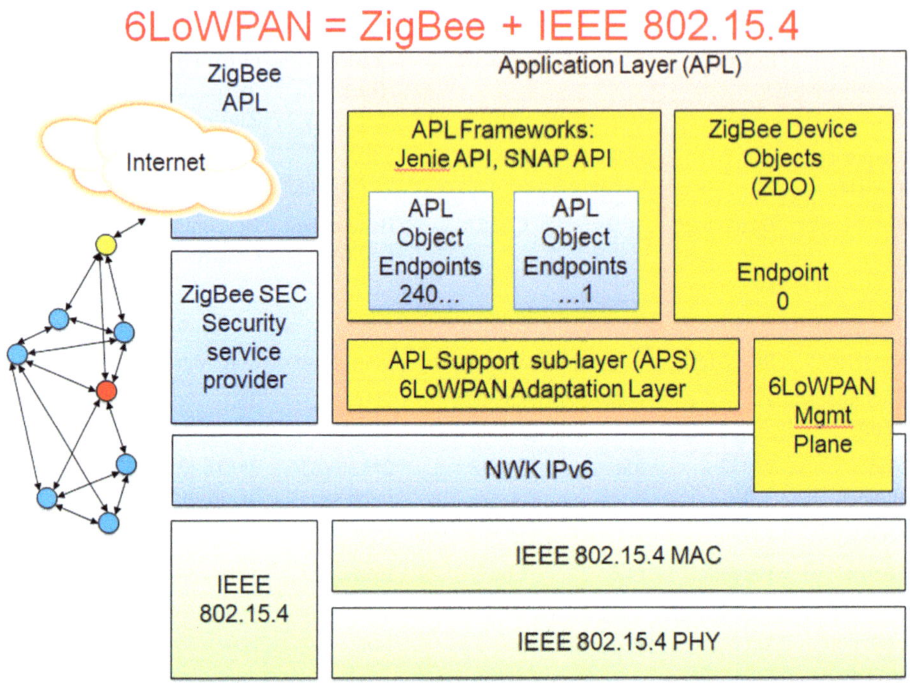

Fig. 9.4 6LoWPAN Architecture: what is inside?

observations, intrusion and fire alarm systems, intelligent robotics, indoor climate control optimisation.

9.2.3 Case Study 1: Energy-Efficient Self-Organizing WSN

As the mostly used technologies for such energy-efficient ad-hoc sensor networks ZigBee (IEEE 802.15.4), EnOcean, 6LoWPAN can be here mentioned (Fig. 9.5). The strength of ZigBee lies in the transport of small amounts of data and is therefore particularly suitable for use in building automation, home automation and medicine. An alternative to the ZigBee is EnOcean with energy harvesting with very low energy consumption and great accumulator durability. As an advanced technology, 6LoWPAN uses IPv6 advantages in addition to the energy autarky. The localisation is possible via the following procedure:

- The nodes can measure the signal field strengths PRx of the wakeup signal along the three spatial axes X, Y, Z.
- This feature is used for an accurate estimation of the position of a sensor node.

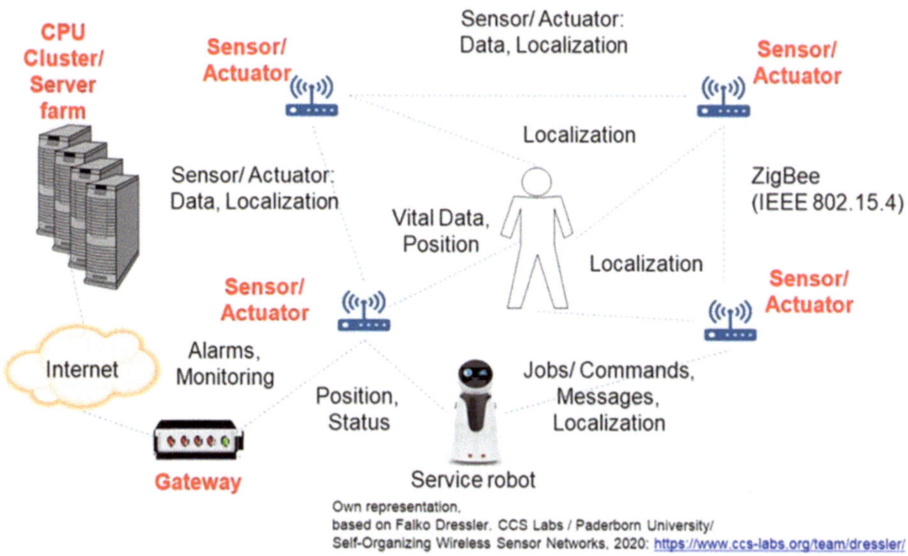

Fig. 9.5 Energy-Efficient Self-Organizing Wireless Sensor Networks with Localisation Functionality [4–6]

Wireless sensor networks often cooperate with mobile communications (LTE and 5G), WLAN and are considered in connection with traffic telemetry, IoT and Smart City.

1. WSN and Internet of Things means, among other things, that PHY objects get a virtual representation. IoT closes the existing gap between real and virtual worlds. The state of real objects is available to digital Internet services.
2. Traffic telemetry in connection with 4G, 5G and WSN understand intelligent solutions for public and private traffic. Increasing mobility and growing vehicle fleets require flexible infrastructures for intelligent interlinking, optimised traffic flows as well as acceptable safety and comfort while reducing pollutants, including a CO_2 footprint.
3. Smart City exists as a 4/5G mobile and WSN-based concept for intelligent cities in the medium term, which result from advances in the IT and telecommunications industry. The concept now contains innovative ideas for attractive urban services.

9.3 Principles for Energy Efficiency in WSN and WPAN

Generally, a WSN consists of the spatially distributed autonomous sensor nodes (SN, Sensor Nodes), which are suitable for cooperative monitoring: physical parameters (industry) and environmental conditions, e.g. temperature, sound, vibration, pressure, movement or contamination, pollutant concentration. The main principles for achieving energy efficiency in WSN.

Aimed to energy efficiency, WPAN are built under use of so-called Holistic Multi-Layered Method. They meet multiple trade-offs (Fig. 9.6) between the predefined requirements for each layer, which are often contradictive [1–3]. Possible application scenarios for wireless sensor networks are as follows:

- Numerous traffic scenarios—tunnels, railways and highways
- Disaster control, e.g. GITEWS (German Indonesian Tsunami Early Warning System) for measuring temperature, pressure, wind speed in tsunami areas with interoperability with satellite radio and GPS
- Military, e.g. MFI (Micromechanical Flying Insects, drones)
- Environment and Biology, e.g. ZebraNET (Princeton University)
- Animal and patients monitoring, healthcare (Smart Pills).

Wireless sensor piconets are in general a subject to the condition of unified development with optimisation of energy consumption in all layers of the OSI reference model up-to-down and across. The uniform approach to energy efficiency through so-called "Low-Duty-Cycle" is followed (Fig. 9.7). To increase energy efficiency in WSN, the combination of the following, often contradictory, optimisation methods is used (refer Fig. 9.6):

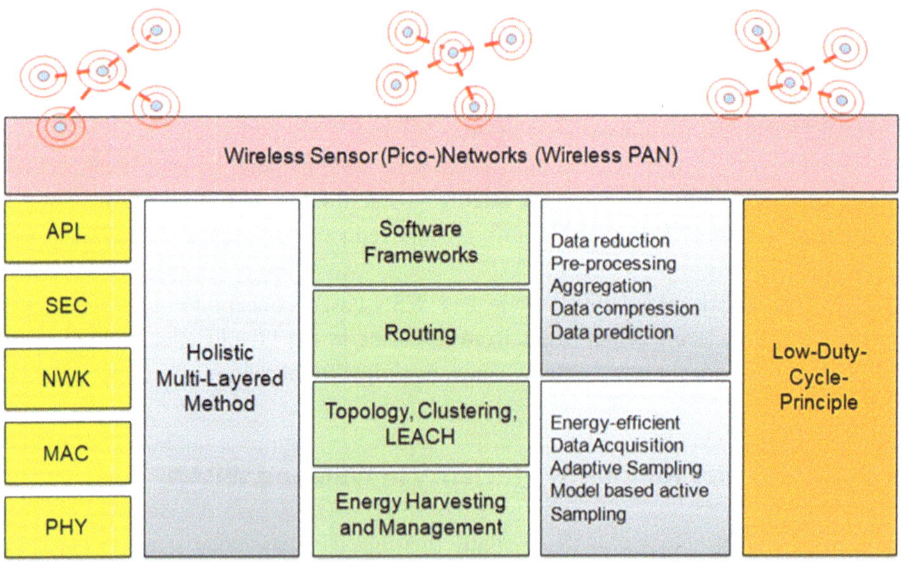

Fig. 9.6 Holistic Multi-Layered Method for Energy Efficiency: Sub-Optimal-Architecture with Trade-offs [1–3]

1. Optimised software components and frameworks with data reduction and improved data acquisition
2. So-called data-based energy efficiency
3. Optimised routing, i.a. under use of IPv6
4. Optimised topology and clusters with dynamic selection of the node and CH (Cluster Heads) or LEACH
5. Energy harvesting and management of power supply (Dynamic Voltage, Scheduling, Event based processing in nodes).

Optimisation principles for energy consumption in WSN (cp. Fig. 9.6) can be only considered as a suboptimal process with several tradeoffs (compromises), for example, like:

- Tradeoff "Transmission Power—Noise"
- Tradeoff "Frequency—Energy"
- Tradeoff "Voltage—Data Rate"
- Tradeoff "Energy—Accuracy"
- Tradeoff "Energy—Latency".

Efficient dynamic cluster formation with dynamic selection of nodes and CH (cluster heads) is known as the LEACH method (refer Fig. 9.8).

The LEACH Principle can be described as follows (refer Figs. 9.8 and 9.9). The nodes, which are already CHs, cannot operate again as CHs for 1/P rounds, where P is a wished part of Cluster Heads within the network is. Furthermore, each node possesses

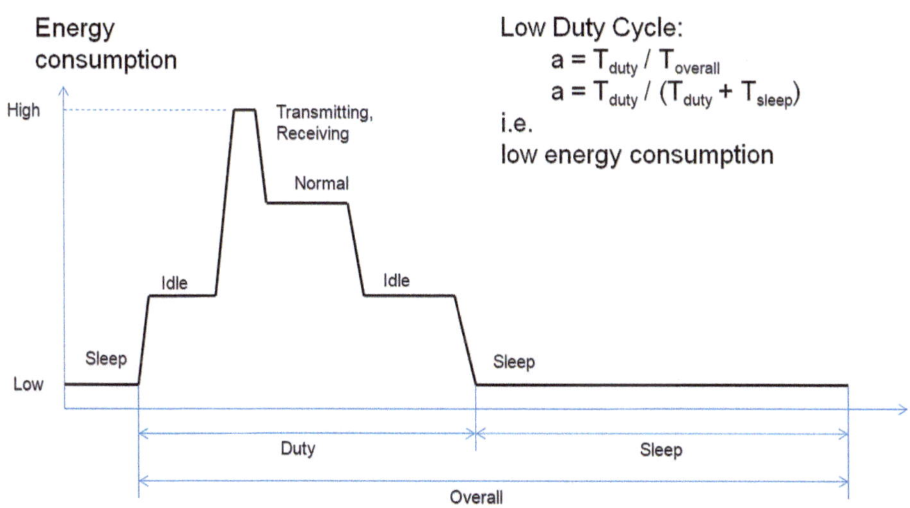

Fig. 9.7 Energy Efficiency via 'Low Duty Cycle'

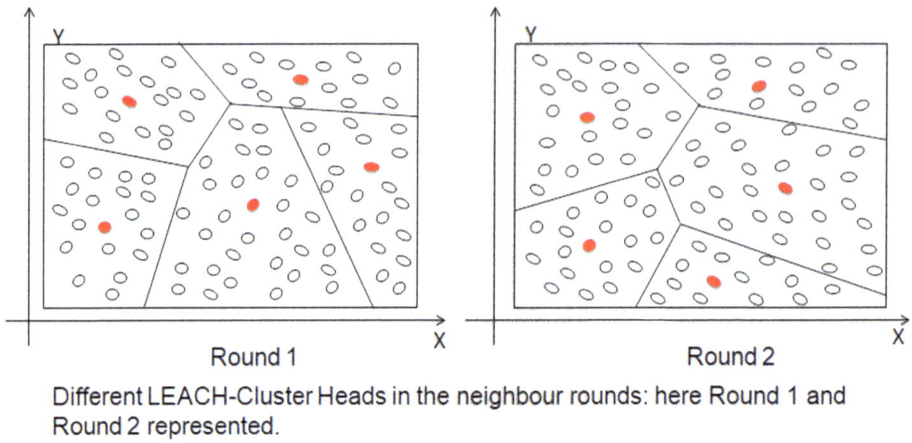

Fig. 9.8 LEACH-Clustering for Energy Efficiency [1–3]

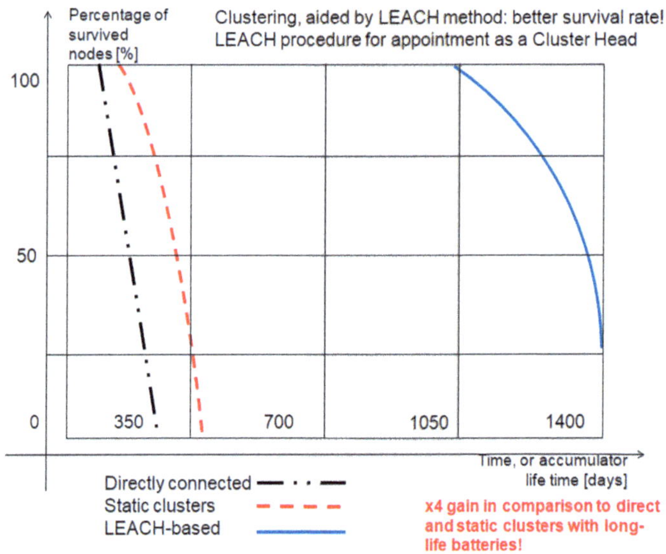

Fig. 9.9 LEACH: more survivability

the average probability Z<T(n) to become to Cluster Head in each new round. Finally, on the end of the round each node, which did not become to CH connects to a next Cluster Head and become to Cluster Member (Join Cluster). Each of the Cluster Heads schedules a slotted communication plan (Cluster Schedule) for each node from the own Cluster aimed to the successful data transfer (Slot 1–Slot m). The procedure extends up

to 4 times the average life expectancy of the WSN (refer Fig. 9.8) compared to static solutions:

1. directly connected sensor nodes and
2. static-coupled nodes and CH in the clusters.

9.3.1 Case Study 2: Annual Cost Calculation for a WSN

Let us illustrate the following calculation example (Table 9.4). Suppose we installed a wireless sensor network that works on the sensor nodes that use ZigBee modules or Wi-Fi modules. The following reference time is considered: 1 year = 365 days per 24 h. It means the energy efficiency is 100 × or, even, 200 × times higher for the WSN.

Let us a ZigBee WSN consists of 10.000 sensors (refer Table 9.4). Depending on the event frequency within this WSN, the event driven, periodically, permanent, and the behavior of the sensors (push, pull), a sensor survey and data accumulation are performed. Thus, if each sensor can transfer a short telegram up to 100 Bits only, then:

- The survey for each sensor is conducted 20 times per hour: 2000 Bits/h
- × 24 h = 48000 Bits/daily = 6000 Bytes/daily for each sensor
- In general, an average sensor wireless network accumulates experimental data for 3 years × 365 days:
- × 365 × 3 = 6.57 MByte for each of the sensors
- The overall-data for the mentioned network:

Table 9.4 Annual Costs Calculation for a WSN

Parameter	ZigBee (IEEE 802.15.4)	Wi-Fi (IEEE 802.11 ax)	
Frequency	900 MHz, 2400 MHz	2.4 GHz	5 GHz
Power consumption	1 mW (0 dBm)	100 mW (20 dBm)	200 mW (23 dBm)
Calculation of electricity costs = 29ct / kWh =	1mW * 24 h * 365 d = 8.76Wh = 8.76 * 10 ** − 3 kWh Amount [ct] = 8.76 * 10 ** − 3 kWh * 29 ct/kWh = 0.254 ct/year	100 × performance, i.e Amount × 100 Amount [ct] = 25.4 ct/year	200 × power, i.e Amount × 200 Amount [ct] = 50.8 ct/year
Annual costs for 2.500 sensors	6,35 EUR	635 EUR	1270 EUR
Annual costs for 10.000 sensors	25,40 EUR	2.540 EUR	5.080 EUR
Note:	Average electricity price in Germany: 29 ct/kWh		

- 6.57 MBytes × 10.000 sensors 66 GB of raw data!

Within three years the electricity costs can only reach 76 EUR 20 Cent (refer Table 9.4). The unit costs per GByte for the acquired data in the given case are tiny and amounts to: 76.2 EUR/66 GByte = 1.15 EUR/GByte only (i.e. 38 ct/GByte yearly)! The system is energy efficient and provides low data unit costs for the collected data.

9.3.2 Security in WSN

Data security in wireless sensor networks (s. Fig. 9.10) can also be structured hierarchically (based on BSI.de). The following techniques such as WPA3, IPv6, TLS and firewalls are used for this multi level concept. Three hierarchy layers (strata) are provided: SN stratum, CH stratum, GW stratum. The mentioned strata offer the following functionality [18]:

- GW—Gateway, GW implements the protocols DHCP, IPv4, IPv6, as well as firewall filtering and key management and WPA3
- SN—Sensor Nodes, SN provide AES encryption and IPsec
- CH—Cluster Heads, theCH links are implemented by IEEE 802.11 and the SN links by IEEE 802.15.4.

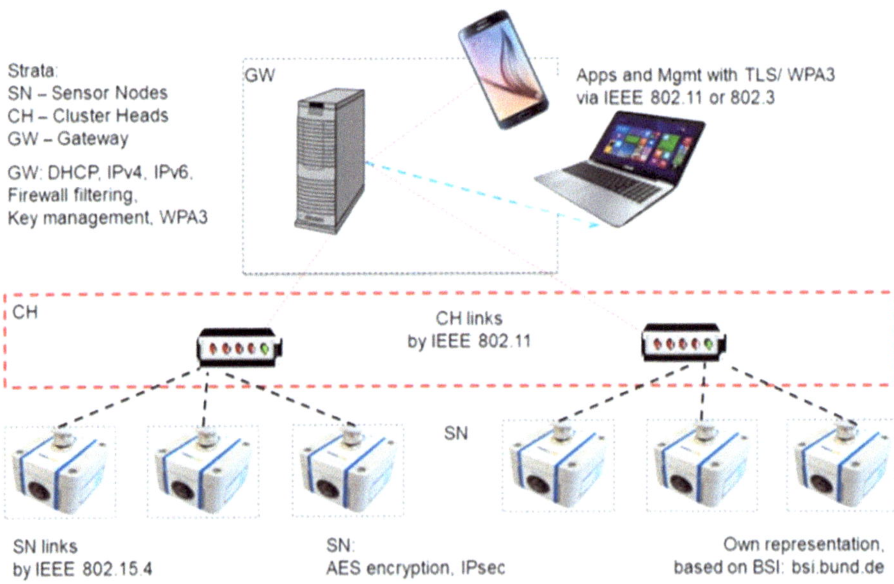

Fig. 9.10 Security in WSN [18]

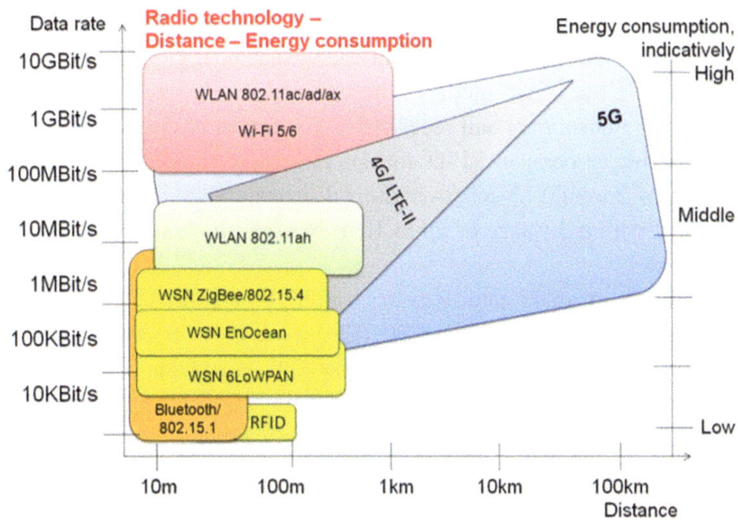

Fig. 9.11 WSN in comparison with different RF technologies [6–17]

9.3.3 Comparison and Evaluation

As a conclusion to this section, radio networks and WSN are offered at a glance (Fig. 9.11). The comparison is carried out according to criteria such as DR, range and energy consumption. The most appropriate energy-efficient RF technologies are WSN and RFID. They provide in combination with Wi-Fi APs and 5G BSs the possibility of development of energy efficient hierarchical IoT solutions for monitoring and control for the industries, digital economy and everyday life over the world and for each institution.

9.4 Energy Efficiency in Contactless Communication via RFID and NFC

With the aim of further optimisation, the inexpensive and energy-efficient RFID and NFC readers as well as their farms can be combined hierarchically to the 4G–5G mobile BS, Wi-Fi APs, WSN SN.

9.4.1 Energy Efficiency via RFID

What is an RFID (Radio Frequency ID) transponder today? This is used for fast, uncomplicated data exchange between transmitter and receiver over a short distance. The primary application is in logistics, trading, retail as well as for contactless payment systems

(Fig. 9.12). Unsatisfied data security is a major disadvantage. However, increasing the data security is possible by reducing the working distance. Otherwise, the encryption can be provided via the active models. The essential construction elements for RFID technology are the transmitters and receivers: a distinction is made as to whether communication is active or passive. RFID enables automatic and contactless identification and localisation of (mobile) objects over short distances. This modus operandi provides energy efficiency with a distance of up to 10 m (passive mode), as well as up to 100 m for active systems.

For applications in trade and logistics, multiple identification is possible (see Fig. 9.13) as well as identification via EPC (Electronic Product Code) with 64, 96, 128 bits. In combination with RFID technology, the EPC is suitable for the detection and tracking of objects (products, logistic units, packaging, transport pallets, systems, service relationships, documents, reusable transport containers, storage locations, locations) that are equipped with a transponder with EPC, possible without visual and touch contact. It is about an internationally used key and code system.

There are multiple constructions of the RFID transponders (refer Figs. 9.13 and 9.14):

(a) RFID Label
(b) A sticker with EPC-RFID for WalMart chain
(c) RFID Chip for Animal Identification
(d) RFID Chip in key ring pendant

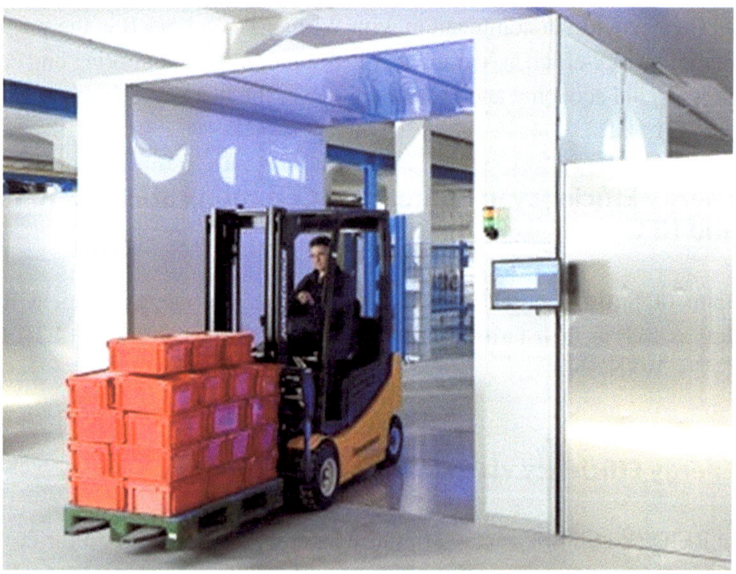

Fig. 9.12 Applications in logistics and trading: a RFID-gate

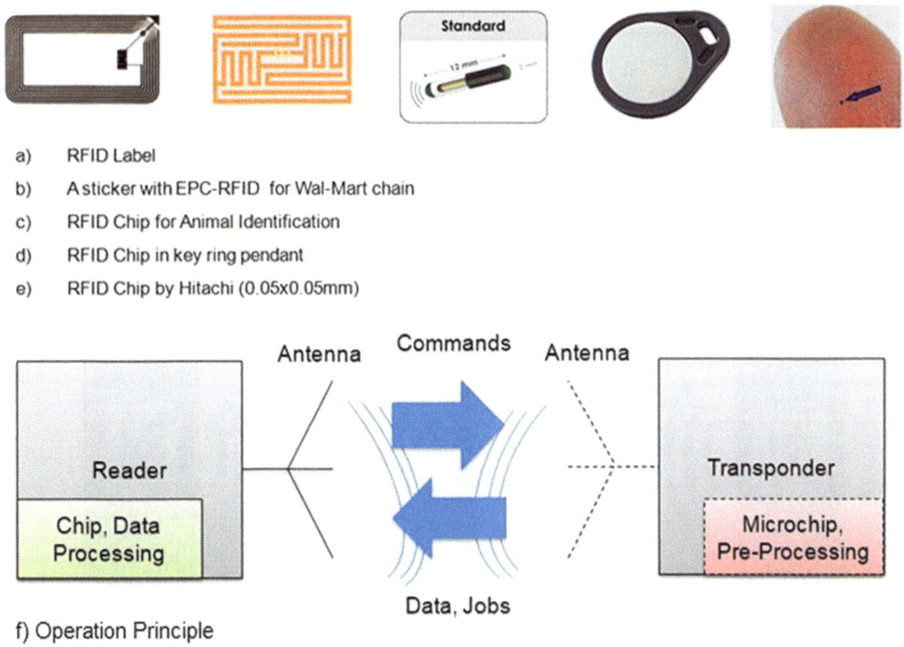

Fig. 9.13 RFID types and communication principle [13–17]

(e) RFID Chip by Hitachi (0.05 × 0.05 mm). Furthermore, inexpensive and loosely coupled RFID transponders with sensors are suitable for the measurement of physical and chemical variables, in particular, like pressure, acceleration, expansion, humidity and electrical conductivity are possible.

Typical RFID frequencies are as follows:

- Low Frequency (LF): 30–500 kHz for motor vehicle immobilisers, animal identification
- High Frequency (HF): 10 MHz up to 15 MHz, for smart labels on goods in the logistics and retail.

What is the general structure and function of RFID transponders and RFID readers? RFID technology has the following components as transmitters and receivers:

1. RFID detection device (reader) can contain a reading and other writing unit. In this case, reader transmits commands to the RFID transponders. Reader can be of various types: handheld, mobile, stationary, can also be combined with a barcode reader. The reader contains a generator for generating an alternating magnetic field. The reader

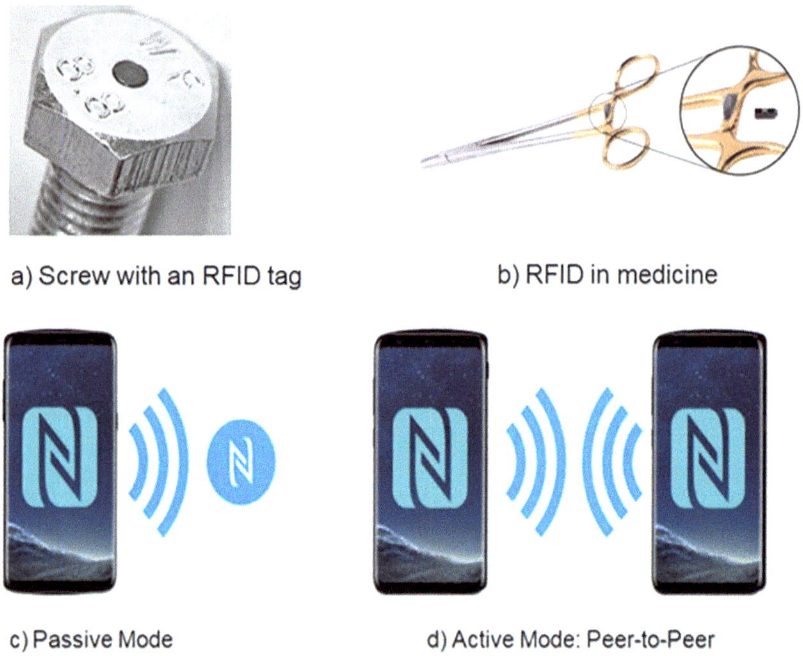

Fig. 9.14 RFID and NFC types [13–17]

generates a magnetic field in the radio frequency range. Reader contains a chip, which is mostly rewritable, and the data recorded can be encrypted optionally. RFID reader is actively equipped with memory, μ-controller and battery: therefore, a slightly shorter life expectancy with programmable and configurable devices.

2. RFID transponders contain an LC resonant circuit with a balanced resonance frequency, which is usually located between two foiletched coil and capacitor plates and whose size varies between cm and μm. The transponder transmits data. If the transponder has a microchip, it is a programmable variant (refer Fig. 9.13f).

The operation principle (refer Fig. 9.13f) is as follows:

1. The reader sends commands via an electromagnetic field (Transmitting)
2. Reception takes place via the antenna of the transponder
3. Probably forwarding to the microchip if any for a Preprocessing
4. Transponder sends the Response via electromagnetic field
5. Data processing in the reader.

For RFID technology, the following ITU-coordinated frequency bands are defined: LW at 125–134 kHz; SW at 13.56 MHz; UHF at 865–869 MHz (EU); UHF at 950 MHz (USA/Asia); SHF at 2.45 and 5.8 GHz.

The following modes are distinguished:

- Passive RFID: only antenna
- Active RFID: with memory, μ-controller and battery: therefore, shorter life expectancy, but programmable, configurable.

RFID transponder is often used together with wireless sensors for measuring of physical and chemical variables. They are, in particular: pressure, acceleration, elongation, humidity and electrical conductivity. As an example (Fig. 9.14a, b), a screw with an RFID tag (technology) and an RFID for medicine are given.

9.4.2 Energy Efficiency via NFC

NFC (Near Field Communication) is a further development and unification of RFID technology. NFC is an international standard based on RFID and oriented to contactless data exchange via EM induction using loosely coupled coils. NFC is designed for a very short range of up to 10 cm. the technology is comparable and interoperable to RFID, WSN and Bluetooth technologies, but there is no direct competition. NFC technology is usually used in IoT and Industry 4.0 due to integrated reading units and transponders. There are the following most important NFC features and specifications (refer Fig. 9.14c, d):

- Frequency band $f = 13.56$ MHz
- Data rate, in Kbit/s: 106, 212, 424
- Operating modes: Read-Write, Peer-2-Peer, Card Simulation
- Specifications: ISO18092, ECMA 340, ETSI TS 102190.

Multiple smartphones provide a NFC interface or, otherwise, it can be integrated via a microSD slot. The NFC devices each have active and passive modes (refer Fig. 9.14c, d):

- Active mode: as a reader or peer-to-peer mode for data exchange, using your own energy source.
- Passive mode: is a smart card or emulates one, the energy is drawn from the RF field of the active site.

NFC is supported by numerous mobile OSs for smartphones, for instance, under use of Windows 10 Mobile or Android:

- connectionless (with passive HF-RFID tags according to ISO 14443 or ISO 15693)
- connection oriented, peer-to-peer (between two equivalent active transmitters Tx).

9.4.3 Case Study 3: Energy-Efficient Monitoring and Management of Farm Animals via RFID and Wi-Fi

The use of RFID technology for energy efficient monitoring and management of farm animals was depicted in Fig. 9.15. The following components are used:

- 2D-Farm with RFID-Readers and Wi-Fi Access Points
- Ethernet-Switch to Intranet and LAN coupling
- VPN-Router and Authentication Server for securing of data transfer
- JINI Server for die Discovery, Lookup and Join Services
- Application Server for support of the apps for analysis of the animal status and stall layout, for animal tracking, feeding and milking, as well the alarms.

The relevant DB are called accordingly: animal condition, steel layout, tracking, feeding, milking, alarms etc. The Monitoring and Management is handled by mobile native and web apps in the smartphones and tablets. Based on the combination of RFID and Wi-Fi this IoT solution is energy efficient and provide low data unit costs.

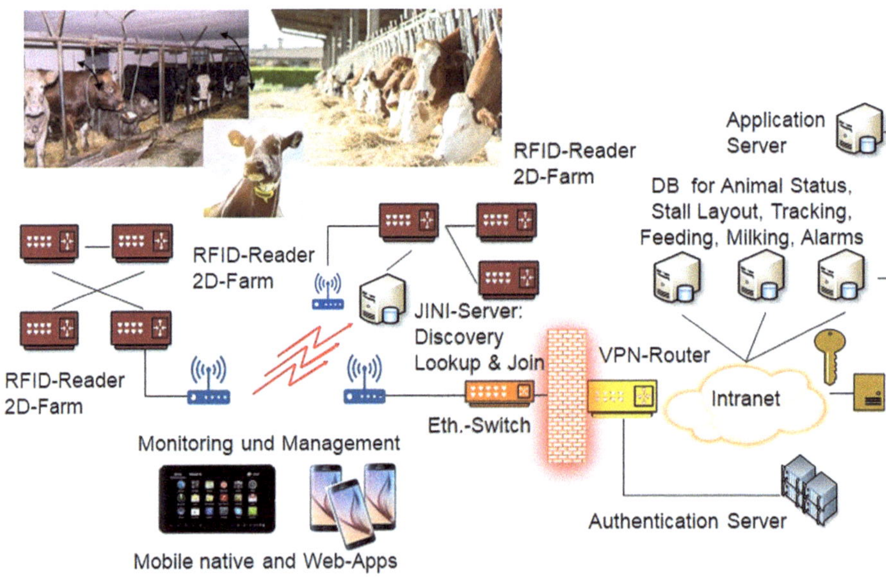

Fig. 9.15 RFID and Wi-Fi based monitoring and management of farm animals

9.4.4 Evaluation

- Origins in music, trading and retail, logistics and warehouse management on shorter ranges (10 cm–10 m)
- very energy-efficient contactless technology with big durability (12–72 months)
- unsafe due to attacks by third parties but encryption is optionally available
- inexpensive, if passive, only antenna
- Heterogeneity, variety of the constructions for RFID vs. NFC unified specifications, which are RFID based
- Unification of RFID-Technology by ISO 14443, ISO 15693
- Networking of the machines and work pieces
- Integrated reader and transponder unit
- Holistic tracing the products from the supplier to the customer
- Automatic entries in the machines' and work pieces' databases.

9.5 Conclusions and Outlook

The given chapter represents a demarcation between IoT, IoS and IoP as well as provides an overview on interoperable IoT technologies: 2-5G, WSN, LoRa WAN, WI-Fi 5/6, RFID/NFC, which promise 'a new technological breakthrough' for widespread modern energy efficient systems. Furthermore, in the given chapter, the energy efficiency is examined as one of the most important issues for IoT. An overview for energy efficient approaches and solutions for hierarchically organised infrastructure WSNs and selforganised ad-hoc WSNs is provided. They are widely interoperable with the 4G/5G base stations and microcells, the backboned Wi-Fi access points. A qualitative comparison of the RF/NFC technologies has been done by the criteria of data rate, distance and energy consumption. With the aim of further optimisation, the inexpensive and energy efficient RFID/NFC reader farms can be used in combination to convenient WSN. In regard to provide the energy efficient WSN protocols, a holistic multilayered method is offered, which is based on Low-Duty-Cycle-Principle, energy harvesting, LEACH Clustering and topology optimisation, efficient OS and software frameworks, enabling data reduction. The chapter possesses a WIP status (Work-in-Process).

The discussed case studies and deployment scenarios are provided. The case studies on energy efficiency of IoT were discussed.

In midterm, the standards for ML and AI will accompany IoP, IoS and clouds, IoT and fog, as well as the industries, digital economy and everyday life over the world and for each institution. Therefore, it is very important for the systems to provide energy efficiency and low data unit costs in view of the increasing demands on the QoS [1–6].

References

1. A.Luntovskyy, B.Shubyn. Energy Efficiency for IoT, based on IEEE Workshop RECI-2020 on "Reliability Engineering and Computational Intelligence", Zilina, Slovakia, 27-29 Oct. 2020, in Springer LNCS series "Reliability Engineering and Computational Intelligence - Studies Comp. Intelligence", Vol. 976, by Coen van Gulijk and Elena Zaitseva (Eds.), ISBN: 978-3-030-74555-4, pp. 199–215 p., on July 2021, 507420_1_En, (Chapter 12).
2. A. Luntovskyy, B. Shubyn, T. Maksymyuk, M. Klymash. 5G Slicing and Handover Scenarios: Compulsoriness and Machine Learning, in "Current Trends in Communication and Information Technologies", in Springer Lecture Notes in Networks and Systems series, Volume 212), by Vorobiyenko, Petro, Ilchenko, Mykhailo, Strelkovska, Iryna (Eds.), based on Conf. IPF-2020 "Infocommunications - Present and Future", Odessa, 16–19 Nov. 2020, pp. 223–255, ISBN: 978-3-030-76342-8, 32 p., June 2021.
3. Andriy Luntovskyy, Tim Zobjack. Secured and Blockchained IoT, IDT-2021 Zilina, IEEE Xplore, 22-24 June 2021, in IEEE Issue "Information and Digital Technologies", ISBN 978-1-6654-3692-2, ISSN 2575-677X, 2021, pp. 1–10.
4. Ralf Gessler, Thomas Krause. Wireless-Netzwerke für den Nahbereich, Vieweg+Teubner, Wiesbaden 2009, ISBN 978-3-8348-0247-7, 342 S.
5. Falko Dressler. CCS Labs / Paderborn University/ Drahtlose Selbstorganisierende Netzwerke (Online): https://www.ccs-labs.org/team/dressler/.
6. 6LoWPAN (Online): http://www.scantec.de/.
7. Virtenio Sensors (Online): https://www.virtenio.com/.
8. EnOcean Alliance (Online): https://www.enocean-alliance.org/.
9. Preon 32 Sensor Nodes (Online): https://www.virtenio.com/.
10. Smarter World Sensors (Online): https://smarterworld.de/.
11. Field Protocol (Online): https://www.computer-automation.de/feldebene/zensoren/die-protokoll-frage.74052.html.
12. Clemens Otte. Industrie 4.0 (Online): https://bdi.eu/artikel/news/was-bedeutetindustrie-40/.
13. RFID Transponders (Online): https://moodle.ruhr-unibochum.de/m/mod/wiki/view.php?pageid=1144#toc-5.
14. RFID in der Industrie 4.0 (Online): https://ihk-industrie40.de/leitfaden-industrie-4-0/technologien/rfid-technologie/.
15. RFID (Online): https://www.elektronik-kompendium.de/sites/kom/0902021.htm.
16. NFC in der Industrie 4.0 (Online): https://industriemagazin.at/a/nfc-alsproduktivitaetsturbo.
17. NFC (Online): https://www.elektronik-kompendium.de/sites/kom/1107181.htm.
18. BSI – Federal Office for Information Security (Online): https://www.bsi.bund.de/.

Secured and Blockchained IoT

10

The given chapter examines multiple case studies and gives the answers to the key questions about secured and blockchained as well as energy efficient IoT solutions. The following IoT features are discussed:

- Small energy consumption and energy efficiency
- Wide interoperability to 5G, 6G, WSN, RFID, NFC, Robotics, Wearable
- New efficient communications models with decentralisation, M2M, fog-based models, P2P instead of convenient C-S, cloud centric, but not only….

Certain advanced demands on security are required. In the following, the conventional and advanced IoT security approaches are considered:

- Conventional
  - Centralized solutions
  - Service orchestration
  - 1:1 and 1: n communication
  - Firewalls, IDS, IPS
  - Public Key Method and PKI
- Advanced
  - Decentralized solutions
  - (Micro-)Service choreography
  - P2P and n: n communication
  - CIDN
  - Blockchain technology.

A. Luntovskyy and D. Gütter, *Highly-Distributed Systems*,
https://doi.org/10.1007/978-3-030-92829-2_10

10.1 Internet of Things: State-of-the-art

In general, a distributed networking PHY system with IT components can by commonly described as IoT nowadays. This includes embedded, sensors, RFID/ NFCs, robots, computers, smartphones, tablets and further "intelligent stuff" with processing algorithms and analytics for PHY objects that interact each other. These systems, or "things", are also clearly identifiable and have to be secured. IoT devices (things) can autonomously acquire then partially process and forward collected data. IoT augments [1–5] the convenient distributed IT and wireless sensor networks (WSN). The common Architecture and Components of an IoT device is given in Fig. 10.1. The classic communication models like C-S and P2P are extended for IoT via the direct development of so-called M2M systems. The IoT devices can operate, in general, with different frequency of data acquiring: periodically, non-periodically and event driven. This means that, depending on the system, communication with a networking participant takes place either via the requests from the network with the corresponding return of the desired information, or on the device side when critical events occur. E.g. messaging (warning), if a temperature sensor detects that a certain limit temperature has been exceeded or fallen below. There is no debt for a permanent transmission of data is due to the energy efficiency requirements for IoT systems. On the one hand, there is a requirement to use as small telegrams and tiny data rates as possible on a sensor network (WSN) by energy savings. Such restrictions cannot be applied to every IoT device; however, as soon as the use of cable systems and fixed networks for efficient power or data transmission no longer makes economic sense, these requirements become even more important. IoT devices are mainly operated by batteries or have their own type of power supply via so-called energy harvesting, e.g.

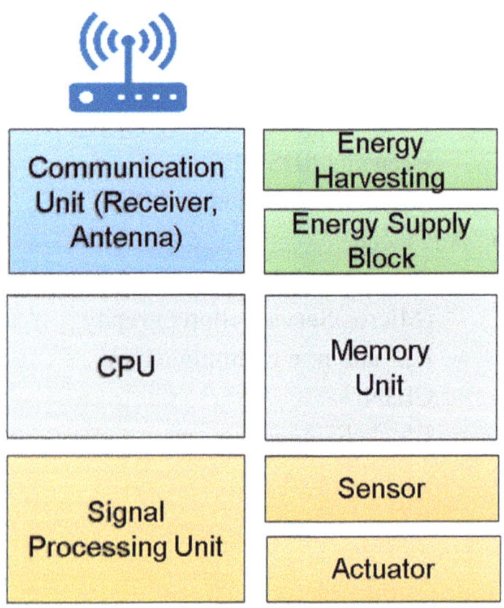

Fig. 10.1 Architecture and Components of an IoT Device

via solar energy, kinetic and thermo energy converters or wind power. The protocol stack with multiplicity of protocols for IoT is depicted in Fig. 10.2. The well-known protocols for enabling networks are used in the PHY and Data Link layers, for example, mobile radio 4G, 5G, BT, RFID, NFC and as well as some IoT-specific protocols.

Most of them use the established combinations with IPv4, IPv6, TCP and UDP for data transfer. An exception is ZigBee, which provides its own protocol for transport and is based on the IEEE 802.15.4 standards (the lowest layers). ZigBee is specified for WSN and corresponds to the basic IoT concept of data economy and energy efficiency. The further development of ZigBee is 6LoWPAN under use of IPv6. The IoT applications have access to the data, provided via various higher-level protocols (HTTP, CoAP, SEP 2.0, MQTT, OneM2M). They are the widely used HTTP can also be used for data queries and data strings transfer. RPL (Routing Protocol for Low power and Lossy Networks) is optimised to use in combination with IPv6 and further WSN. There are several initiatives in industry to promote and spread IoT. The aim is to increase the productivity and flexibility of the producers, while reducing costs and energy consumption. The mostly known are as follows:

- Industry 4.0 strategy paper of the Federal Republic of Germany
- "Industrial Internet" from General Electric
- "Factories of the Future" from the EU.

These approaches are summarised under the term "Industrial IoT" (IIoT), which is sub area of general IoT. The particular challenge is the integration of two core technologies,

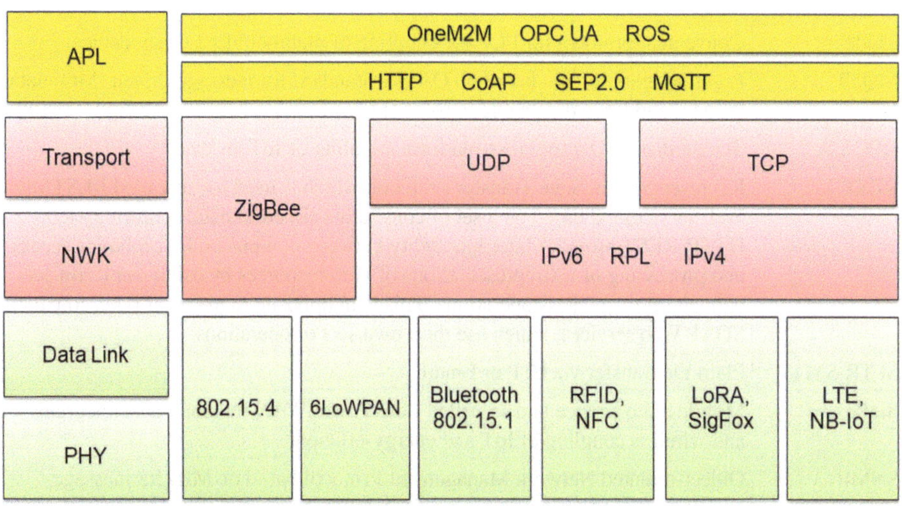

Fig. 10.2 Protocol Stack for IoT

the programmable logic controller (PLC) and a system for monitoring and data acquisition (SCADA). Together with a connected network, they form what is known as operational technology (OT) and LON (Local Operation network). The focus here is on continuous real time operation and security. The OT is parallel to IT and must be considered separately. At the same time, the challenge is to build interoperability between the two levels. Operational technologies are particularly widespread in so-called critical infrastructure. IIoT can refer to two reference architectures:

- ITU-T Y.2060 of the International Telecommunication Union (ITU) as a general reference architecture for IoT since 2012
- Industrial Internet Reference Architecture V1.7 of the Industrial Internet Consortium (IIC) from 2017 (this reference architecture of the IIC can be seen as a detailed extension of the more generalised architectural description of the ITU).

The available application protocols (refer Fig. 10.2) for IoT are summarised in Table 10.1. These are separated by their universality and the supported communication protocols (refer Table 10.1): from the mostly simple tools and frameworks up to the whole integration platforms.

Table 10.1 Application Protocols for IoT

Protocol	Description
AMQP	Advanced Message Queuing Protocol is an open standard application layer protocol for message-oriented middleware (MOM)
CoAP	REST-based protocol for connecting devices with small QoS. Mainly used for M2M communication
Modbus	Convenient protocol for PLC for coupling of mainly older Legacy devices
MQTT	Light-weight protocol based on OASIS-Standard for message-driven (MQ) data transfer and limited data rate in IoT filed
OPC UA	Standardized SOA for cross-platform coupling of IoTdevices
REST	Representational State Transfer Protocol, which is used for so-called RESTful Web servicesand based on a set of constraints aimed to max. conformity to HTTP. REST provides interoperability between computers, IoT, robotics,access and processing of textrepresentations of Web resources by using a uniform predefined set of stateless operations. This is in opposite to the other type, so-called SOAP Web services, which use their own sets of operations
SFTP, SMTP	Plain file transfer via FTP or Email
SigFox	Mobile radio service and so-called Low-Power WAN for small data telegrams and wireless coupling of IoT and energy-efficient
SNMPv3	Object-oriented Network Management Protocol based on MIB for message exchange in LAN or correspondingly in IoT scenarios

There is nowadays a large number of platforms from well-known software manufacturers for the management and data processing of IoT devices [4–8]:

- IBM Watson IoT Platform
- Microsoft Azure IoT Hub and Win for IoT
- Google Cloud IoT
- Amazon AWS IoT
- SAP Internet of Things, SAP Leonardo IoT, SAP Edge Service.

On the other hand, multiple open-source SW solutions and platforms for IoT device integration can be mentioned like Robot OS, OPC UA, RabbitMQ, MosQuitto, AutomationML tools, which are based the both on the above listed application protocols (refer Table 10.1). These can be taxonomised by their universality and the supported communication protocols: from the mostly simple tools and frameworks up to the whole integration platforms.

10.1.1 Motivation and example for Energy-Efficient IoT with LoRa WAN

LoRa WAN devices are used for so-called low power wide area Long Range communication (LPWAN). The Semtech Corporation acts as s pioneer of LoRA Alliance. The LoRa protocol is based on chirp spread spectrum modulation techniques (CSSM). The LPWAN is operated in license free frequency areas like SRD and ISM: 2.4 GHz, 868/915 MHz, 433 MHz, 169 MHz. The basic modules are available as open source software (OSS). LoRa WAN provides asymmetrical links and energy efficiency for M2M scenarios (Fig. 10.3).

The UL ranges are d= 10 km (from the end device to the network). However, low DR by approx. 292 bps to 50 kbps is only available. The middle ranges for LoRa extend from 2 km in urban areas up to 40 km in rural areas. Multiple LoRaWAN networks are deployed Europe- and worldwide: they provide IoT coverage in the Netherlands, Switzerland, South Korea etc. These countries are the first countries with area-wide LoRaWAN coverage. As an example, Swisscom LPN can be mentioned:

- Frequency band from 863 to 870 MHz (SRD Band for Europe).
- Gateway PTx = 500 mW (27 dBm).

10.1.2 Case Study 1: OPC UA Platform for IoT Integration

As appropriate examples OPC UA (OPC Unified Architecture) and programming framework, standardized by IEC 62,541-2015 [6], as well as ROS (Robot OS) and

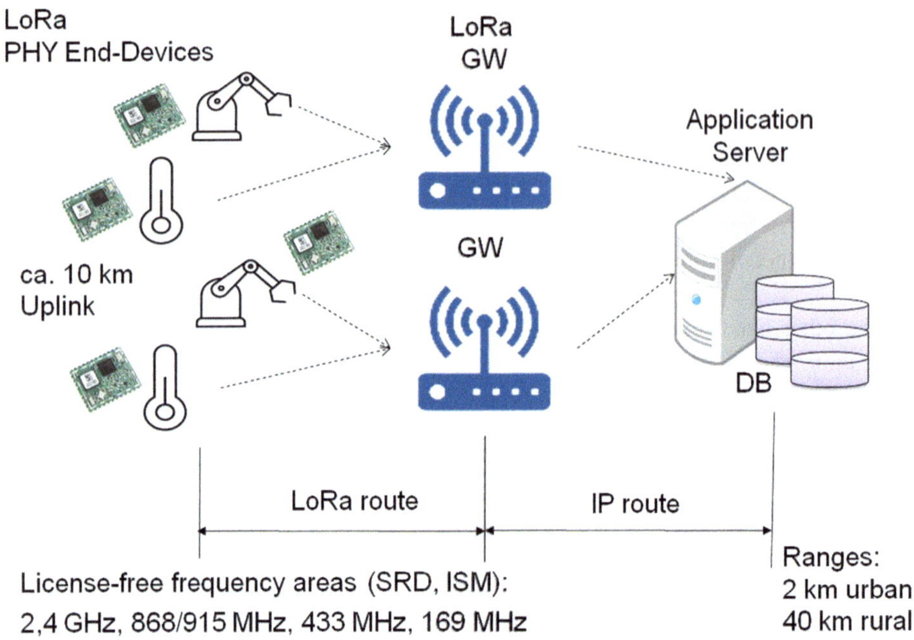

Fig. 10.3 LoRA Architecture

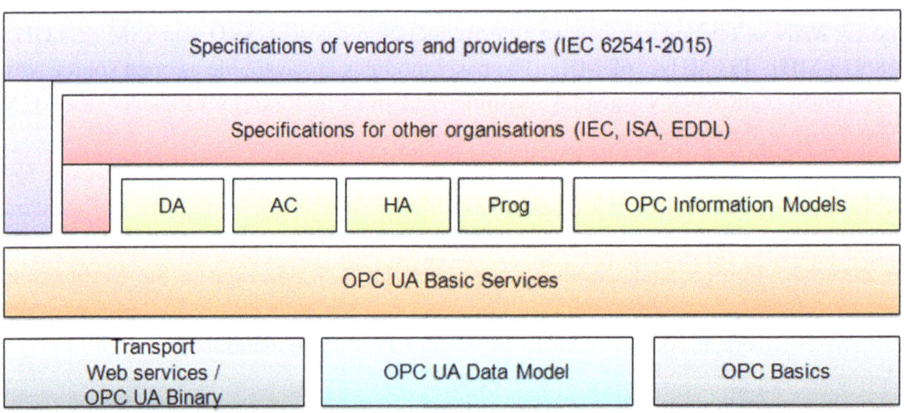

Fig. 10.4 OPC UA Architecture

programming framework can be considered. The OPC UA specification as well as the based framework for IoT and Embedded is a multi-part specification and consists of the following parts, layers and tiers as it was depicted in Fig. 10.4:

(1) Concepts, Security Model, Address Space Model;
(2) Services, Information Model, Mappings;
(3) Profiles, DA (Data Access), AC (Alarms and Conditions), Prog (Programs), HA (Historical Access);
(4) Discovery, Aggregates, PubSub.

The following bus formats can be used: EDDL (Electronic Device Description Language); ISA Bus (Industry Standard Architecture); Profibus as well as further fieldbus formats. The advantages of this OS are as the follows: fault-tolerance, reduced development complexity, availability of the examples on functionality and usage, documents and DB, video streams with tutorials, navigation via way-point navigation and location tracking [1–6].

10.1.3 Case Study 2: IoT Application integration via the SAP Platform

The IoT application integration via the SAP platform [7, 8] is shown in Fig. 10.5. The following basic principles are used for such integration:

1) "Out-of-the-box" integration means standardised technologies and applications, as well as preconfigured scenarios for the simple linking of services.
2) "Open integration" means that the connection of non SAP services is also supported and promoted, based on public APIs.
3) "Holistic integration" means that the type of sender and the receiver is irrelevant, any integration is supported.
4) "AI-driven integration" describes the use of artificial intelligence (AI), which supports the creation of integration scenarios with suggestions based on experience and crowdsourcing.

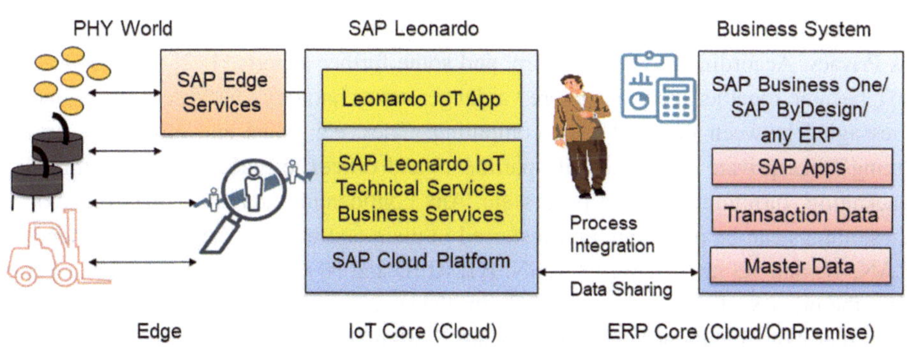

Fig. 10.5 SAP based IoT Integration. (Source: https://blogs.sap.com/2019/05/06/sap-leonardo-iot-for-smbs/)

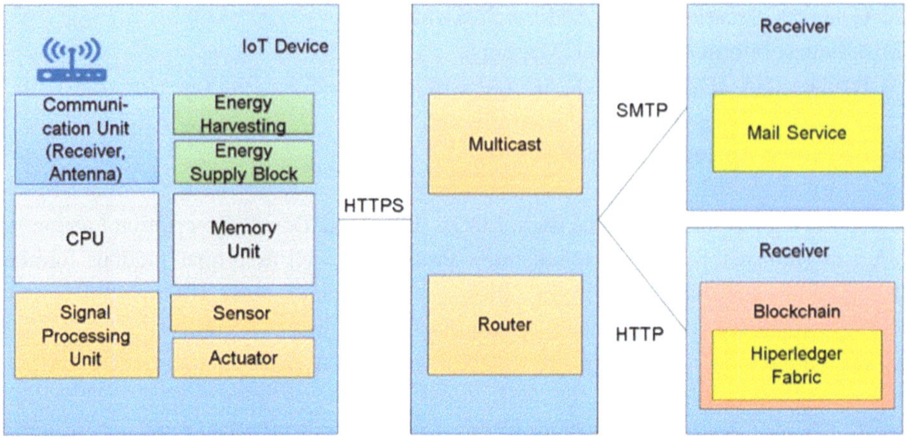

Fig. 10.6 IoT Integration Scenario

5) Crowdsourcing is the term for outsourcing internal tasks to external, voluntary participants, in this case the users of the Integration Suite. The integration scenarios created by these users are used anonymously to improve AI.

An integration scenario is given in Fig. 10.6. An IoT energy efficient device can be represented via DHT11, DHT22, AM2302 sensors for temperature & humidity. The further components like CPU, memory, energy supply, communication units with antenna are provided via a small single-board computer: e.g. Arduino MKR WiFi 1010 or Raspberry Pi 4 Model B; 4 GB, ARM-Cortex-A72 4x, 1,50 GHz, 4 GB RAM, WLAN 11ac, Bluetooth 5, LAN, 4 × USB, 2 × Micro-HDMI.

10.2 Security Aspects for IoT

The aspects of security can be specially divided into Data Security and Safety, as well as Privacy. According to Gartner Corp. and some further experts [1, 2] the robotics and IoT became together more and more "a dangerous technology" because of the standing messaging between the intelligent computing nodes, e.g. usual and ordinary everyday 'things' like private cars, walls of apartments and offices, electrical goods, commodity packaging, furniture, valuable papers etc. Any confidential data about the contactor of these things can be immediately transferred in plain text through the available channels: LAN, 4G and 5G mobile, Wi-Fi, WSN, BT, RFID and NFC. Such technologies can concern the privacy of citizens and, even, harm the state security interests. Therefore, nowadays the above-mentioned technologies are thoroughly studied by the leading political and security structures worldwide, in particular, by the EU commissions and in USA. The solution is possible through the use of the corresponding cryptographic protocols, firewalls, intrusion detection systems, that provide reliable protection, but â€¦ not only of them. Therefore, let us discuss the security aspect for up-to-date IoT and platforms more in details. The IoT platforms and apps have to consider the security aspect too and provide the necessary functionality.

The following approaches for can be used in this way [1–3, 9–19]:

1) The partition of the interacting robots and "things" to the smaller clusters (piconets).
2) LEACH method for voting for the wireless or mobile cluster heads.
3) Choice of the secured gateways (application level firewalls with antimalware) to linking to the secured clouds for the secured analytic processing.
4) Security risks monitoring (virus, intrusions) on the edges of the robotic clusters and between the clusters.
5) Use for the robotic nodes of the Intrusion Detection Systems (IDS) which operate cooperatively each other.
6) CIDN technology deployment for the interacting IoT and robotic nodes.
7) The traffic between the interacting robotic nodes can be encrypted and authenticated via cryptographic protocols like VPN, IPsec, TLS/SSL and/or HTTPs.
8) Sensor systems detecting possible threats using different sensor modalities, such as radio, acoustics, electro-optics, infrared and radar.
9) Use of the private and public Blockchained technologies.
10) Interoperability of IoT and WSN and Multi-Layered Monitoring.

10.2.1 Case Study 3: Multi-Layered Monitoring and Control for infrastructure WSN

Monitoring of wireless sensor networks can be organised hierarchically (Fig. 10.7). The SN are coupled into the clusters with single hop to the Wi-Fi AP. The stratum of the sensors is thus formed. The highest Gateway Stratum is based on the middle backbone stratum of the summarised Wi-Fi Access Points. The Gateway Stratum enables holistic management and monitoring and control of the wireless sensor networks based on 4G/5G.

IoT and WSN operate together under considering of energy efficiency (energy harvesting, hierarchy and dynamic clusters, data aware SW solutions) as well as of hierarchical security. Therefore, data security in wireless sensor networks (s. Fig. 10.7) can also be structured hierarchically (based on foundations of BSI.de). As an example for "best practices", the Infotec BRAIN [23] for Monitoring of Wireless Sensor Networks can be considered. The system provides energy efficiency and low data unit costs.

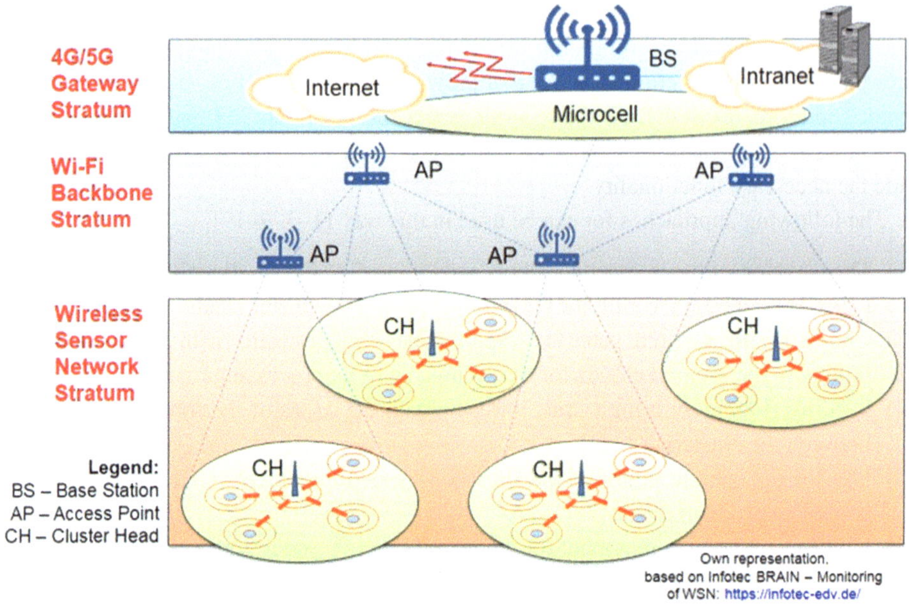

Fig. 10.7 Multi-Layered Monitoring and Control based on an infrastructure WSN

The following techniques such as WPA3, IPv6, TLS and firewalls are used for this multilevel concept. Three hierarchy layers (strata) are provided: WSN stratum, Wi-Fi-BB-stratum, GW stratum. The mentioned strata offer the following functionality:

- BS – Base Station, BS builds a mobile radio microcell with the coupled private cloud as well as Internet and Intranet
- AP – Access Points, AP implement the protocols DHCP, IPv4, IPv6, as well as firewall filtering and key management and WPA3
- CH – Cluster Heads, the CH links are implemented by IEEE 802.11 and the SN links by IEEE 802.15.4.
- SN – Sensor Nodes, SN provide AES encryption and IPsec.

10.2.2 Case Study 4: Hierarchical Security in IoT: Flying Drones

Drones are becoming increasingly easier and less expensive to access and can thus offer criminals and terrorists a new means of attack (s. Fig. 10.8). The next Multi-Layered Combined Solution for Anti-Drone Defence is discussed below [24].

„BOS – Behörden und Organisationen mit Sicherheitsaufgaben“ (in German) means “Authorities and organisations with security tasks” and is a collective term for

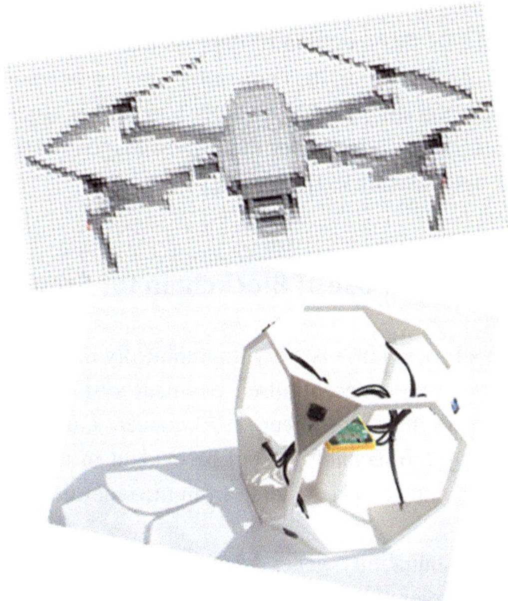

Fig. 10.8 Anti-Drone Defense System AMBOS [24]

institutions that are entrusted with the prevention of dangers for safety and security (emergency services). So-called system "AM-BOS" for defence against unmanned aerial objects for BOS (Authorities and organisations with security tasks), developed by the IDMT Fraunhofer Institute for Digital Media Technology, enables the security forces to support the detection of potential threats from drones. The system features possesses hierarchical architecture.

AMBOS sensor system came from the field of audio signal processing and can detect possible threats using four different sensor modalities such as radio, acoustics, electro-optics, infrared and radar. The system AMBOS is developed via the partners from science and industry, which are working since 2017 on information support for the detection and localisation as well as an assessment of flying drones.

10.2.3 Case Study 5: CIDN for IoT, Sensor Piconets and Robots

The widespread modern Intrusion Detection Systems (IDS) evaluate and prohibit the potential hackers' attacks that are directed against a computer systems or a network. IDS increase data security significantly in opportune to the classical firewalls which lonely deployment is not satisfying. Intrusion Prevention Systems (IPS) are the enhanced IDS which provide the additional functionality aimed to discovering and avoiding of the potential attacks [1, 2]. Nevertheless, as a rule the classical IDS/IPS are operated autonomously. They are not able to detect temporary unknown hackers' threats, which became more sophisticated and complex year by year. Those dangerous threats can serve to disorder the operation of data centres, IoT and robotic clusters round the clock in 24/7-mode. Therefore, the cooperation and collaboration of the IDS within a network is of the great meaning [1, 2]. A CIDN is a further concept for a collaborative IDS/IPS network intended to bridge over the disadvantage of the standalone defence against the unknown dangerous attacks (refer Chap. 8).

10.3 On Use of Blockchain for Secured IoT Applications

Blockchain (BC) is a cryptographically distributed computer network application [9–21] supporting a decentralised payment system and decentralised financial online transactions in the peer-to-peer (P2P) concept (since 2009). However, the economic success of this crypto technology will be evident in the next 10 up to 20 years. Potential areas of application can be resulted from the renewed consideration of the main properties of BC:

- Sustainability, general transparency and commitment
- Acceleration of economic workflows, value chains and digitalization processes (so-called "IT in the digital age")
- Completeness and accountability protection in digital Supply-Chain-Management
- Scalability, decentralised network
- Possibility to eliminate the need for an intermediary in financial transactions by making all necessary documents via a BC accessible and fake proof manner for all involved parties
- Compulsoriness, accountability and liability, especially, for IoT solutions.

The deployment of Blockchain technology speaks for a decentralised financial system. The advantages of such a solution are obvious. BC crypto technology is also well suited to supporting current crypto currencies (such as Bitcoin, Ethereum, Ripple, Litecoin, ZCash, Monero, Stellar etc.). In the Smart Contracting (SC), in the logistics industry, entire supply chains, from the production of raw materials to the ready-on-market products for the end user can be represented via an appropriate BC. Every participant within the logistics chain have insight into the good conditions and WF steps. Subsequent manipulation of data afterwards is completely excluded. Therefore, the trustiness within a supply chain grows between the participants. There are some pilot projects of a kind of micro grids, in which the residents can sell on excess self-generated solar power directly to their neighbors. The use of BC technology can also bring some advantages in this area via the SC. Therefore, there is also multiple BC types, which can be used in public space, by enterprises, for communities or privately (communities, crowdfunding). Table 10.2 offers the BC types, ordered by their access model with some provided BC implementation examples.

Table 10.2 BC types, ordered by access model

Blockchain Types		Read OP	Write OP	BC example
Open	No access limitations	Everyone	Everyone	Bitcoin, Ethereum
	Limited access	Everyone	Authorized	Sovrin
Limited	Consortium-limited access	Authorized	Authorized	Corda, Hyperledger, Quorum
	Private limited access	Completely private or authorized	Network Operator	

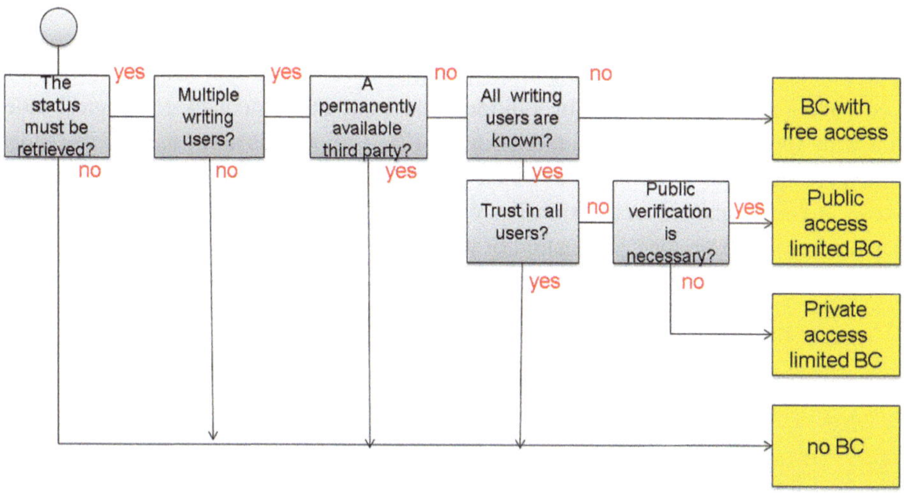

Fig. 10.9 Wuest-Gervais Flowchart as Decision-Making Support for BC: own representation, based on [15]

10.3.1 Constructing a BC

A flowchart as a decision-making support for BC by Wuest-Gervais [15] is depicted in Fig. 10.9. Totally, the five following FAQ can help with decision making to choose the correct BC type:

(1) BC with free access;
(2) public access limited BC;
(3) private access limited BC, or
(4) none.

A main disadvantage of the BC technology is resource intensity and energy consumption. Especially, with a view to the large energy and consumption for BC it makes sense for further research and integration consultants to develop further skills in this area. Multiple projects are still in a pilot phase, but with increasing of maturity degree can be expected so that Blockchained IoT can already be provided in mid term.

The general "chaining principle" for BC construction is depicted in Fig. 10.10. Hash and signature algorithms like SHA-256 and RSA are commonly used for chaining, authentication and integrity verification.

The financial transactions are used as a rule as blocked transactions. Therefore, the BCs possess in the practice the following detailed structures (Fig. 10.11). The central element of a BC is a block, which contains the transactions that are carried out between the participants within a P2P network. According to Distributed Ledger Technology, the transactions are saved in chronological order. A block has only a limited storage capacity at a time; when the maximum number of transactions is reached, the current block is closed and a new one is created. In addition to the transactions, a block contains a header area with various attributes (see refer Fig. 10.10). The attributes are usually the block number, a time stamp, the previous block key value (hash) and, depending on the technology, a sample and a nonce optionally. When a block is completed, a unique key value (hash) is created from the combination of the transactions and the header attributes. This hash is used for validation and is reused in the following block as part of the header area.

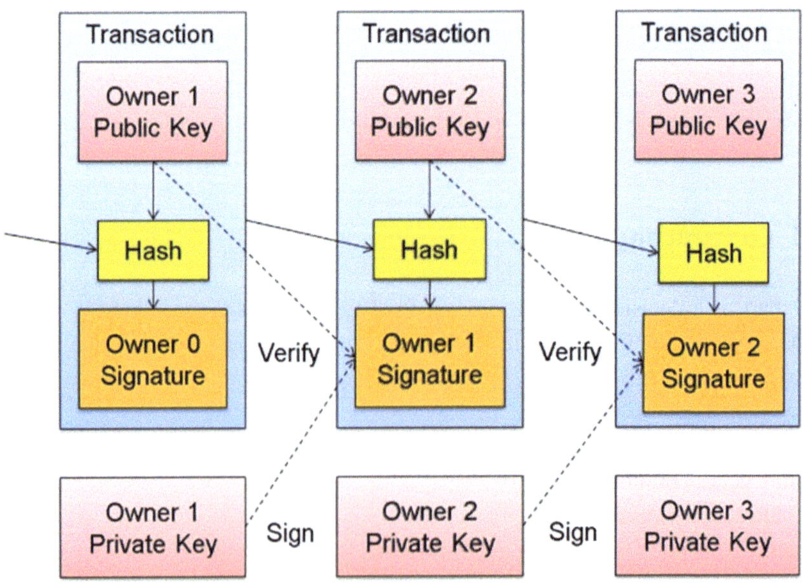

Fig. 10.10 General principle of BC

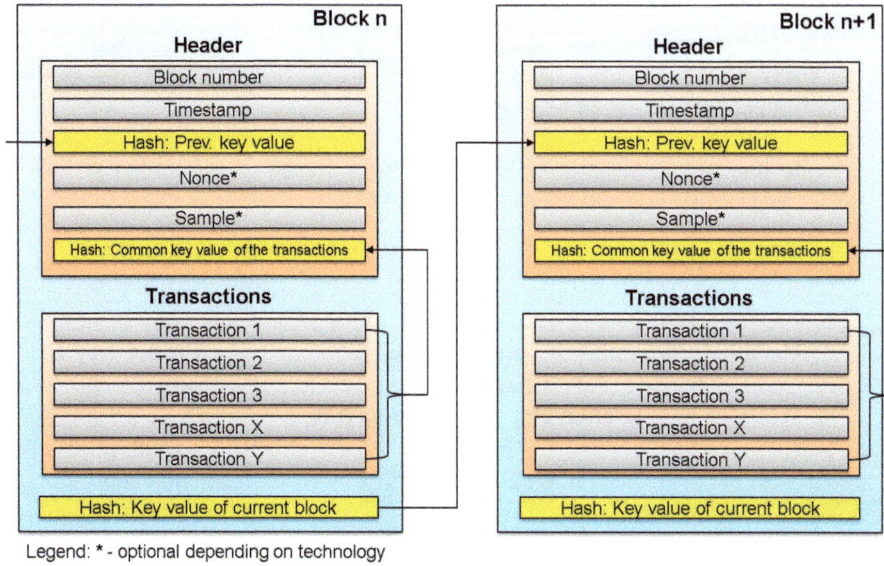

Fig. 10.11 General BC Structure (detailed representation)

The individual key values (hashes) of the blocks are interdependent as well as the blocks are cryptographically chained.

The key values for the block: previous key, common key of the transactions and current key (refer Fig. 10.11) are generated from the considered block and called the "Hash Values" according to the SHA-256 algorithm from NIST. By definition, so-called SHA-256 hash algorithm delivers a hash, i.e. an output with a fixed size, from an input of any size and shape. The hash functions are unidirectional, so that conclusions about the origin are not available.

It means that each change for each character in a string in the initial data (i.e. transactions) creates a completely new hash value. In addition to the SHA-256 hash algorithm, the public key method (Fig. 10.12) is another important component (e.g. RSA) for integrity control of the data (transactions) within a BC. The method can be used both to verify the sender and to encrypt a transaction. A transaction is the sending of data via the BC network, for example, the digital payment of an invoice under use of crypto-currency (BTC, ETH, ZEC, XMR).

10.3.2 Recent BC Types

Aimed to the distributed transactions several consensus methods and algorithms for the BC block creation must be used. The most important of these methods are listed in Table 10.3 below [9–21].

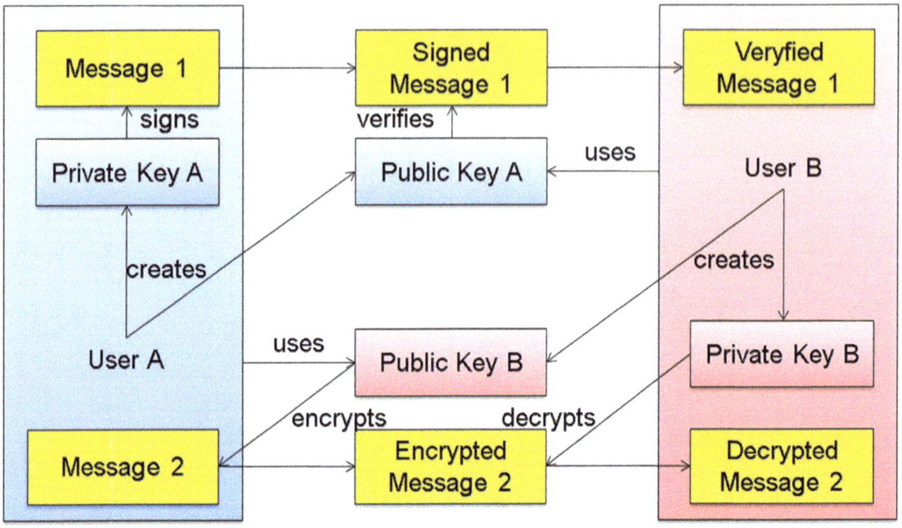

Fig. 10.12 Public Key Method for BC

Table 10.3 Different Consensus Methods and Algorithms

Method	Description of the algorithm	Disadvantages	System examples
Proof-of-Work	Participants (so called miners) provide their computing power for a fee in order to calculate the hash values of new blocks. The difficulty of the calculation is adjusted so that blocks are created at regular time intervals	Very computationally and energy-consuming	Bitcoin
Proof-of-Authority	Only certain nodes within the P2P network can create the blocks, queuing-based choice	Less robust to manipulation	Hyperledger Fabric, Ripple
Proof-of-Stake	The participants possess the P2P network's crypto-currency themselves. The higher the percentage of participation, the more likely it is to choose to create a block for a reward, which is measured in this crypto-currency	Capital accumulation can put off other participants	Stratis, Reddcoin

The Blockchains in general is presented in Fig. 10.13. BC can be separated into two big classes. Generally, the public (open) Blockchains are accessible for everyone, although a distinction is made here between:

- A completely open access, for instance, Bitcoin.
- A limited access, to which the participants must first register themselves, for example, by Sovrin.

Multiple crypto-platforms can be used for BC integration within different applications: Ethereum Classic, Codius for Ripple, Hyperledger Sawtooth [11, 14, 17, 18] etc. The following APIs are used nowadays for these multiple BC types (s. Table 10.4).

A detailed comparison between some recent open source BCs [9–21] and their advantages are given in Table 10.5.

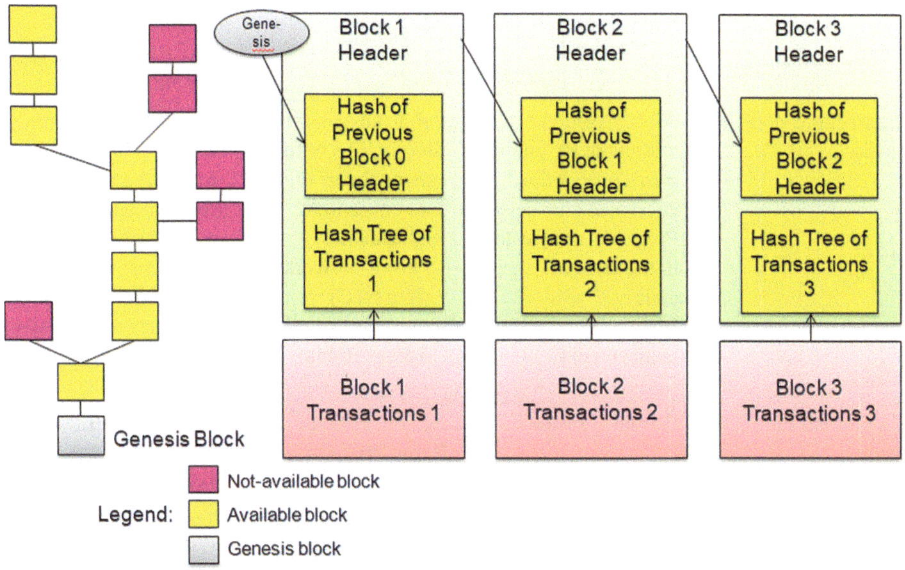

Fig. 10.13 BC as Distributed Public Ledger: Headers and Blocks within a P2P Hash Tree

Table 10.4 BC APIs in Comparison

Blockchains	Available API
Hyperledger Fabric	REST
Quorum/Ethereum/ETH	JSON-RPC
Corda	RPC
MultiChain/BTC	JSON-RPC
Ripple	REST

Table 10.5 Comparison of Recent BC Types

Blockchain Type	Hyperledger Fabric 2016	MultiChain 2015–2019	Quorum 2016
BaaS in a cloud	Suitable	Suitable	Suitable
Private or public P2P network?	Private, participation is only by prior invitation, subject to approval	Designed for private networks, subject to approval, both publicly and privately operated	Both publicly and privately operated
Open source	Open source	Open source	Open source
Modularity	Available	Available	Available
Suitability as a solution in the corporate environment	Suitable	Suitable, a modified BC version of BTC	Quorum is the opposite of MultiChain: Like MultiChain is an enterprise version of the public BTC-BC, Is Quorum as an enterprise version of the public ETH
Transactions und blocks	All transactions within the BC run via previously defined, encrypted channels Programs are created in the form of "chain code" Chain code management	Public key encryption Since v2.0 SW for so-called "Smart Filters" Two consensus mechanisms: – "Proof-of-Work" (like BTC) – resource optimized Round-Robin	Based on Ethereum/ETH Multiple alternative consensus mechanisms Programs, which are based on Quorum-BC are called analogously the ETH like "Smart Contracts"
Support via PL	Go, Java, JavaScript	Python, JavaScript, Ruby, PHP, C#	Proprietary PL: Solidity as JavaScript-Dialect
System examples	Collection of container weights by Kuehne + Nagel and Capgemini Platform: monitoring and transparency of the container weights for all parties involved in a supply chain	"Air-Quality-Chain" for measured environmental data of the city of Vienna are saved in a MultiChain and made publicly available	This technology has not yet been used productively Bank and Business BC in mid-term

However, as the main disadvantages of the BC-based tools and APIs for IoT, the performance reduction by real time services as well as enormous energy consumption can be mentioned as a critical position. A further important requirement to IoT is the data security and trust especially by risks of ransomware and other crypto Trojans like Petya, WannaCry or GandCrab. In opposite to standard solutions, based on PKI and combined symmetric-asymmetric encryption (RSA) and digital signatures, BC provides its own security decentralised infrastructure, which distinguish from centralised PKI or bilateral Web-of-Trust incorporated in convenient distributed apps. In Germany, the BC cloud initiatives came from the Telekom.de and Siemens too. So-called augmented BaaS (BC as a Service) BaaS among other XaaS (Everything as a Service) from the cloud can reduce the investment risks for the enterprises in the digital age.

10.4 BC Based Applications and Platforms for IoT

IoT can be successfully combined with a correct digitisation solution as a BC is. MindSphere from Siemens as well as further Siemens solutions for Blockchained IoT can be used for food and beverage industry aimed to complete ensure traceability from the farm and field up to the consumers. Furthermore, Siemens relies on the BC, which secures the tool management during repair of power plants Siemens brings transparency to the supply chain. The increasing recognition of BC technology in the enterprise cloud environments by "the digital age" has also prompted multiple giants of IT industry, i.e. companies such as MS and SAP to offer their own BaaS among other XaaS. The following kinds of BC are integrated within these software platforms [9–21]: Hyperledger Fabric, MultiChain, Quorum, and Ripple.

A further successful BC example is based on the use of Hyperledger Fabric BC from the Hyperledger Foundation and concerns to the collection of container weights. The worldwide known logistics company Kuehne+Nagel and the consulting company Capgemini have developed a platform for monitoring and transparency of the container weights for all parties involved in a supply chain. MultiChain from CoinSciences Ltd. is a modified version of the BTC BC, which is specially designed for private P2P networks and can be adapted to the respective requirements too. The proven public key method is used. Quorum by J. P. Morgan is an open source BC based on Ethereum. It can be operated both publicly and privately and offers several alternative consensus procedures. The software on the Quorum BC offers analogously Ethereum SC network. Thus, the software has some legal restrictions and cannot yet be directly deployed; probably, in mid-term as bank and business BC.

Three BC frameworks are currently available in SAP Cloud Platform [7, 8, 22]: Hyperledger Fabric, MultiChain and Quorum. These services enable the creation and operation of own nodes in the corresponding BC P2P networks as well as integration of external networks. A BC business service is specially designed for cross-company communication via these BCs of the types: Hyperledger Fabric, MultiChain and Quorum.

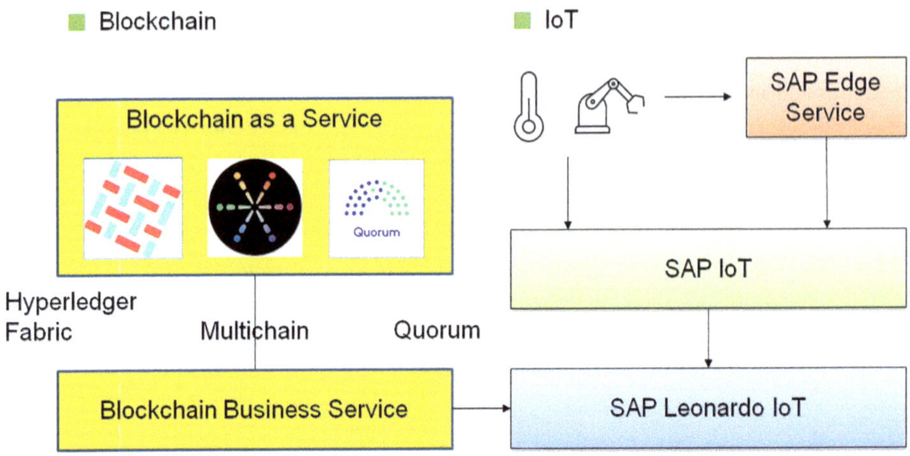

Fig. 10.14 Blockchained IoT with SAP Platform [22]

The services support the software, i. a. under IoT, which are based on BCs and provide compulsoriness, accountability and liability for their WF steps and IoT communication. The following BC components can be provided:

- A timestamp service for the BC.
- A proof-of-state service that can be used to write an object in JSON format to the BC.
- A proof-of-history service for displaying the version history of an object in the BC.
- All of these components can be used via standardised REST APIs.

An example of the integration of BC with IoT within the SAP Architecture is given below via the Fig. 10.14.

10.5 Conclusions and Evaluation on Use of BC with IoT Scenarios

1) Based on the statistics and predictions of leading consulting firms like Gartner's, it can be concluded that IoT and BC became the subjects of intensive research and development in recent years. Multiple IT companies have started own pilot projects to test them as new opportunities for the digital age.
2) This work examined the architectures, protocols and SW platforms for IoT devices as well as the important issue for data security and safety.
3) A BC is a digital ledger hosted by a P2P network that stores a digital record of transactions as discrete secured blocks.

4) Secured and Blockchained IoT can be provided only under use of the appropriate combination of the existed approaches possible like crypto protocols, FW, IDS/ IPS and CIDN. As advanced approach, the BC technology can be systematic deployed with the aim to compulsoriness, accountability and liability for IoT solutions.
5) As one of the known BC problems is use of existing computing and resources in a responsible manner. The trade-off between energy efficient IoT and BC must be found as well as the BC deployment must be appropriate and acceptable in terms of energy consumption.
6) The case studies on the Secured and Blockchained IoT devices are discussed. The focus is shifted on the integration using the platforms. However, it remains to be expected that these integration platforms for IoT and BC will be equipped in the future with additional functions, which improve handling and increase their acceptance.

References

1. A. Luntovskyy, J. Spillner. Architectural Transformations in Network Services and Distributed Systems: Service Vision. Case Studies, Springer Nature, 2017, 344p. (Monograph, ISBN: 9-783-6581-484-09).
2. Luntovskyy, A., Guetter, D. Moderne Rechnernetze - Lehrbuch, Modern Computer Networks, Handbook, Springer (2020), 481 p. + 265 pict., ISBN: 978-3-658-25616-0 (in German, 11 July 2020), vol. 1.
3. Luntovskyy, A., Guetter, D. Moderne Rechnernetze - Übungsbuch, Modern Computer Networks - Exercises, Springer (2020), 145 p. + 44 pict., ISBN: 978-3-658-25618-0 (in German, 11 July 2020), vol. 2.
4. Mahmood Zaigham (Ed.): Fog Computing: Concepts, Frameworks and Technologies, Springer 2017, London, ISBN 978-3-319-94890-4.
5. Jamil Y. Khan, Mehmet R. Yuce (Eds.). Internet of Things (IoT): Systems and Applications, 2019, New York, Jenny Stanford Publishing, ISBN 9780429399084, 366 p.
6. Lueth, Knud Lasse. The 25 best IoT Platforms 2019 - Based on Customer Review (Online):https://iot-analytics.com/the-25-best-iot-platforms-2019/
7. SAP Cloud Integration (Online): https://www.sap.com/.
8. SAP Leonardo for IoT (Online): https://blogs.sap.com/2019/05/06/sap-leonardo-iot-for-smbs/.
9. A. Antonopoulos, G. Wood. Mastering Ethereum: Building Smart Contracts and Dapps, 2019, O'Reilly Media, 345p. ISBN: 978-1491971-949.
10. Smart Contracts (Online): http://www.icertis.com/.
11. Survey on Crypto-platforms 2019 (Online): https://hackernoon.com/top-blockchain-platforms-to-watch-out-in-2019-aa80e336a426/.
12. MS Bletchley (Release & Roadmap), Cryplets Deep Dive (Online): https://azure.microsoft.com/.
13. A. Luntovskyy, D. Guetter. Cryptographic Technology Blockchain and its Applications, in "Advances in Information and Communication Technologies", Springer (ISBN: 978-3-030-16769-1), LNC "Processing and Control in Information and Communication Systems (Int. Conf. UkrMiCo-2019)" (eds.: M. Ilchenko, L. Globa et al.), 2019, pp. 14–33 (https://link.springer.com/book/10.1007/978-3-030-16770-7).

14. MIT Blockchain Course (Online): https://executive.mit.edu/course/blockchain-technologies/a056g00000URaa7AAD.html.
15. Karl Wuest, Arthur Gervais. Do you need a Blockchain? ETH Zurich & Imperial College London (Online): https://eprint.iacr.org/2017/375.pdf.
16. Leske, Christophe, Göbel, Andreas, Joswig, Steffen (2020): Blockchain mit SAP, 1. Aufl.
17. Codius: Open-source Hosting Platform for Smart Programs (Online): https://codius.org/.
18. Hyperledger Sawtooth (Online): https://www.hyperledger.org/projects/sawtooth/.
19. Blockchain as a Service: Teamwork ohne Daten-Grenzen? (Online): https://www.t-systems.com/blockchain/.
20. Siemens Blockchain IoT (Online): https://new.siemens.com/.
21. Siemens Blockchain (Online): https://www.plm.automation.siemens.com/.
22. Andriy Luntovskyy, Tim Zobjack. Secured and Blockchained IoT, IDT-2021 Zilina, IEEE Xplore, 22–24 June 2021, in IEEE Issue "Information and Digital Technologies", ISBN 978-1-6654-3692-2, ISSN 2575-677X, 2021, pp. 1–10.
23. Infotec BRAIN – Monitoring of Wireless Sensor Networks (Online): https://infotec-edv.de/
24. AMBOS (Online): https://www.fraunhofer-innovisions.de/oeffentliche-sicherheit/ambos/.

Part III

CONCLUSIONS AND OUTLOOK, LEARNING TASKS AND APPENDIXES

Conclusions and Outlook 11

11.1 HDS Construction Paradigms

The following HDS Construction Paradigms P1–P4 can be formulated considering the above-discussed positions within this monograph with a handbook function:

- Paradigm P1. The first paradigm is the use of advanced architecture models (M3–M5) in addition to (M1, M2) for efficient communication, providing advanced QoE and energy autarky within HDS Apps (IoT, Robotics, mobile Apps based on well-known platforms), **refer Chap.** 12 **(Models M1–M5).**
- Paradigm P2. The next one is the deployment of modern SWT process models (like e.g. Scrum) and use of flexible Micro Services, leading to the efficient decentralised highly distributed systems (so called HDS), which provide better scalability, reliability and reconfiguration.
- Paradigm P3. The further paradigm regards to security, privacy, authentication and compulsoriness of HDS workflow steps, modules and services under use of Blockchain technology. However, as the main disadvantage of the Blockchained IoT, the performance reduction by real time services as well as energy consumption become to a critical position.
- Paradigm P4. In addition to the classic algorithms, ML must be also integrated within HDS applications imperatively to overcome Big Data problematics.

Thus, there is, probably, a lot of further innovative techniques, methods, models and unclear options, which are to be accurately handled: energy autarky and harvesting, combined privacy, safety and security, Blockchaining, Micro Services, basic platforms, mastering of development tools and frameworks for HDS (IoT, Robotics, mobile Apps).

A. Luntovskyy and D. Gütter, *Highly-Distributed Systems*,
https://doi.org/10.1007/978-3-030-92829-2_11

11.2 Final Considerations

1. HDS generally and IoT-scenarios especially need nowadays the efficient access and management models under considering of the QoS parameters and low energy criteria.
2. Some classical as well as the advanced approaches to the optimization of Performance, Reliability and Scalability for IoT Scenarios were analysed.
3. An important role advanced SWT solutions play for choreographed decentralised micro-services.
4. The extended Internet of the future needs these solutions based on the cooperation between fog and clouds with efficient delegating of the analytic blocks via agents, adaptive interfaces and protocols.
5. The aspects of Security and Privacy in such potentially dangerous IoT-scenarios need a special considering for risks evaluation and cooperative intrusion detection with a separated research.
6. To provide the compulsoriness within the used workflow the Blockchain technology can be used and Blockchained IoT has to be created. Using Blockchain we can win in terms of security and productivity even unless more complicated and resource consuming character of hash computing.
7. This work can be qualified as a "work-in-progress". The authors try to find new efficient communication models and method combinations to overcome the above-mentioned Big Data problem as well as provide decentralised compulsoriness via Blockchained architectures.

12 Learning Exercises and Solutions

The following learning exercises and solution examples are prepared here which are referred with the work contents and aimed to the deepening of the material comprehension.

These tasks unify and practically illustrate the above-discussed subjects within the given book. They are divided into three essential topics:

1) Basics and Performance in HDS.
2) Efficiency and Security in HDS.
3) Internet of Things.

1) Tasks on Basics and Performance in HDS
    - Task 1. Definition and Comparison of Distributed and Highly-Distributed Systems
    - Task 2. Comparison of Distributed and Highly-Distributed Systems
    - Task 3. Basics: Web Services and SOA. Micro-Services
    - Task 4. Service Composition in Highly-Distributed Systems
    - Task 5. Performance Optimisation in Highly-Distributed Systems
    - Task 6. Heterogeneity Elimination via Virtualisation
    - Task 7. Big Data in Highly-Distributed Systems
    - Task 8. Vitero: Videoconferencing and Online-Tutorials for High Schools and Universities
2) Tasks on Efficiency and Security in HDS
    - Task 9. Clustering and Energy Efficiency
    - Task 10. Clustering: Performance, Speedup and Parallelism Grade
    - Task 11. Green IT: PUE and ERE
    - Task 12. Advanced Security in Highly-Distributed Systems: Firewalls, IPS and CIDN
    - Task 13. Passwords in Highly-Distributed Systems

A. Luntovskyy and D. Gütter, *Highly-Distributed Systems*,
https://doi.org/10.1007/978-3-030-92829-2_12

- Task 14. Regular Data Backup
- Task 15. Blockchain and Highly-Distributed Systems
- Task 16. IoT Efficiency: Electricity and Data Unit Costs for a WSN

3) Tasks on Internet of Things
   - Task 17. Introduction to IoT
   - Task 18. Data Acquisition for IoT
   - Task 19. Traffic Telemetric
   - Task 20. Smart Energy
   - Task 21. Sensor Networks.

12.1 Tasks on Basics and Performance in HDS

12.1.1 Task 1. Definition und Comparison of Distributed and Highly-Distributed Systems

1. Which components and which distinguishing features identify a Distributed System? Which kind of functionality must be realised herewith?
2. Specify five essential distinguishing features of a distributed system!
3. Decide, which of the following examples belong to the so-called Distributed Systems? Prove and justify the made decision!
   - Decentralised office environment, based on workstation net
   - Multiprocessor system
   - Grid system
   - Transputer Network for Parallel Processing (since 1980)
   - Central mainframe of an airline with worldwide 10,000 star-coupled terminals
   - Public cloud
   - Infotainment in the city trams and underground railways
   - Container
   - Skype
   - Blockchain
4. Distributed Systems are deployed, in general, to the following main aims:
   - Data sharing (shared use of data)
   - Device sharing (shared use of the devices)
   - Computing power sharing (shared use of computing power)
   - Communication between the users within a Distributed System

Provide at least one example for each above listed aim!

Can such systems also run in background mode (for instance, like network protocols or parts of an OS)? Provide your examples!

Solution to (1):
The components of a distributed system are as follows:

- Client application
- Server application
- Network infrastructure
- Data base or distributed DB
- GUI (for pre-processing)
- Business logics
- RPC or RMI as communication mechanism
- Low-Level-Interface—Sockets (as extension to RPC)
- Directory Service: Search for the dedicated servers
- Security Service: Cryptographic securing for services
- Time Service: Time reference/Synchronisation for services
- Distributed File System: Standardization of resource access, file transparency.

Solution to (2):
Five distinguishing features to identify a Distributed System:

- Coupling of spatially distributed computers by via a network infrastructure
- Cooperation with the aim of providing a specific application functionality
- No shared physical storage, no strictly synchronized clocks
- Decentralized organization and administration
- Often also fault tolerance and load balancing through replication.

Solution to (3):
A distributed system:

yes or no?

System example	Decision	Reasons
1. Decentral organised office environment	Yes	Decentral, cooperative task, groupware (GIT, CVS, CMS)
2. Multiprocessor system	No	Centralized managed, no special cooperation aim
3. Grid system	Yes	Set of heterogeneous computers with the common task of cooperation
4. Transputer network for parallel processing (from 1980 er)	No	Common storage, no special cooperation aim, no practical relevance
5. Central mainframe of an airline with worldwide 10,000 star-coupled terminals	No	Central and synchronized, in spite of common task of cooperation

System example	Decision	Reasons
6. Public cloud	Yes	Common network storage, a special cooperation aim, high practical relevance
7. Infotainment in the city trams and underground railways	No	There is a central administration point!
8. Container	No	Common physical storage, central management instance is available (Host OS)
9. Skype	Yes	No central administration instance, decentralized, no synchronization, cooperation task available: VoIP, teleconferencing over IP, instant messaging, chat
10. Blockchain	Yes	Distributed transactions are cryptographically secured, multilateral communication and compulsory as a cooperative aim, distributed DB and replication

Solution to (4):

Distributed Systems are deployed, in general, to the following main aims. Sometimes they are used in a background mode:

Aim	Example	Comment
• Shared data use	Cloud storage, network group, Office 365, Overleaf LaTeX CVS-Editor and Cloud, Groupware (GIT, further Multiuser-Editors)	Users can access the data at the same time and at any time, depending on the authorization
• Shared device use	Webcam, network printer, network scanner, automated sensor radio network	Users can use the shared devices at the same time and at all times depending on the authorization
• Shared computing power	Cluster, grids	Concurrency of processes or transactions
• User communication in a distributed system	Chats, WhatsApp, Web, P2P exchange, File Sharing, Skype, Gaming Clouds	Users can send the messages over available protocols on each other (N:M) simultaneously
• Background mode	In addition to those distributed applications that are visible to the end users, there are also those that are used internally (in background mode) by other components and remain hidden from the end users. For instance, to such type the applications belong for calculating routing tables (OSPF, RIP, BGP etc.), domain name services (DNS), configuration services (DHCP) as well as network management services (SNMP)	These systems do not necessarily perform their functions explicitly on the application layer!

12.1.2 Task 2. Comparison of Distributed and Highly-Distributed Systems

1. How do the highly-distributed systems differ from conventional distributed systems?
2. Give some brief characteristics and discuss the most important properties of highly-distributed systems!
3. Specify the most important construction paradigms for the highly distributed systems!

Solution to (1) + (2):
Distributed Systems
The term “Distributed Systems” has been used for many years for applications, which operate in modern combined wired-wireless-mobile networks under clear coperation goals, as well as have no centralization in memory access or synchronization in the clocking. The distributed applications are constructed on the sample n-tier and often possess redundancy in form of server and database replications. They follow to established SOA (service-oriented architecture) concept and can be often organized as cloud-centric structures. Significant architectural transformations in network services and distributed systems characterize an ongoing trend nowadays. The clouds, clusters with explicit cooperation goal (e.g. parallelized computing) as well as grids belong to the above-mentioned systems.

Highly-Distributed Systems
The term “Highly-distributed systems” include all modern combined, wireless, and mobile networks and have a complex internal structure. This technology should be well protected, maintaining the maximum possible value of the QoS parameter (higher DR and availability, low latency).

HDS allows us to deploy flexible structures based on SOA and micro services, as well as deploys efficient communication models like P2P, cloud fog or M2M, which will allow us to resolve distribution conflicts in a short time and support fast access to data analytics. Since 2005, P2P systems, the C-S communication model, as well as serverless structures (SLMA, robotics) have gained popularity. In 2011, cloud solutions became popular too. With the predominant use of load balanced “thin clients” with the delegation of functionality to the cloud. So what do we mean by the term “highly distributed systems”? The distinguishing features of HDS are as follows (Fig. 12.1):

- Advanced communication models (C-S with Clouds, Fog, P2P, M2M)
- Advanced methods for performance management and optimization as well as for QoE (Quality of Experience increasing.
- Advanced SWT (agile approaches like XP, DevOps, Kanban, Scrum, Micro-Services).
- Advanced Data Analytics regarding to solving of “Big Data” shortcomings [1, 2] via Machine Learning (ML).

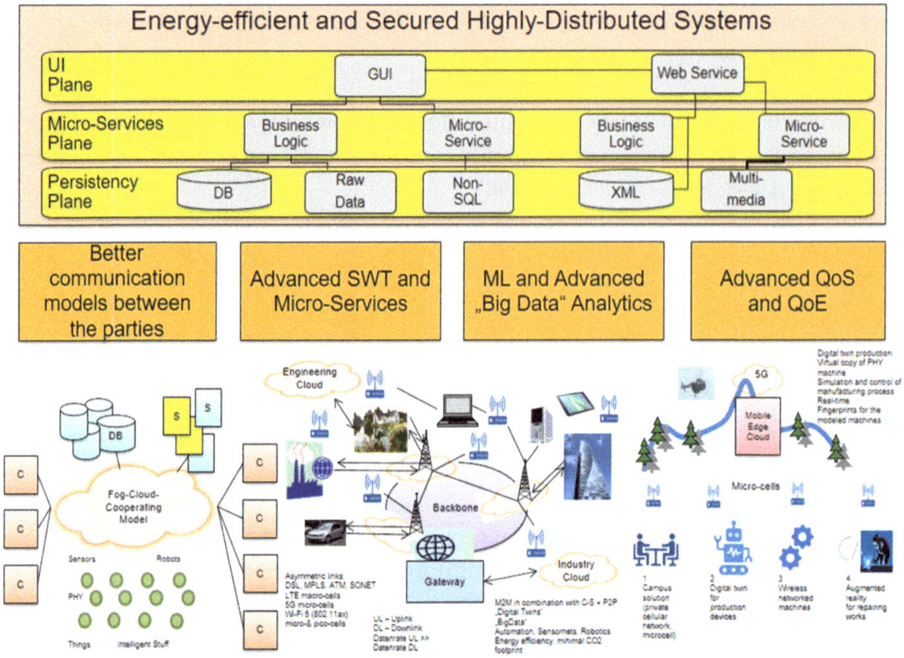

Fig. 12.1 To the motivation on HDS [1–4]

HDS systems can be efficiently used and provide powerful modern apps only if they achieve needed QoS requirements and parameters (refer Fig. 12.1), such as: Performance; Reliability; Scalability as well as support necessary security and privacy. HDS cooperate with basic modern technologies and/or offer the apps for such technologies on existing digital platforms, e.g. such subject like: Blockchain (Decentralized Transactions, Data Mining, Smart Contracting); Machine Learning and Neural Networks; IoT, 5G networks and Robotics. For rapid development of efficient HDS the novel SWT approaches and process models are widely used, like Scrum and DevOps.

The appropriate construction basis for HDS are often SOA and Micro-Services, which combine development process agility with flexible communication models [1, 3, 4]. Systematically, HDS are involved to the creation of AI (Artificial Intelligence). The neural networks and ML (Machine Learning) techniques are widely integrated into internal structures of HDS. The main purpose of ML for HDS is a special deployment of some workflow steps and learning algorithms, which are performed without human intervention. The time factor and amount of data, which are required for training, belong to the most important indicators of performance and QoS for apps and HDS.

Solution to (3):
HDS Construction Paradigms P1–P4
The following HDS construction paradigms P1–P4 can be formulated taking into account the above discussed positions:

- Paradigm P1. The first paradigm is the use of advanced architecture models (M3–M5) for efficient communication, providing advanced QoE and energy autarky within IoT Apps.
- Paradigm P2. The next one is the deployment of modern SWT process models (like Scrum) and use of flexible Micro-Services, leading to the efficient decentralized highly distributed systems (so called HDS), which provide better scalability, reliability and reconfiguration.
- Paradigm P3. The further paradigm regards to security, privacy, authentication and compulsoriness of HDS workflow steps, modules and services under use of Blockchain technology. However, as the main disadvantage of the Blockchained IoT, the performance reduction by real time services as well as energy consumption become to a critical position.
- Paradigm P4. In addition to the classic algorithms, ML must be also integrated within HDS applications imperatively to overcome Big Data problematic.

The SOA per Web Services and Micro-Services are frequently the regular construction materials for apps in the world of IBN, i.e. Distributed and Highly Distributed Systems (HDS) like e.g. so-called "Lego pieces". The providers of such kind basic elements are often called as SOA providers, which follow non-monolithic software modules concepts.

Due to Service Composition, rules the modern distributed apps can be loosely coupled or constructed as well as EAI (Enterprise Application Integration based on Services) can be provided, inter alia for the clouds.

12.1.3 Task 3. Basics: Web Services and SOA. Micro-Services

Web Services and Micro-Services are frequently regular construction material for apps in the world of Distributed and Highly Distributed Systems like e.g. Lego pieces. The providers of such kind basic elements are called SOA often. Due to Service Composition modern distributed apps can be loosely coupled or constructed. Furthermore, Enterprise Application Integration (EAI) based on Services can be provided, i.e. coupling of the app functionality to the clouds.

1. What does it mean "Web Service"? Why the provision of a WS does support the cooperation and integration between different applications that are operated on different platforms?

2. Specify three necessary building components for Web Services (a so-called WS trio)! What is the function of the UDDI registry? What does SOAP do? Why was WSDL implemented on the basis of XML?

Solution to (1):

- A Web Service (WS) is a software application that can be uniquely identified with a Uniform Resource Identifier (URI) and whose interface can be defined, described and found as an XML artifact. A web service supports direct interaction with other software agents using XML-based messages by exchanging them via Internet based protocols. About the functionality and structure of web services (s. next question). Internally, a WS can be implemented in any programming language (comparable to middleware components).
- The next example (Fig. 12.2) illustrates the thesis that WS support cooperation and integration between different applications. The depicted applications for E-Commerce are titled "Online Shop" and "Supplier".
- They belong both to different enterprises, are both the complex n-tier applications that are based on several application servers (operated on different platforms), but provide a unified communication interface for Goods, Customer Data and Store tables for support E-Commerce functionality.

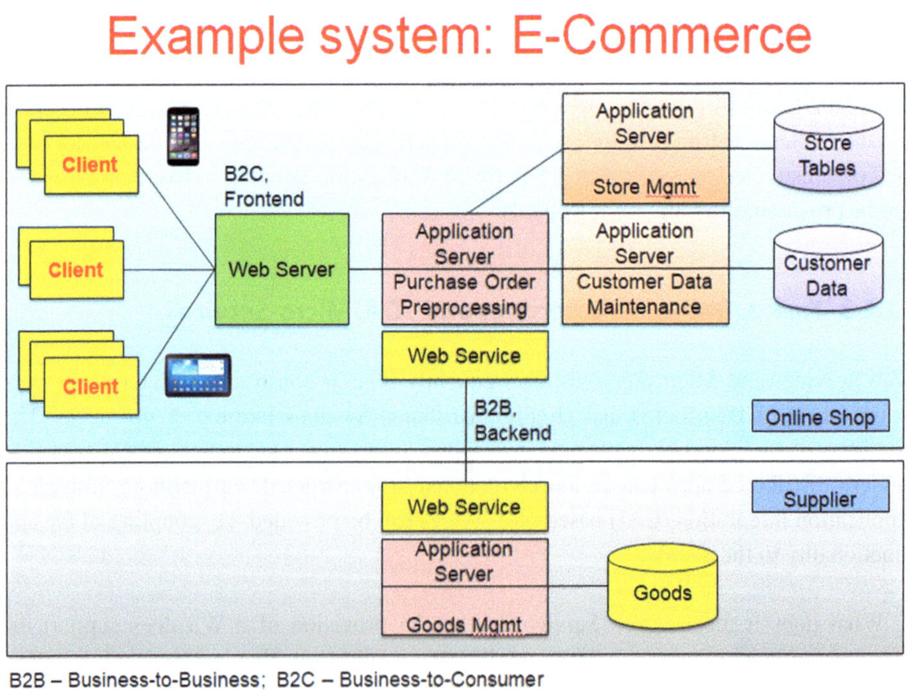

Fig. 12.2 Example system on E-commerce

Solution to (2):
The terms referred to as "WS trio" play a major role in WS use:

- UDDI (Universal Description, Discovery and Integration) is information and directory service (service registry) simultaneously. UDDI is itself implemented as a WS. UDDI Specification is used for describing and searching for a web service. Finding web services in "Business Registries" like e.g. MS, IBM and Ariba is provided. UDDI services can be queried using the service name (so-called "white pages") or by specifying attributes (yellow pages; a kind of "business directory") or technical details (so-called "green pages").
- WSDL (Web Service Description Service) is description of the user interface of WS (execution is transparent). WSDL describes method calls based on XML schema including the associated input/output parameters. WSDL can contain links to SOAP and HTTP GET/POST messages.
- SOAP (Simple Object Access Protocol) is a transfer protocol for XML-oriented messages. SOAP is based on underlying application and transport protocols (e.g. HTTP/TCP and SMTP/UDP). The protocol is more flexible than conventional HTTP GET/POST mechanisms for more complex content and its better structuring. A distinction is also made between SOAP web services and REST (Representational State Transfer) web services. REST protocol for web services represent a lightweight alternative to SOAP.

3. Outline the processes of publishing and deploying a WS!
4. How can we prove broad interoperability between WS based apps? How do the web services enable access to the clouds?
5. Clarify the differences and demarcation between WS and SOA. How is EAI (Enterprise Application Integration) provided?

Solution to (3) + (4) + (5):
Refer Fig. 12.3 with the outlined Find-Bind-Execute Paradigm for Web Services:

As an example (Fig. 12.4) we consider the heterogeneous apps from different "software worlds" and development circles. Use of WS guarantees dismantling heterogeneity for apps and applications on different platforms, as well as provide application integration.

Cloud computing use WS indispensable to provide the access between the thin clients to the server part (a cloud).

Differences and demarcation between WS and SOA:

- SOA is not Web Services because SOA describes an architecture paradigm that is detached from specific implementation technologies.

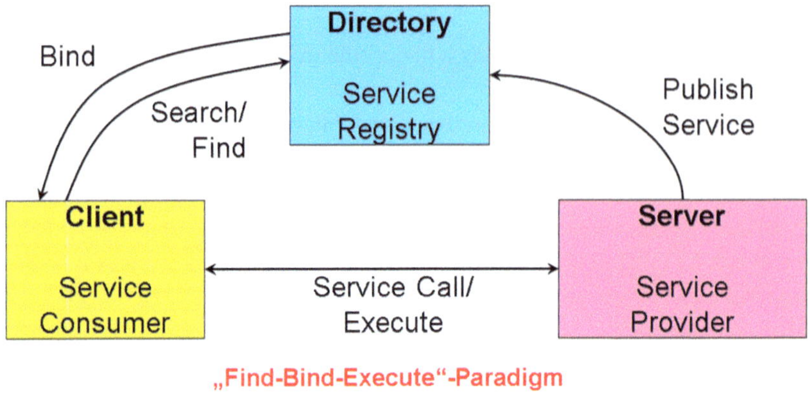

Fig. 12.3 Find-bind-execute paradigm for web services

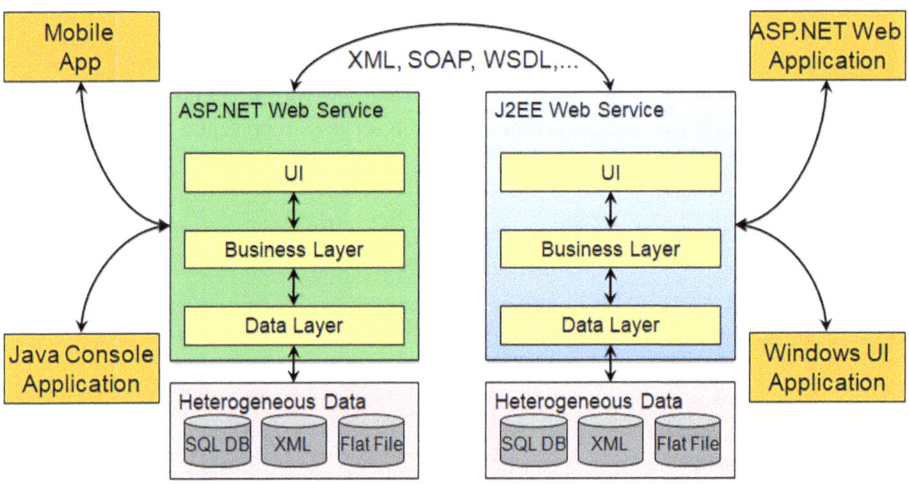

Fig. 12.4 Dismantling heterogeneity with WS for Apps and Application Integration

- The SOA approach could already be implemented in 1990s with the technologies available at the time and was used with so-called "CORBA middleware", among other things (former CORBA—Common Object Request Brokering Architecture).
- SOA is not a solution for business problems. Furthermore, as an architecture paradigm, SOA does not give any recommendations for dealing with business problems.
- SOA is individual, and there is no so-called "Standard SOA". A company always has to tailor an SOA to its own needs.

Further brief statements about web services:

1) Web services are based on a service-oriented architecture
2) Time, costs and complexity of integration projects can be significantly reduced by using an SOA
3) Use of web services does not necessarily lead to an SOA. The most important innovations in SOA are proven Web Services Standards.

6. Why and in which cases is REST is considered as an alternative to SOAP in web services?

Solution to (6):
So-called REST communication is given in Fig. 12.5. The protocol REST provide a simpler mode for WS deployment. What for the main advantage is the use of HTTP semantics in comparison to SOAP Web Services.

Further advances on this area are signified by so-called WS-* (WS Star) that can optimize "classic Web Services" with SOAP. However, these coexist together with so-called RESTful web services for EAI and app implementation in Distributed Systems.

7. Specify the main platforms on which web services can be run?

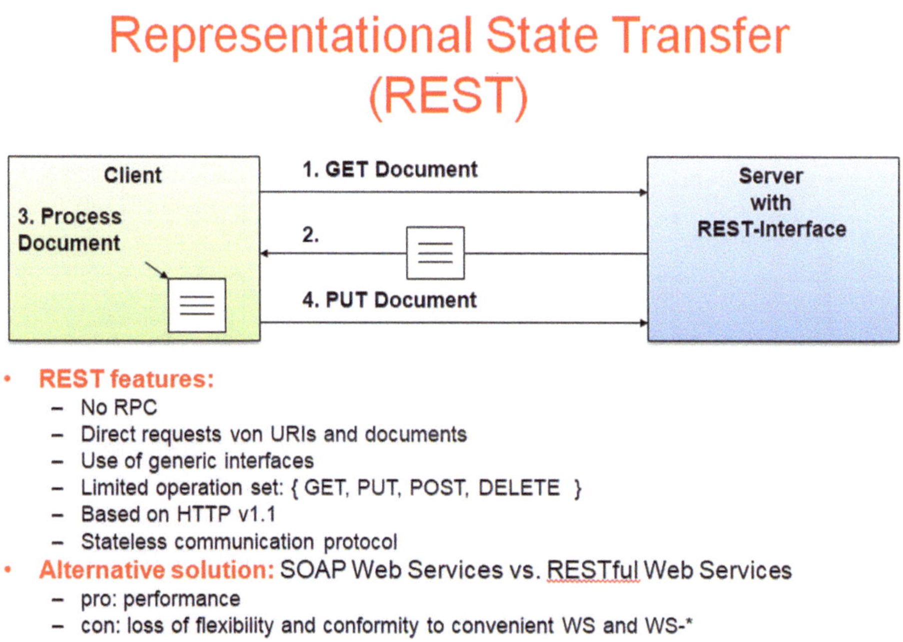

Fig. 12.5 REST communication in detail

Solution to (7):

The platforms to WS are used in the same manner like the platforms for MW components. The Web Services can be placed in the same order on the application servers:

- Apache Axis 1, 2 and Tomcat
- Novell Mono Development Platform
- Microsoft.NET Server
- SAP Web Application Server (SAP NetWeaver)
- Java Web Services Development Pack (JWSDP) on Oracle (based on Jakarta Tomcat)
- Oracle Application Server (in Oracle Fusion Middleware)
- IBM WebSphere Application Server (based on Apache and J2EE Platform)
- DotGnu/GNU Project.
- WebLogic von BEA Systems
- Macromedia ColdFusion
- Cordys WS-AppServer etc.
- Cordys WS-AppServer etc.
- Demarcate the terms "Web Services" and "Micro-Services"!

8. To illustrate and explain the evolution for app implementation in distributed systems, please use the sketch below (Fig. 12.6)!

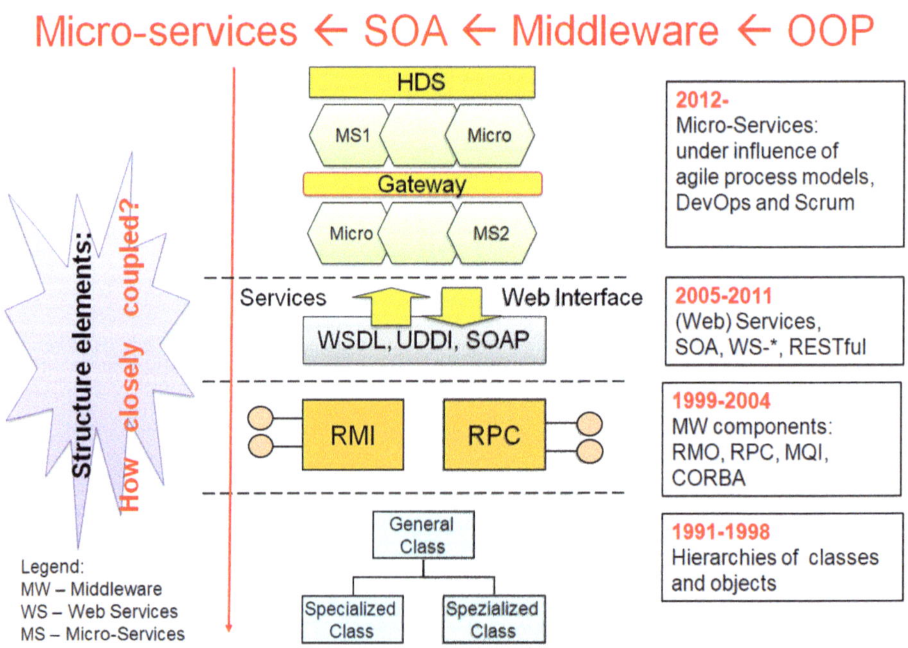

Fig. 12.6 Periodization for changing app implementation basis in DS and HDS

Solution to (8):
A comparison of SOA and Micro-Services is given in Table 12.1.

The distinguishing features of the both terms "Web Services" and "Micro-Services" are examined.

The sketch presents the periodization in the app implementation in distributed systems or, correspondingly, in advanced HDS (refer Fig. 12.6). Since the 1990s, the implementation of apps in distributed systems (DS) has been striving forever more loose coupling of the components with one another or convenient adaptation or EAI under heterogeneity conditions. The expenditure factor is also reduced through source code unification and increased reusability. The evolution extends from OOP to MW (Middleware) components to WS, SOA and micro services. Increasingly loose coupling is being pursued as an approach.

9. Which SWT (Software Technology) process models are based on the expansion of micro services? Which architecture components use micro services? Can you name the main platforms on which micro services can be run?

Solution to (9):
Refer Fig. 12.7 for conventional and agile SWT methods and Process Models.

A process model organizes a workflow for creative production (SW development) into various, structured sections and phases, which are assigned to corresponding technologies, tools, languages, protocols and methods of the organization (refer Fig. 12.7).

The following concepts are used in the given comparison:

- AOP/Aspect-Oriented Programming (Xerox PARK)
- BPMN/Business Process Model and Notation (OMG)

Table 12.1 Comparison SOA and micro-services

SOA	Micro-services
Both SOA and micro-services use services as architectural elements	
SOA uses the services for integration of different apps (EAI) Frequently external integration, as internal also available	Micro services bring a structure to an App under use of the services Better internal integration
So-called quasi-monolithic structure	n-tier structure
The combination of the services (Service Composition) can be "orchestrated" or "choreographed" (so-called Orchestration or Choreography), and the portals can provide a common GUI for all services (central or decentralized composition)	Each micro service can include a GUI and implement the business processes in similarity to a SOA with Orchestration (centralization)
Conclusion for the comparison: loosely coupled, more flexibility by micro-services	

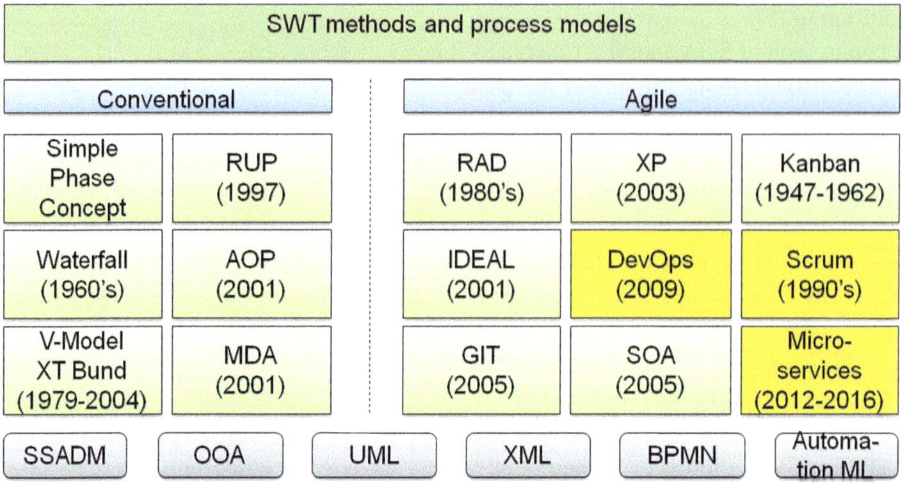

Fig. 12.7 Conventional and agile SWT methods and process models

- IDEAL/SW process of Carnegie Mellon University
- MDA/Model-Driven Architecture (OMG)
- OOA/Object-Oriented Analysis (P. Coad, E. Yourdon)
- RAD/Rapid Application Development (J. Martin, B. Boehm)
- RUP/Rational Unified Process (P. Kruchten, Rational)
- SOA/Service-Oriented Architecture (Web Services)
- SSADM/Structured systems analysis and design method (T.de Marco, E. Yourdon)
- UML/Unified Modeling Language (G. Booch)
- V-Model/SW process federative model (Germany)
- XP/eXtreme Programming (M. Fowler).

For instance, SCRUM as process model (Fig. 12.8) was inspired by a popular sport.

- Scrum is a virtual analogy to rugby sports and describes the successful teams, which are working on the development of a project or SW product together and involved into close human communication.
- As an iterative SWT process model, Scrum does not describe classic project phases, but instead of attaches the values of so-called continuously usable results (artefacts) right from the beginning of the project.
- The Scrum is oriented on three basic principles for three people (Roles), four meetings (Sprint) and six results (Artefacts).
- Scrum is nowadays a "de facto" agile SWT standard and is interoperable with advanced SOA as well as micro-service architectures.

Fig. 12.8 What is really SCRUM?

- Based on Conway's Law, the micro-services are involved to flexible SW development.
- For unified definition of a SW project, roles, events and artefacts (i.e. obtained handmade results) must be assigned. The defined roles belong the Scrum workflow: Development Team, Scrum Master, Product Owner. Furthermore, a so-called Sprint Process with regular meetings with validation and review of project results must be processed.

Typical slogans are used for software development under deployment of Micro-Services with Scrum:

- Slogan 1: "One for all and all for one" (a musketeer slogan).
- Slogan 2: "Do twice in half time" (s. XP).
- Slogan 3: Unix-Philosophy ("Do One Thing and Do It Well!")
- Slogan 4: Two pizzas teams (6–8 people can be satisfied with two big pizzas), refer Fig. 12.9.

Fig. 12.9 Two pizzas teams

Use of Micro-Services: What does it mean?

- Concept for modularization, specific organization and SWT approach
- Component: a single micro-service, independent and technically oriented
- Architecture: Micro-service architecture is represented as a large entity across all individual micro-services

The mostly popular citations about:

- James Lewis:
  "Micro-service is a small application that can be deployed, scaled and tested independently. This application and its code should be easy to understand and do exactly one job."
- Sam Newman:
  "Micro-service is a self executable software component that collaborates with other software components within an integrated application. Micro-services should take on a role within a specialised business domain."

10. Which architectures and architecture components use Micro-Services for construction of the apps that distinguish from so-called quasi-monolithic concept?

Solution to (10):
Using, for example, Scrum, a significantly more flexible MS-based structure can be derived from a WS-based 3-tier structure (use of so-called Macro-Services). Using the application for a supermarket as an example, a software engineer can design simpler and more casual coupling of the services, saving on access times to the involved layers, in particular, on the data layer (see Fig. 12.10) with individual DB control.

The main platforms for implantation with micro-services are as follows: Netflix, Spring, Kubernetes.

An appropriate example of MS use is given below (cp. Fig. 12.11):

A gateway with MS acts as a service provider and can deliver the required codes from the content specific MS1, MS2, MS3 under use of widespread protocols and interfaces like HTTP/REST, AMQP/JSON-RPC, WebSockets/WAMP to several apps like Browser App, Native App, Server App.

12.1.4 Task 4. Service Composition in Highly-Distributed Systems

1. What does it mean "Service Composition"?
Are there differences in service composition for WS and micro-services?

Solution to (1):

- Service composition is as a rule process- and service-oriented

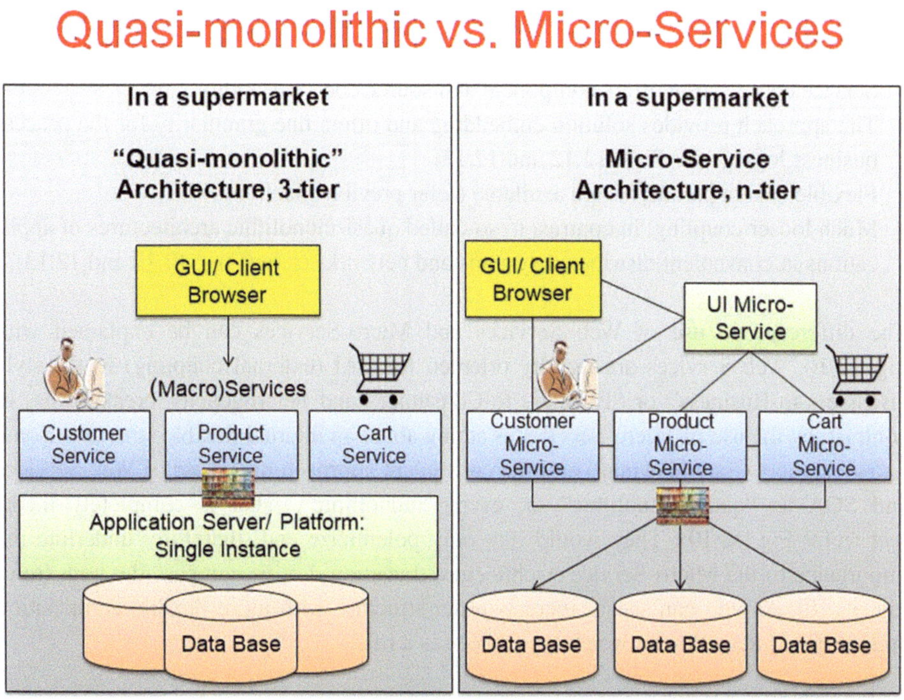

Fig. 12.10 Quasi-monolithic or micro-services?

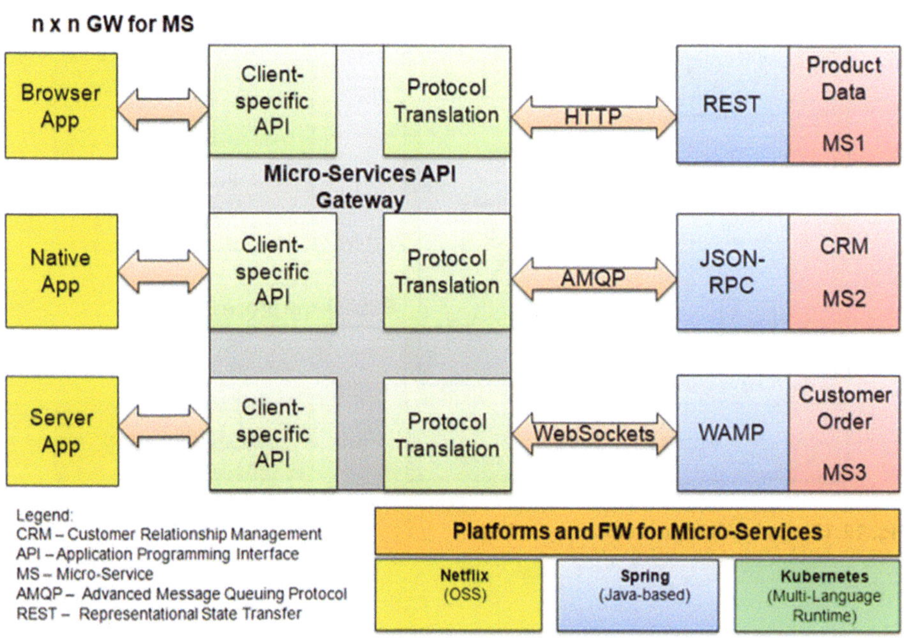

Fig. 12.11 Micro-services as architecture components

- Web services, Micro-Services and their platforms are very suitable as an implementation basis
- Service composition offers component and source code reusability
- The approach provides solution embedding and offers fine granularity for the process business logic (refer Figs. 12.12 and 12.13)
- Flexible design paradigms are available (refer previous Sections)
- Much looser coupling, in contrast to so-called quasi-monolithic architectures of applications in convenient distributed systems and networks (refer Figs. 12.12 and 12.13).

The difference in use of Web Services and Micro-Services can be explained with Fig. 12.10. Web Services are usually oriented for EAI (external coupling) in the style "Business-to-Business" or "Business-to-Consumer" and heterogeneity breakdowns. In contrast, by the use of Micro-Services is acting about an internal flexible structure. Some known authors (Sam Newman) call the convenient approach under use of Web Services and SOA as "quasi-monolithic" or, even, "monolithic", what is completely incorrect (refer Fig. 12.10). They would like only polemicize and, therefore, underline the importance to the Micro-Service Architecture transition. Let us compare the both (refer Fig. 12.10). As you can see it, there is n-tier-structure with more flexible composition and intrinsic interaction between the services as a rule.

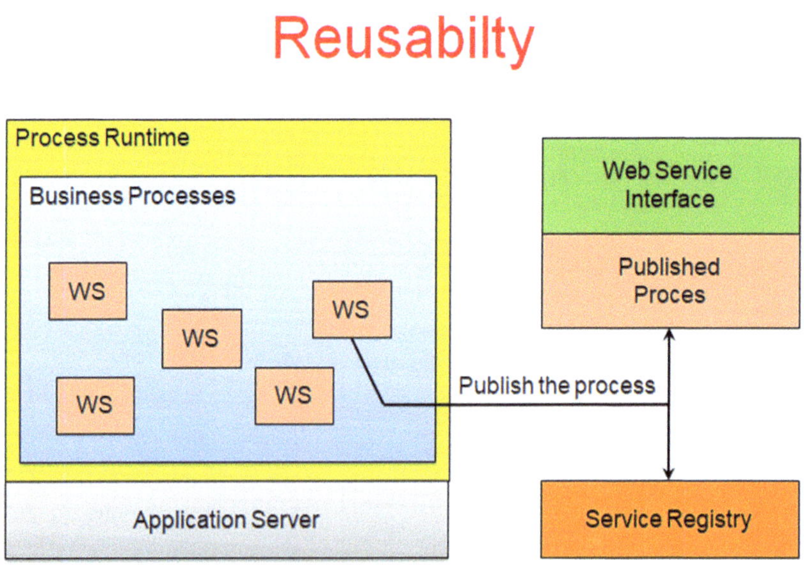

Fig. 12.12 Embedding and reusability of WS

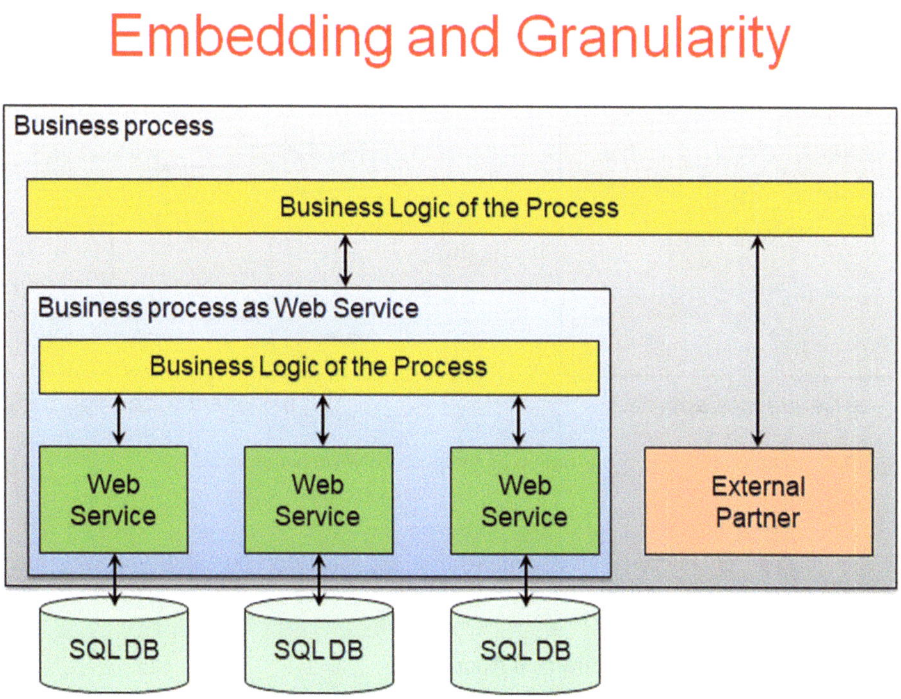

Fig. 12.13 Embedding and granularity of WS

Table 12.2 Comparison SOA and micro-services in regard to service composition

Centralized approach	Decentralized approach
Service orchestration represents a single centralized executable business process (the orchestrator) that coordinates the interaction among different services. The orchestrator is responsible for invoking and combining the services	Service choreography is a global description of the participating services, which is defined by exchange of messages, rules of interaction and agreements between two or more endpoints
The relationship between all the participating services are described by a single endpoint (i.e., the composite service)	Choreography employs a decentralized approach for service composition
The orchestration includes the management of transactions between individual services	The choreography describes the interactions between multiple services, where as orchestration represents control from one party's perspective
Orchestration employs a centralized approach for service composition	Choreography employs decentralized approach for service composition

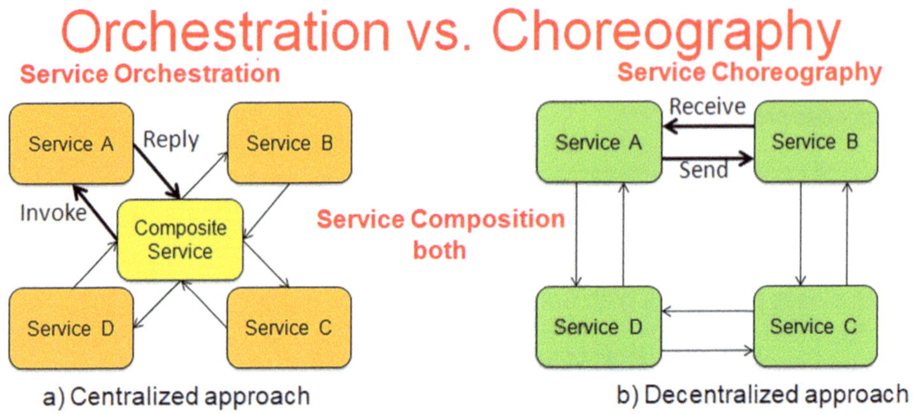

Fig. 12.14 Composition via orchestration or composition via choreography?

The macro- or micro-service composition is generally possible via two main approaches (refer Table 12.2):

(a) Service Orchestration (Centralized Approach);
(b) Service Choreography (Decentralized Approach).

2. What are the advantages of centralized or decentralized composition of services (Web Services or Micro-Services)? Clarify the differences between the both concepts: orchestration and choreography (Fig. 12.14).

Solution to (2):
The service composition can be divided into Centralised Approach and Decentralised Approach.

Centralised Approach: Service orchestration represents a single centralised executable business process (the orchestrator) that coordinates the interaction among different services. The orchestrator is responsible for invoking and combining the services. The relationship between all the participating services are described by a single endpoint (i.e., the composite service). The orchestration includes the management of transactions between individual services. Orchestration employs a centralised approach for service composition.

Decentralised Approach: Service choreography is a global description of the participating services, which is defined by exchange of messages, rules of interaction and agreements between two or more endpoints. Choreography employs a decentralised approach for service composition. The choreography describes the interactions between multiple services, whereas orchestration represents control from one party's perspective. This means that a choreography differs from an orchestration with respect to where the logic that controls the interactions between the services involved should reside.

12.1.5 Task 5. Performance Optimization in Highly-Distributed Systems

1. Which performance optimization methods for highly distributed systems do you know? Specify them.

Solution to (1):
The performance optimisation methods for HDS (refer Fig. 12.15) are as follows:

1) Targeted server or DB replication (for load balancing, min. response time)
2) Concurrent server or client-side processing through multithreading (lightweight sub-processes with their own program counter and stack)
3) Caching on the client side (result cache with e.g. names)
4) Targeted further distribution to n-tier or P2P structures (load balancing, scalability)
5) Use of suitable communication models (refer Figs. 12.15 and 12.16)
6) Further empirical rules.

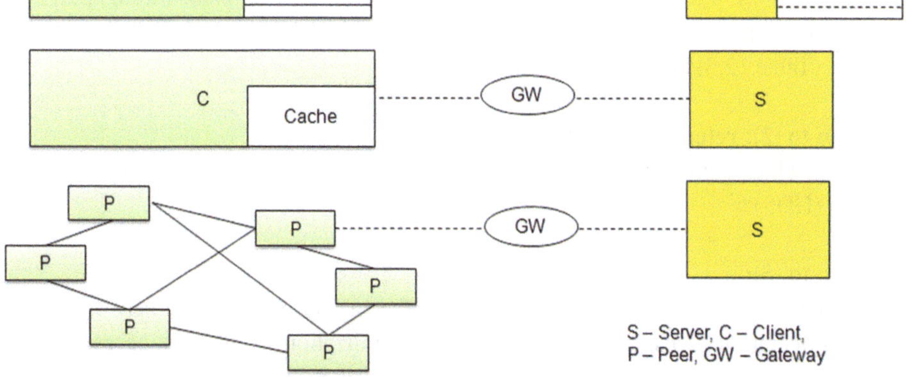

Fig. 12.15 Performance optimization methods for HDS

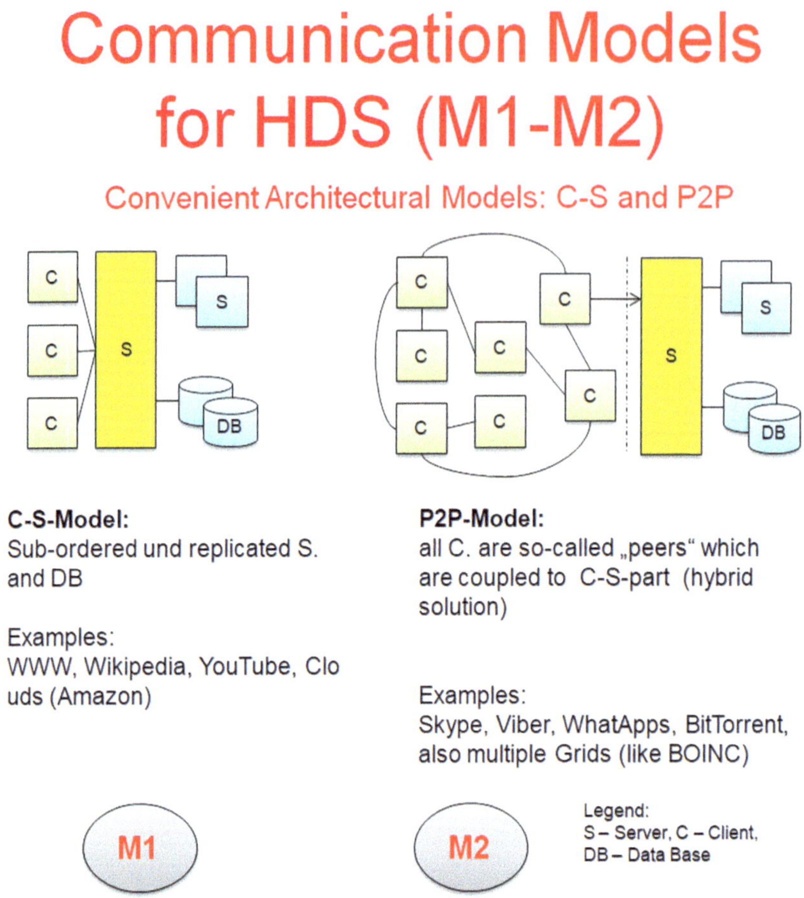

Fig. 12.16 HDS: convenient architectural models M1–M2

2. How do they influence the following QoS criteria such as max. throughput, min. response time, probability of failure, availability, reliability etc.? Use the prepared table (Table 12.3) below for your analysis:

Solution to (2): refer the table above.

3. Use of which communication models is mostly suitable for the highly-distributed systems (HDS)?

Solution to (3):

Convenient Distributed Systems M1-M2 use efficient and performance networks under clear co-operation goals, as well as have no centralization in memory access or synchronization in the clocking (Fig. 12.16). Distributed Systems provide efficient communication under use of convenient Architectural Models M1 and M2 (cp. figures).

Table 12.3 Performance optimization methods and QoS criteria for highly distributed systems

Approaches for QoS	Max. throughput	Minimal response time	Reduced downtime	Better availability	Better scalability	More reliable
Multi-threading	x	o		o	x	
Replication			x	x	x	x
Caching on client		x		o	x	o
n-tier	o	o		o	x	x
Use of proxies and gateways	o			x	x	x
Peer-2-Peer		x		x	x	
Further empiric rules	x		o	x	o	o
Notice:	o—partially compliant			x—fully meets		

Additionally, they offer replications due to redundancy, use of flexible P2P structures, cloud and fog services. A very important role the energy efficiency and autarky plays.

The Advanced architectural models for communication are represented via M3-M4 both (Fig. 12.17). They are cloud-centric and fog-cloud correspondently as well as more relevant for HDS use. The focus lies on the placement of analytics blocks as a mostly intensive part of software due to resources workload and energy consumption:

- M3 with the feature for less extern performance for Analytics, consolidated servers, DB and thin clients
- M4 with the feature for more extern performance for Analytics, which is partially delegated to the fog. Therefore, there are five main types of communications models, which are widely deployed in HDS: Client-server (C-S) communication model, Peer-to-peer (P2P) communication model, cloud-centric and fog-cloud communication models and, finally, Machine-to-Machine (M2M) communication model.
- Machine-to-Machine (M2M) communication is a communication model that involves one or more instances that do not necessarily require human interaction or intervention in the process of communication (cp. Fig. 12.18). The applications are used in automated and half-automated modes. The typical scenarios are sensor networks, telemetry, IoT and robotics. The main differences to C-S and P2P models (refer the previous sections) are as follows: M2M applications generate short 'bursts' on periodic data packets (so-called short telegrams); The communication is realised typically

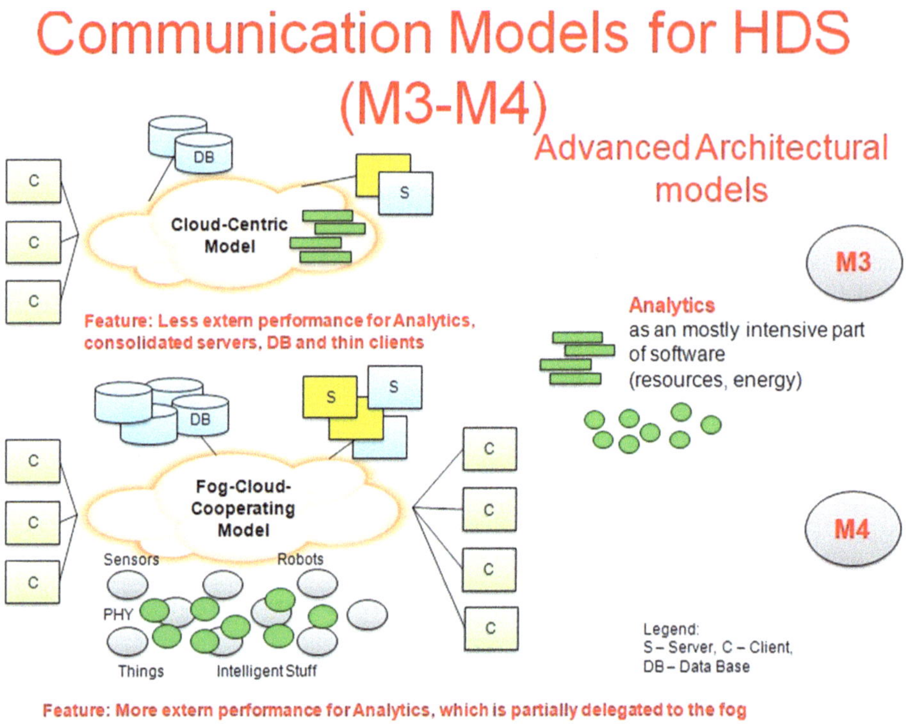

Fig. 12.17 HDS: advanced architectural models M3–M4

via asymmetric links; M2M communications are established through uplink channels (UL); Uplink traffic is bigger than downlink (DL) traffic (direct channels from providers). Usually, M2M traffic is machine generated and does not require any human intervention: no human-to-human (so-called H2H). The QoS requirements for M2M communication model differ significantly from regular applications such as in the mobility, delay tolerance on offline and data volumes, on priorities. The automated M2M applications have, furthermore, low power consumption and significant requirements on security and privacy.

12.1.6 Task 6. Heterogeneity Elimination via Virtualisation

1. What are the advantages of virtualization from an operational and technical point of view? Specify at least four other advantages:
   - Heterogeneity Elimination
   - Better utilization of virtual machines

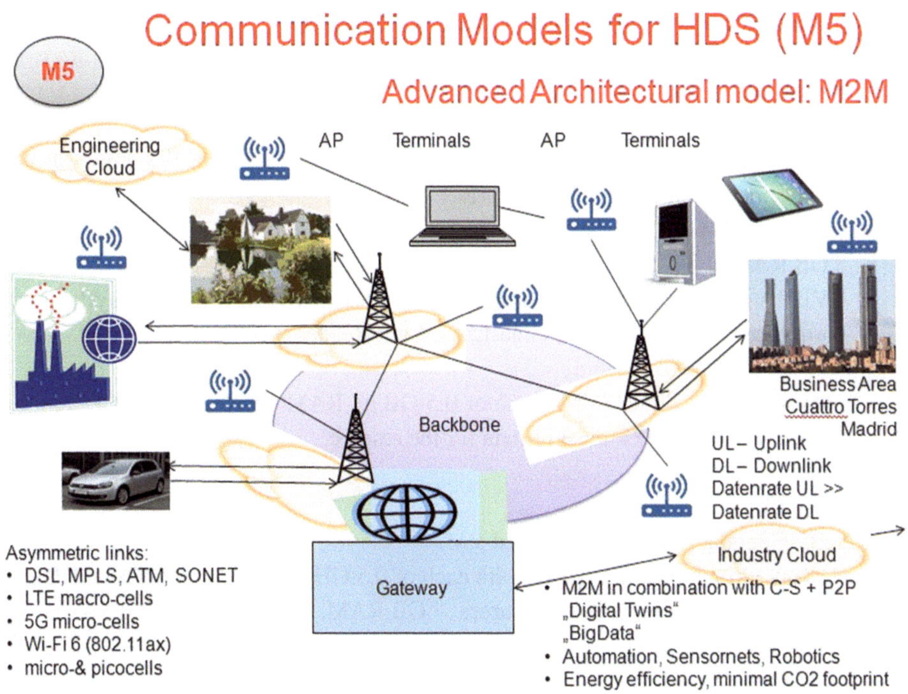

Fig. 12.18 HDS: advanced architectural model M5, M2M

- __________________________
- __________________________
- __________________________
- __________________________

Solution to (1):

Virtualisation advantages from an operational and technical point of view are as follows:

1) Heterogeneity Elimination, abstraction from HW/SW, independence from HW/SW, reduction in heterogeneity
2) Better utilization of virtual machines than for single PHY servers (functions such as emulator + simulator, i.e. these simulate environments for certain OS as well as properties of certain OS)
3) Flexible resource consolidation
4) Flexible resource partition if necessary from a big PHY machine
5) Downtime reduction
6) Minimizing of investments (CAPEX)
7) Minimal amortization
8) Saving of personal and maintenance costs, OPEX reduction

9) Dynamic deployment of a hot reserve at needed place
10) Electricity costs reduction
11) Reliability increasing.

2. A computer company sells you a powerful desktop PC with the OS Windows 10. This computer provides the following features:
   - Intel Core i5-8400 CPU 4.0 GHz, Hexa-Core
   - 8 GB RAM, 1 TB SSD, 128 TB SATA, DVD DL burner, USB Type C
   - NVIDIA Ge Force GTX 1050 graphics (2048 MB), HDMI, DVI, DisplayPort
   - Form factor: Tower, WLAN, Giga-LAN, Card Reader etc.

How many virtual machines (VM) each of 0.5 GByte RAM and 6 TByte HDD can this PC virtualize, if at least 20% must remain for the reserve!

Solution to (2):

Notice: OS-Container serves as a limited VM.

How many virtual machines (VM) with each of 0.5 GByte RAM and 6 TByte HDD?
We consider the following host resources: 8 GB RAM, 1 TB SSD, 128 TB SATA
Searched: n = min (x, y)

RAM: 8 GB	x = 0.8 * 8 GByte = 6.4 GByte	20% reserved
HDD: 128 TB	y = 0.8 * 128 TByte = 102.4 TByte	20% reserved
SSD: 1 TB		100% reserved

$$x = 6{,}4\,\text{GByte}: 0{,}5\,\text{GByte} = 12\ \text{VM}$$
$$y = 102{,}4\,\text{TByte}:\ 6\ \text{TByte} = 17\ \text{VM}$$
$$\text{VM number}$$
$$n = \min\,(x,\ y) = \min\,(12{,}17) = 12$$

if n = 12 VM are deployed, then they possess the following overall resources:

$$\text{overall RAM} = 6\ \text{GB of}\ 8\ \text{GB}$$
$$\text{overall HDD} = 72\ \text{TB of}\ 128\ \text{TB}$$

Attention: overall SSD = 1TB (100% is dedicated only to the host OS)!
Check list:

Remaining RAM	= 2 GB ca. 20–25%	→ OK
Remaining HDD	= 56 TB reserve > 20%	→ OK
Remaining SSD	= 1 TB (100%)	→ OK

12.1.7 Task 7. Big Data in Highly-Distributed Systems

1. Define the term "Big Data"! What kind of problems can appear for the Highly-Distributed Systems due to the weak structured and unmanageable data (refer Fig. 12.19)?
2. A WSN consists of 15,000 sensor nodes (refer Fig. 12.21). Based on 'Low Duty Cycle' rule each sensor must transmit only an economised short telegram with up to 100 bits. Herewith the request is processed for each sensor node 20 times/hour via the gateway (base station). The average life expectancy for such kind of WSN lies between 3 and 9 years. Compute please the overall acquired data volumes for the discussed period under given conditions DVoverall!
3. Which kind of sources do so-called "Big Data" have? Clarify please the principle of "6 V for Big Data" (refer Fig. 12.22).

Solution to (1):

"Big Data" is a term for the current and medium-term collections, whereby the data volumes are so large and immobile (growth from ~100 PByte to 100 EByte) that their management and access are only hardly possible without special new procedures and under

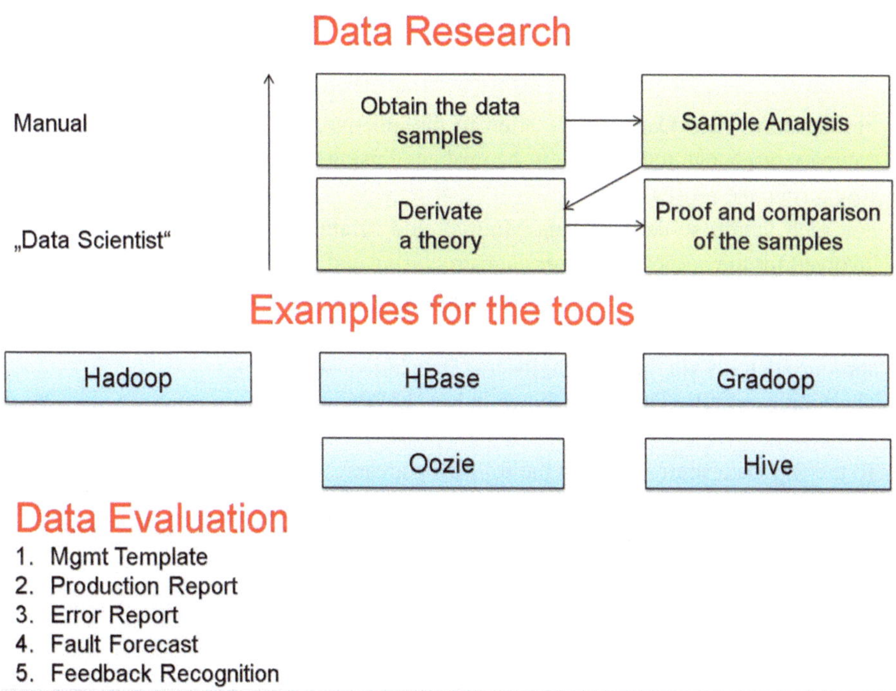

Fig. 12.19 Big data and appropriate tools

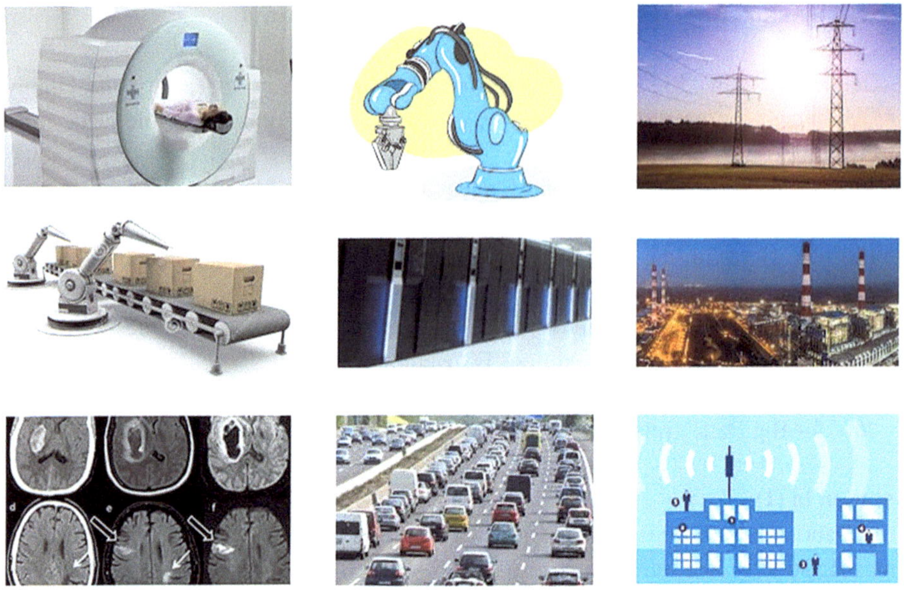

Fig. 12.20 Sources of "Big Data"

use of the conventional storages. The retrieval and analysis procedures become frequently inadequate under existing conditions.

- Furthermore, "Big Data" refers often to the amounts of data that are too large, too complex or change too quickly to be evaluated using classic manual methods of data structuring and processing.
- For such target, so-called "Data Mining" and "Data Pattern recognition" are often involved for the purpose of better data structuring and processing.
- Refer to Fig. 12.19, where use of new smart methods for "Big Data" is shown. From convenient "manual methods" to "smart methods", i.e. under deployment of the appropriate tools via the data engineers (Data Scientist).
- The sources of "Big Data" are shown in Fig. 12.20:

  To the Big Data sources belong the following areas:

- Power Grids
- Patient Data (CT, MRT)
- Highway Traffic
- Telco and Mobile Provider Data
- Robotic, Logistic and Automated Industry
- Cluster Computing.

Solution to (2):
Refer Fig. 12.21

Given:
N = 15,000 sensors
FL = 100 Bit as a short telegram
Frequency = 20 1/h

(a) T = 3 years—average life expectancy
(b) T = 9 years—average life expectancy

Searched:
K—Number of GW requests? (Gateway/Base Station)
Min/Max. Data Volumes DV1, DV2—?

During to 3 years (without maintenance):
K = 20 1/h * 24 h* 365 d * 3 y = 20 * 24 * 365 * 3 = 525,600 requests per sensor per average life expectancy period
Herewith the following data volumes can be acquired

As a minimal value for the discussed WSN:

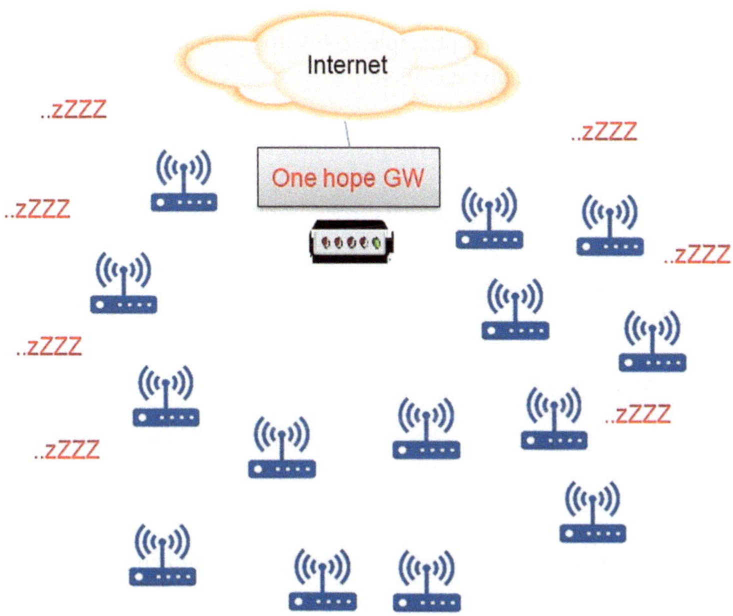

Fig. 12.21 Wireless sensor network

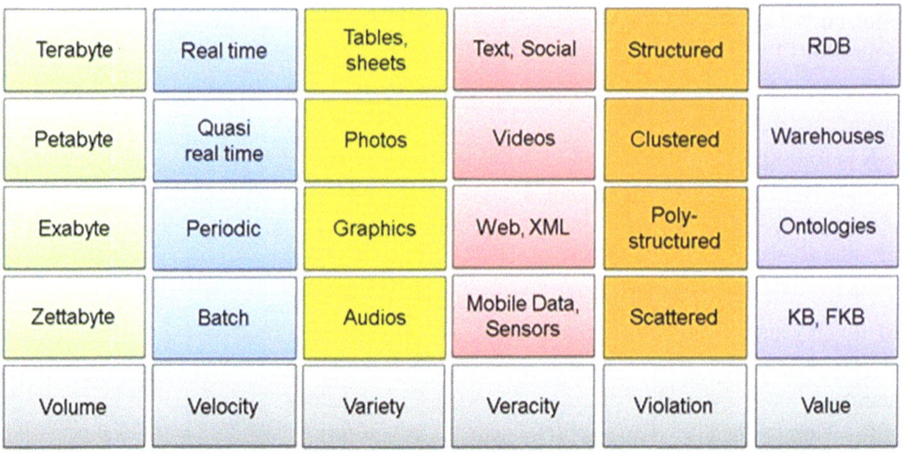

Fig. 12.22 6 V for Big Data

DV1 = FL * K * N = 100 Bit * 525,600 requests * 15,000 sensors = 788,400,000,000 Bit = 788 TBit = 98.5 TByte ~ 100 TByte

As a maximal value for the discussed WSN:

$$DV2 = 3DV1 = 300 \text{ TByte}$$

Solution to (3):

To clarifying the above-mentioned principle of 6 V for "Big Data" (refer Fig. 12.22) the following common sources can be specified:

- Data from sensors and production systems (Smart Manufacturing)
- Traffic data and telematics (IoT)
- Robot data, QoS controls and traffic data from mobile providers
- Power plants and renewable energy sources (smart grid, smart metering)
- Simulation data (clusters, grids)
- CT and MRT patient data, recorded personal data (Fog).

The following 6 V for "Big Data" are known:

Volume/Velocity/Variety/Veracity/Violation/Value = 6V

These six V-factors are formulated as follows (Fig. 12.22):

1) Volume: The amount of generated and stored data is moving from 100 PB up to 100 EByte nowadays.

2) Velocity: It means the frequency and speed of data acquisition and processing (from slow batch techniques to real time with strong limitation on reaction time or latencies).
3) Variety: Multiple codec types of multimedia data. This helps for more effective the resulting insights.
4) Veracity: The quality of captured data can vary greatly, affecting accurate analysis.
5) Violation: The discussed data are often scattered, without any clustering or any structuring.
6) Value: Inconsistency of the data sets can negatively affect the processes of data handling and management.

12.1.8 Task 8. Vitero: Videoconferencing and Online-Tutorials for High Schools and Universities

Nowadays, at universities and high schools the following video conference tools for the online classes and tutorials are widely used: Zoom, MS Teams, Cisco Webex Meetings, BigBlueButton etc.

Herewith we discuss Vitero software (https://www.vitero.de/) as a reference tool for virtual classroom creation. The tool offers effective collaboration for its users/participants under use of "Virtual Telco Room" concept through usual web conferences.

The decisive advantage of Vitero compared to traditional teleconferencing (by phone, VoIP, video, web) results from its ergonomics. Vitero makes the following services available to users:

- Audio (per phone and VoIP)
- Photos and live pictures (per web cam)
- Document and application Sharing as well as Groupware
- Visualisation and interactions (slides, online commitment, subgroups for the users).

The following scenario can be examined (see Fig. 12.23). You want to set up a video conference with several partners via Vitero. You use a multipoint conference system with a star architecture and a central MCU (Multipoint Control Unit). Vitero sends a video with the resolutions A1, A2, A3 correspondently (see below) with a colour depth FT of 24 bit/pixel and a frame rate FR of 25 fps:

- Video playback resolution $A_1 = 480 \times 270$ Pixel
- Video playback resolution $A_2 = 640 \times 360$ Pixel
- Video playback resolution $A_3 = 1280 \times 720$ Pixel

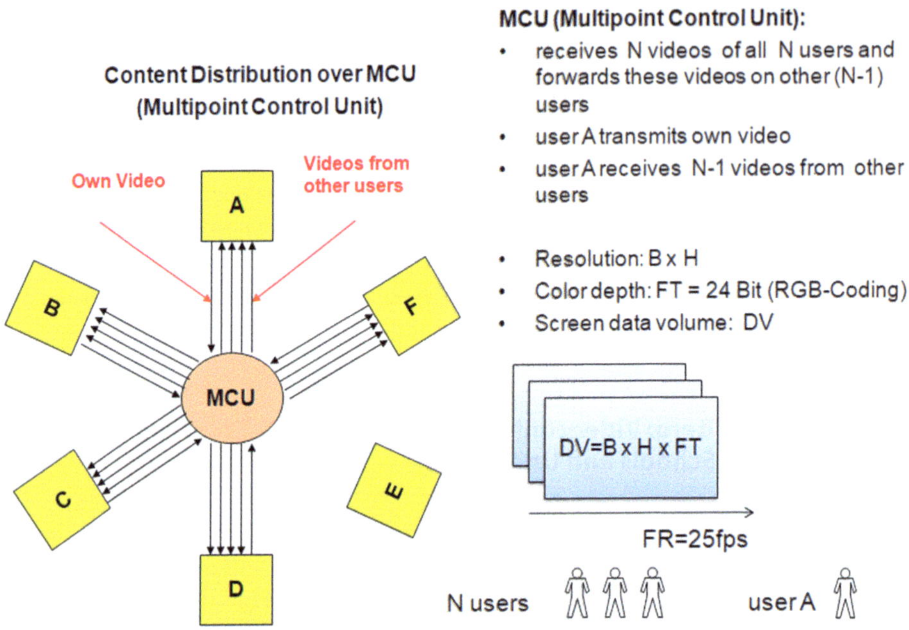

Fig. 12.23 Videoconferences via Vitero for N users by the given DV and FR parameters

Answer please the following questions:

1. With which compression factor do you have to compress the video to be sent via Vitero if you are using a DSL connection with upstream of 5 MBit/s and downstream of 100 MBit/s?
2. With how many users N can you set up a video conference if all users send their videos of the same quality with the compression factor fk = 111: 1 and with 20% overhead, which is caused by protocol headers?
3. However, as a rule, you should only use a maximum of 60% of the available data rate DR for video transmission. How can the amount of data to be transferred be adjusted so that you can continue to communicate with all partners? How does this affect the quality of the videos?
4. What is the maximum data volume V that can be transferred per user and per hour T = 1 h with the maximum transfer efficiency of k = 0.6 (60%)?
5. What other video conferencing systems can you specify? Please give 2–3 examples of the systems and specify their main features!

Solution to (1):

With which compression factor do you have to compress the video to be sent via Vitero if you are using a DSL connection with upstream of 5MBit/s and downstream of 100MBit/s?

Given:

$$\text{Colour depth FT} = 24 \text{ bit/Pixel, frame rate FR} = 25 \text{ fps}$$

By diverse resolution factor:

- Video playback resolution $A_1 = 480 \times 270$ Pixel
- Video playback resolution $A_2 = 640 \times 360$ Pixel
- Video playback resolution $A_3 = 1280 \times 720$ Pixel

DSL connection: Upstream: 5 MBit/s, Downstream: 100 MBit/s

Searched: compression factor fk

Screen data volume for a uncompressed video signal: $\mathbf{DV = A_i * FT\ (i = 1,2,3)}$
Data rate for a uncompressed video signal: $\mathbf{DR = DV * FR}$
Video playback resolution A_1 (480×270 Pixel): $480 * 270 * 24 * 25 = 77.76$ Mbit/s
Video playback resolution A_2 (640×360 Pixel): $640 * 360 * 24 * 25 = 138.24$ Mbit/s
Video playback resolution A_3 (1280×720 Pixel): $1280 * 720 * 24 * 25 = 552.96$ Mbit/s

The transmitting channel possess the data rate **DR = 5 MBit/s** and operates under compression factor $\mathbf{fk = DR_{Videoplayback}/DR}$

$$\text{fk} = 78 \text{ Mbit/s}/5 \text{ Mbit/s} = 16$$

$$\text{fk} = 139 \text{ Mbit/s}/5 \text{ Mbit/s} = 28$$

$$\text{fk} = 553 \text{ Mbit/s}/5 \text{ Mbit/s} = 111$$

The provided compression is as follows: 16:1 or 28:1 or 111:1 depending on the resolution A_1, A_2, A_3.

Solution to (2):
With how many users N can you set up a video conference if all users send their videos of the same quality with the compression factor fk = 111:1 and with 20% overhead, which is caused by protocol headers?

Given:
Video playback with A_3 (1280×720 Pixel): $1280 * 720 * 24 * 25 = 552.96$ Mbit/s
Compression factor fk = 111: 1
DSL connection: Upstream: 5MBit/s, Downstream: 100 MBit/s

Searched:
Number of users/video conference partners N

Required data rate:

$$DR_{netto} = 552.96 \text{ Mbit/s}/111 = 5\text{Mbit/s}$$
$$+ 20\% \text{ overhead:}$$
$$DR_{brutto} = 120\% * 5\text{Mbit/s} = 6 \text{ Mbit/s}$$

The receiving channel (DSL-Downstream): 100 MBit/s

Number of users/video conference partners **N = 100 MBit/s: 6 MBit/s = 16 User** available.

Communication with at least N = 13 partners is provided by Vitero-Tool accordingly the documentation (the tool supports the classes/groups up to 13 users).

Solution to (3):

However, as a rule, you should only use a maximum of 60% of the available data rate DR for video transmission. How can the amount of data to be transferred be adjusted so that you can continue to communicate with all partners?

How does this affect the quality of the videos? You can find the answer in Table 12.4:

Solution to (4):

What is the maximum data volume V that can be transferred per user and per hour T = 1 h with the maximum transfer efficiency of k = 0.6 (60%)?

Given:

DRbrutto = 6 MBit/s, k = 0.6, T = 1 h

Table 12.4 Optimization methods for video transmission/video conferences

Reduction of the amount of data through	Effects on the quality of the videos
Higher compression	Reduction in image quality ("pixeling")
Smaller frame rate	Fewer images are sent per second. The participants' movements appear jerky
Reducing the picture screen size DV = B * H (screen data volume)	Smaller picture is transferred (e.g. 480 × 270 pixels). There are video conferencing systems that only show the current speaker larger, all other participants smaller. Disadvantage: Large computing effort for monitoring, too fast change in discussions between several people, only suitable for one-sided presentation
Reducing the color depth	The picture looks pale
Reducing the audio sample rate	Unsatisfied sound quality (sounds "dull", "tinny")

The possible configuration/requirements:
Audio via VoIP, slide transmitting, video via web cam, application sharing

Searched:
Data volume V

$$\mathbf{V = k \;\; T \;\; DR_{brutto}}$$
$$V = 1/8 \text{ Bit/Byte} * 0.6 * 3600\,\text{s} * 6\,\text{MBit/s}$$
$$= 1/8 * 0{,}6 * 3600 * 6 = 1620 \text{ MByte}$$

The calculations agree with the documentation. The values documented at Vitero are:
Model: Audio over VoIP, no slides, an avatar webcam, application sharing
Number of users N = 13, resolution $A_3 = 1280 * 720$ pixels
"In a session of this kind, you get a total of approx. 225 MB of data volume per hour and participant." (s. the references to Vitero below).

Solution to (5):
What other video conferencing systems can you specify?
Please give 2–3 examples of the systems and specify their main features!

The further world-known video conferencing systems are:

1) Skype for Business, Skype Enterprise
2) Google Meet
3) GoToMeeting
4) Blizz
5) Discord.

Their main features are as follows:

- HD-Video, Shared Screen
- Online-Chat, Business Telephone Systems
- Distance Learning, Social Media, IP-Telephony
- Instant Messaging, Peer-to-Peer, E2E-Encryption.

Availability:
Windows, macOS, Linux, Android, iOS

Target devices:
PC, Tablets, Smartphones

Unfortunately, some of the cited tools that provide the mostly sophisticated solutions are managed not from the EU directly and are not compliant with data protection requirements of the EU GDPR.

References:

1) Vitero University Tool (Online): https://www.vitero.de/
2) Andre M. Chang. Führende Videokonferenzsysteme fallen bei Kurztest durch, ZUMA Wire/dpa (Online 6.7.2020): https://www.internetworld.de/technik/corona-krise/fuehrende-videokonferenzsysteme-fallen-kurztest-2551712.html.

12.2 Tasks on Efficiency and Security in HDS

12.2.1 Task 9. Clustering and Energy Efficiency

Massive energy efficient computing power is required nowadays due to deployment of novel applications (simulation in pharmaceutical industry, civic craft, analysis of financial markets, mining of crypto currencies etc.).

1. What does describe the unit FLOPS?
2. What does it mean “the energy efficient service execution”?
3. What does it mean “Power Usage Effectiveness” (PUE factor) and how can be it calculated? Specify the meaning of the plain variables belong to.
4. Use the criterion PW for the cluster efficiency:

$$PW = \text{Performance}/\text{Power}$$

   Which physical sense does have the criterion PW?
5. Fitful the represented below table with the (calculated) appropriate data (use as a reference so-called TOP500 Clusters):

Clusters and (super) computers from the TOP500 list	Performance, in …	Power consumption (electrical power, MW)	PW = performance/power
Fugaku, Japan	514	28	
Summit, USA	200	10	
Tianhe-2, China	33.9		~2 GFLOPS/W
Titan, USA	17.6		
Regular PC	100 GFLOPS	0.2 kW = 0.0002 MW = 200 W	

Solution to (1):

FLOPS as a performance measure unit means:
FLOPS is a measure of the performance of clusters, computers or processors (CPU, GPU) and describes the number of floating-point operations (additions, multiplications) that they can perform per second.

The number of floating-point operations FLOP is not necessarily directly proportional to the clock speed of the CPU, since—depending on the implementation—floating point operations require a different number of clock cycles.

The FLOPS parameter takes into account the entire computer architecture, consisting of processors, RAM, machine instruction set, multipurpose register set, bus system and protocols, L1/L2/L3-caches, as well as from the usual compilers. But there is not the pure CPU speed to consider. The FLOPS of a computer or cluster can be defined by special program packages, so-called Benchmarks. A special mixture of FLOP in the benchmark is considered. Such programs are, for example, LINPACK, Livermore Benchmark etc.

Exemplarily FLOPS calculation can be done herewith as follows:

Example 1:

- ***Two-Socket-Server with Intel X5675 (3.06 GHz, 6 cores, 4 instructions per tact)***
  Performance = 3.06 x 6 x 4 x 2 = 146.88 GFLOPS
  Example 2:
- ***Two-Socket-Server with Intel E5-2670 (2.6 GHz, 8 cores, 8 instructions per tact)***
  Performance = 2.6 x 8 x 8 x 2 = 332.8 GFLOPS

Solution to (2):
The energy efficient service execution within a cluster can be formulated as optimization problem, for instance:

$$\text{Max } P_N[\text{PFLOPS}] \;|\text{Max } A_N|\; \text{PUE}{-}{>}1{,}0 \;|\; \text{Max PW[GFLOPS/W]}$$

Solution to (3):
PUE factor means "Power Usage Effectiveness":

$$\text{PUE} = (P_{IT} + P_A)/P_{IT}$$

where P_{IT} —peak electrical power of the cluster, P_A —waste power.

Solution to (4):
PW is a criterion for the cluster efficiency. It means Performance per Watt, or PW factor:

$$\text{PW} = \text{Performance/Power} = P_N/P_{IT}$$

Solution to (5):

The represented below table with the (calculated) appropriate data uses as a reference so-called TOP500 Clusters. The table can be fulfilled under use of SI conform units or prefixes:

Clusters and (super) computers from the TOP500 list	Performance, in …	Power consumption (electrical power, MW)	PW = performance/ power [GFLOPS/W]
Fugaku, Japan	514 PFLOPS	28	18.36 GFLOPS/W
Summit, USA	200 PFLOPS	10	20 GFLOPS/W
Tianhe-2, China	33.9 PFLOPS	17.808	1.9 GFLOPS/W
Titan, USA	17.6 PFLOPS	8.209	2.14 GFLOPS/W
Regular PC	100 GFLOPS	0.2 kW = 0.0002 MW = 200 W	0.5 GFLOPS/W

12.2.2 Task 10. Clustering: Performance, Speedup and Parallelism Grade

A powerful cluster with N = 1000 servers is deployed for parallel computing by pharmaceutical industry, where overall peak performance P_N of 66 TFLOPS is reached and the peak performance of a single server P_1 = 100 GFLOPS can be calculated:

1. How large is the parallelism grade p for the above-mentioned simulation task? Calculate the value p and $e = 1 - p$ under use of Karp-Flatt metrics for the both Speedup-Models:
   - Gustafson-Barsis Model
   - Amdahl Model
2. Discuss and justify the results of the analysis!
   To what extent can be Sun-Ni-Model used for the given scenario?

Solution to (1):
Karp-Flatt metrics (1990)

Given:

N=1000—cluster size
P_N = 66 TFLOPS—peak performance of the cluster
P_1 = 100 GFLOPS—peak performance of a single server
$A = P_N/P_1$ = 66 TFLOPS/0.1 TFLOPS = 660 times as Speedup

Searched:

$$\text{Parallelism grade p } -?$$
$$e = 1-p \quad \text{as sequential part}$$

for the both Speedup models:

- Gustafson-Barsis Model
- Amdahl Model

Under use of Karp-Flatt metrics:

$e = [1/A - 1/N]/[1 - 1/N] = 1 - p$
$e = [1/660 - 1/1000]/[1 - 1/1000] = [1/660 - 0.001]/0.999 = 0{,}05\ \%$
$p = 100\% - 0{,}05\% = 99.95$ as parallelism grade

Solution to (2): Sun-Ni Model (1990–1993)
For the above-mentioned scenario, Sun-Ni Model can be also used.

The Sun-Ni Law was developed in 1990–1993. The discussed law refers the memory bounds and increasing workload within a cluster and then leads to a generalization of Amdahl's and Barsis-Gustafson models both. Let us consider that:

$$1 = e + p, p = 1 - e,$$

where p—parallelism grade. If n CPUs are available within a cluster for a workload W, then the parallel portion of the workload Wnewwill be increased in G(n) times as a reflection of the increasing of the memory consumption:

$$W_{new} = eW + (1 - e)G(n)W,$$

where W_{new} is the parallel portion of the workload W. Then the speedup, which considers the memory bounds can be calculated as follows:

$$An = [eW + (1 - e)G(n)W] : [eW/1 + (1 - e)G(n)W/n],$$
$$An = [e + (1 - e)G(n)]/[e + (1 - e)G(n)/n]$$

where $p = 1 - e$, p—parallelism grade. Depending on the growth factor G(n), there are three possible cases:

1) $G(n) = 1$, like Amdahl's:

$$An = [e + (1 - e)]/[e + (1 - e)/n] = 1/(e + (1 - e)/n)$$
$$= 1/((1 - p) + p/n)$$

2) G(n) = n, like Barsis-Gustafson's:

$$An = [e + (1 - e)n]/[e + (1 - e)n/n] = e + (1 - e)n$$
$$= (1 - p) + pn = 1 + (n - 1)p$$

3) G(n) = m > n, for the so-called Sun-Ni Law:

$$An = [e + (1 - e)m]/[e + (1 - e)m/n]$$
$$= [e + (1 - e)m]/[e + (1 - e)m/n]$$
$$= [1 - p + pm]/[1 - p + pm/n]$$

The formula (9) for Sun-Ni Law is used for the case that the workload grows faster than memory. The speedup is slightly higher than Barsis-Gustafson's (4), i.e. the ratio (m/n) = 1.1 … 1.5.

12.2.3 Task 11. Green IT: PUE and ERE

A powerful cluster integrated 1000 CPUs and reached the peak overall performance by 80TFLOPS. The overall electric power P_{IT} amounts 400KW.

1. How large is the PW factor [GFLOPS/W]?
2. How large is the cluster speedup in comparison to a convenient PC?
3. How large is the waste power P_A [KW], if the PUE factor amounts the value of 1.5?
4. Via the cooling process optimization within the cluster the power P_R = 100 KW was "recycled" based on ambience heating and warm water delivering around the cluster building. How large is the ERE factor herewith?

Solution (1)+(2)+(3)+(4):

Given:

N=1000—cluster size
P_N = 80 TFLOPS—peak performance of the cluster
P_1 = 100 GFLOPS—peak performance of a single server as a reference value
P_{IT} = 400 KW as overall electrical power
P_R = 100 KW as "recycled" power
PUE = 1.5

Searched:

PW factor |Speedup A| Waste power P_A| ERE factor?

PW factor (Performance per Watt):

$$PW = P_N/P_{IT} = 80\,TFLOPS/400\,KW = 0.5\,GFLOPS/W$$

i.e. at less 4 times smaller than by TOP500 reference solutions.

E.g. for the worldwide known clusters Tianhe-2 and Titan

(cp. the table with the specifications to the performance and energy consumption):

- ***for the cluster "Tianhe-2" (People Republic China):***
  PW = 33.9 PFLOPS/17.808 MW = 1.9 PFLOPS/MW = 1.90 GFLOPS/W;
- ***for the cluster "Titan" (USA):***
  PW = 17.6 PFLOPS/8.209 MW = 2.14 PFLOPS/MW = 2.14 GFLOPS/W

Speedup A:

$$A = P_N/P_1 = 80 \text{ TFLOPS } /0{,}1 \text{ TFLOPS} = 800 \text{ times}$$

Power Usage Effectiveness:

$$PUE = (P_{IT} + P_A)/P_{IT} =>$$

$$1.5 = (400 \text{ KW} + P_A)/400 \text{ KW} => P_A = 200 \text{ KW as waste power}$$

Energy Reuse Efficiency:

$$ERE = (P_{IT} + P_A - P_R)/P_{IT} = (400\,KW + 200\text{ KW} - 100\,KW)/400\,W$$
$$= 1.25 < PUE = 1.5$$

12.2.4 Task 12. Advanced Security in Highly-Distributed Systems: Firewalls, IPS and CIDN

A firewall is a security system that protects a computer network or an individual computer from unwanted network access and is also an important part of a network. Compare the IDS/IPS modules (Intrusion Detection and Intrusion Prevention Systems) with "classic" firewalls! Make the differences clear!

1. Give examples where the use of a Packet Filter firewall is sufficient or where application-related filtering is required!
2. What is a Circuit Relay used for?
3. Complete the following table (see below) of the filter rules or functionalities for firewall systems. Assign the filter options in the first column of the following table to the correct FW concepts (columns 2–5). In some cases, a term can be assigned to multiple patterns/columns:

Filter rules or functionalities	PF (packet filter)	CR (circuit relay)	AG (application gateway)	Advanced FW-system
Time window control				
IP addresses and DMZ (demilitarized zone)				
Intrusion detection and intrusion prevention systems: IDS/IPS				
Approved/prohibited protocols				
Malware blocking, SPAM-filtering and antiphishing				
Application related authentication and encryption				
Proxy for certain services, proxy server				
Executable scripts, applets, web services and microservices				
Web application firewall				
TCP ports and blocking of DDoS				
Please "X" if applicable!				

4. What is CIDN?

   What types of attacks do they prevent? Discuss using the sketch below (Fig. 12.24):

Solution to (1) + (2) + (3):
The filtering functionalities (rules) of firewalls and DMZ are as follows:
A firewall system is connected as a barrier between the network to be protected (Trusted NW or internal network, LAN/Intranet) and the insecure network (WAN/Internet) so that all data traffic between the two networks is only possible via the firewall system is.

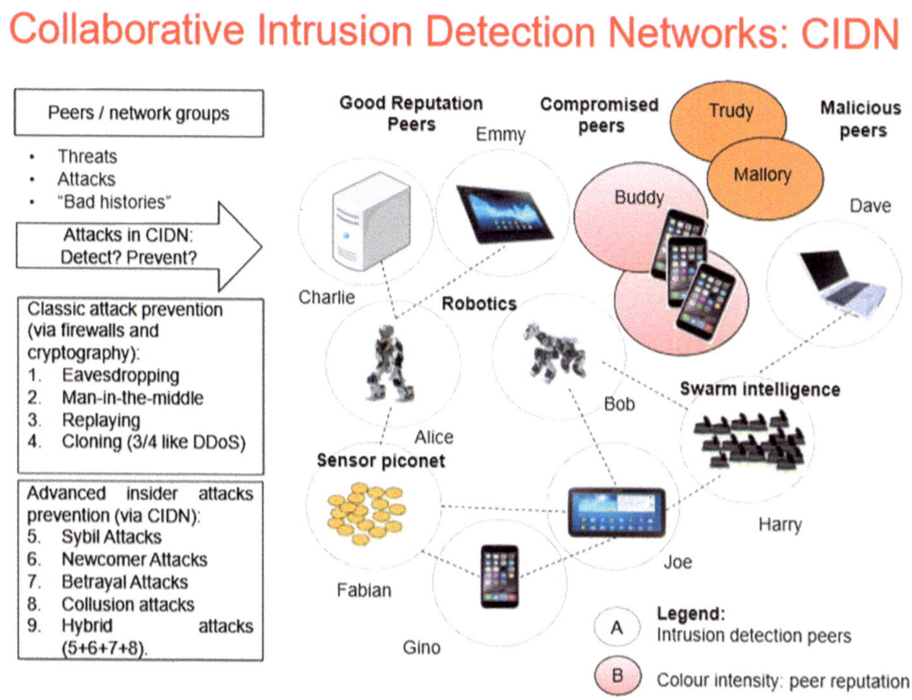

Fig. 12.24 CIDN functionality

However, the function of a firewall is not satisfied to detect the threats/attacks from a network. The classic FW should only implement filtering rules for network communication.

So-called ID/IPS modules (Intrusion Detection and Intrusion Prevention Systems) are responsible for tracking down attacks and are definitely based on the firewall functionality. Together with the firewall modules, these build the advanced firewalls.

So-called multi-level firewall strategy with DMZ includes:

External PF:

- Separation of the insecure Internet with various risks (attacks, viruses, malware).

Internal PF:

- Separation/protection of the internal area or the internal servers: backup server, DNS server, file server, WWW server, a cloud etc.
- Only indirect communication with authorization (authentication and assignment of rights).

Between the firewalls:

- DMZ (Demilitarized Zone), i.e. a zone with publicly available services that are easily accessible from the outside (see direct communication), but not allowed to be manipulated!

The fulfilled table is represented below:

Filter rules or functionalities	PF (Packet Filter)	CR (Circuit Relay)	AG (Application Gateway)	Advanced FW-System
Time window control	x	x	x	x
IP addresses and DMZ (demilitarized zone)	x			x
Intrusion detection and intrusion prevention systems: IDS/IPS				x
Approved/prohibited protocols	x		x	x
Malware blocking, SPAM-filtering and antiphishing			x	x
Application related authentication and encryption			x	x
Proxy for certain services, proxy server			x	x
executable scripts, applets, web services and microservices			x	x
Web application firewall			x	x
TCP ports and blocking of DDoS		x		x
Please "X" if applicable!				

So-called NG-FW can unify and integrate multiple features of the classical FW concepts have been discussed above. Furthermore, they can be divided into some generations (Fig. 12.25). Herewith the NG-FW classification from Check Point is presented.

Firewall Generations by Check Point

Generation	Gen I	Gen II	Gen III	Gen IV	Gen V
Periods	1990	2000	2005	2010	2018
Risks and Treatments	-	-	-	-	CIDN
	-	-	-	-	Sandbox, Sandblast
	-	-	-	Phishing	Antibot
	-	-	IPS	WAF	Anti-Ransom-ware
	-	NG-FW	IDS	*	*
	PF, CR, AG	SIF	*	*	*
	Antivirus	Anti-Malware	*	*	*

Source: https://www.checkpoint.com/

Fig. 12.25 FW generations by check point

Solution to (4):

What is really a CIDN?

The attacks (threats) can disrupt the operation of modern networked data and computing centres in 24/7 mode (refer Fig. 12.24).

Traditional IDS work in isolation and are, therefore, not effective indeed in detecting unknown threats/attacks that are becoming more and more sophisticated!

CIDN is a cooperation network of individual IDS who exchange their knowledge of attacks in pairs, thereby improving the overall accuracy and efficiency, which individual IDS guarantee.

The CIDN (Collaborative Intrusion Detection Networks) can prevent the following types of attacks:

(a) Eavesdropping
(b) Man-in-the-middle (dressing-up attacks, manipulation)
(c) Replay (forwarding attacks, repetition and multiplication of attacks, including multiplication of "malware"—malicious software—in the form of viruses, Trojans, worms, hoaxes, grayware)
(d) Cloning (such as DDoS, Distributed Denial of Service Attack)
(e) Sybil attacks (a large number of pseudonyms, fake identities are generated)

(f) Newcomer attacks (the peers with "bad reputation" delete their bad histories, so-called "X files" with other peers in the network)
(g) Betrayal attacks (the trust mechanism is robust and complies with the social norm: "It takes a long time and consistently good behaviour to build high trust, while just a few bad actions can completely ruin it" ! If a trustworthy peer acts dishonestly, his trustworthiness will fall catastrophically. Therefore, it is difficult for this peer to deceive others or to regain his previous trust within a short period!)
(h) Collusion attacks occur when a group of compromised or malicious peers acts to compromise the CIDN.

12.2.5 Task 13. Passwords in Highly-Distributed Systems

Modern guidelines for passwords (PW) selection or changing have, among other things, the following requirements:

- PW should have at least eight ASCII characters, of which at least one capital letter, at least one digit and a special character/punctuation mark should be included.
- The password is changed regularly depending on the security requirements of the system.

Your task:

Only the alphabet A with a subset of ASCII characters is available. The codes for the digits 0–9 are 48–57; capital letters: 65–90; lowercase letters: 97–122.

1. How much time T1 does an attacker have to spend on average for the brute force password determination if he only knows that a 7-digit PW is present and trying out a password takes 0.1 ms?
2. How does the attacker's effort T2 increase with an 8-digit PW?
   Which measures can significantly minimize the effort of the potential attacker? Explain the principle.
3. How often should the password be changed taking into account (1) and (2)?
   Please give your suggestion (if necessary, select the correct information)!
   - 1 × per 30 days
   - 1 × per 60 days
   - 1 × per year

What speaks against the "compulsion" to change passwords very often?

Solution to (1) + (2):

Given: Alphabet A includes all ASCII codes for the

- Digits 0–9: 48–57
- Capital letters: 65–90
- Lowercase letter: 97–122

$\Delta T = 0.1$ ms for a password test

N = ||A|| = 10 + 26 + 26 = 62 characters are to be used

Searched:

T1, T2 as average duration of the Brute-Force-Attack

Key space SR1 = 62**7 = 3,521,614,606,208 for a 7-position PW

Key space SR2 = 62**8 = 62 * SR1 = 62 * 3,521,614,606,208 = 218,340,105,584,896 for a 8-position PW (**Modern guidelines conformity**)

Duration (PW robustness):

T1 = 0.5 * SR1 * ΔT = 5.5 years as average duration of the Brute-Force-Attack

T2 = 0,5 * SR2 * ΔT = 5.5 * 62 = 346 years as average duration of the Brute-Force-Attack.

Practically safe!

Attention:

The average duration can be significantly shortened due to a wellknown Dictionary Attack!

Theoretically, a secured (crypto-)method is better!

A good example for a PW: ***e12_Jk-X***

A bad example for a PW: ***Timon1977***

Solution to (3):

The password change in the IT system usually takes place once every 90 days to once every 180 days.

Attention:

- So-called "compulsion" to change passwords very frequently leads to negligence when selecting a password or to writing it down on yellow paper slips on screen surface!
- And then, when a user's imagination could be exhausted, passwords that are too simple become easy prey for the hackers (dictionary attack)!

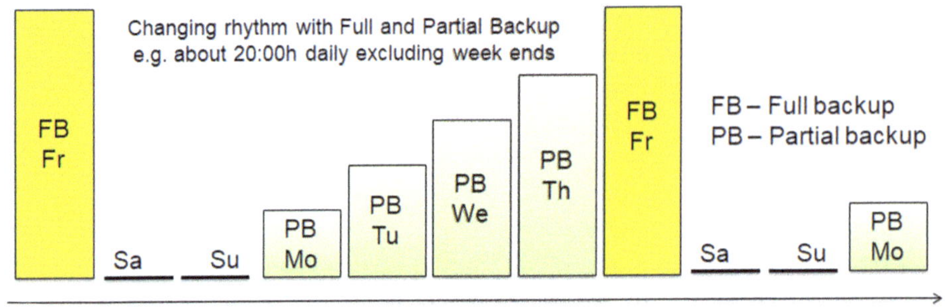

Fig. 12.26 Backup: a regular redundant data securing

12.2.6 Task 14. Regular Data Backup

Regular data backups are an elementary approach for data security. The data are saved redundantly on a storage medium and then are referred as a Backup (Fig. 12.26). The reconstruction of the original data from a Backup copy is called a Restore. The below mentioned example is intended to sensitize the readers to recent backup problems and show how important regular data backups are in the practise, e.g. in own data centers or in the cloud.

1. How many storage media do you need if a full backup is made once a week on Fridays at 20:00 and the partial backups during a week (excluding weekends)? What does a cron job look like?
2. Which backup strategies do you know: refer Fig. 12.27?
   Compare differential and incremental methods!
3. For what purposes and under what conditions can be favored a cloud backup solution?

Solution to (1):
The presented sketch (Fig. 12.25) shows us that five storage media (DVDs, tape drives/streamers, hard disks or virtual storages from the cloud) are required in the given case: four for PB on weekdays as well as one for FB on Fridays.

A tape drive, or streamer, is a device used to store data on magnetic tapes of higher capacity and reliability.

So-called Cronjobs are the typical formats (originally by Unixoids) for so-called Cron-Daemons, the automatic programs, which start regularly the processes (jobs) for full and partial data backup on a given time moment or for a time span (s. Table 12.5).

For the considered case:

Minute	Hour	Day	Month	Week day	Meaning for backup process
0	20	*	*	1–5	PB or FB is processed daily at 20:00, excluding week ends

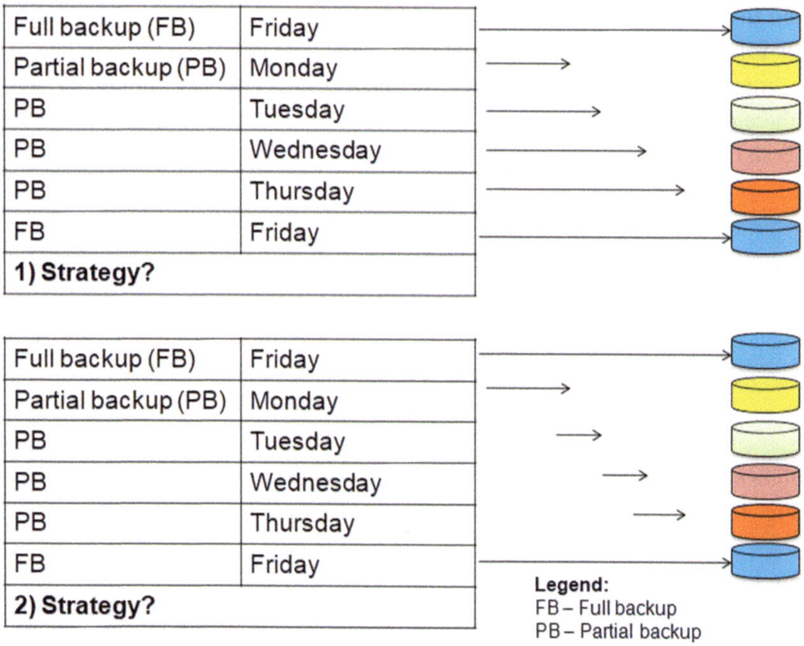

Fig. 12.27 Differential or incremental method?

Table 12.5 Typically examples for the Cronjobs formats for Cron-Daemons, used for full and partial backup

Minute	Hour	Day	Month	Week day	Meaning for backup process
0	0	*	*	*	Daily 00:00 h
5	*	*	*	*	5 Min after each whole hour
*/5	*	*	*	*	Each 5 Min
59	23	*	*	0	Every Sunday (=0 or = 7) at 23:59 h
20.30	1	*	*	1–5	Every Monday till Friday at 01:20 h and 01:30 h
0	1	1–7	12	1	On December 1 till 7 as well as every Monday in December at 01:00 h
5	9–20	*	*	*	5 Min after each whole hour between 9 and 20 h (09:05 h, 10:05 h …20:05 h)
*/10	*	*	*	*	All 10 min
55	23	*	*	0.4	Every Sunday (=0 or = 7) or Thursday (=4) at 23:55 h
50	22	*	*	7.3	Every Sunday (=0 or = 7) or Wednesday (=3) at 22:50 h

Solution to (2):
There is FB as well as PB, which can be used as the both main backup strategies. The PB can be implemented in differential and incremental method. The completed table contains, therefore, two advantages and two disadvantages respectively for the comparison of the both methods:

Backup method	Incremental—IB	Differential—DB
Advantages	1. Fast daily backup available (+)	1. Failure reliability (+)
	2. Less media required (+)	2. Faster restore available (+)
Disadvantages	1. Failure reliability (−)	1. More time required for daily backup (−)
	2. More time required for restore (−)	2. More storage media required for daily backup (−)

The following assignment of the terms IB and DB to the given below figures can be proceed (Fig. 12.28):

Solution to (3):
The cloud backup solutions are nowadays largely favoured by SMEs (small and medium sized companies).

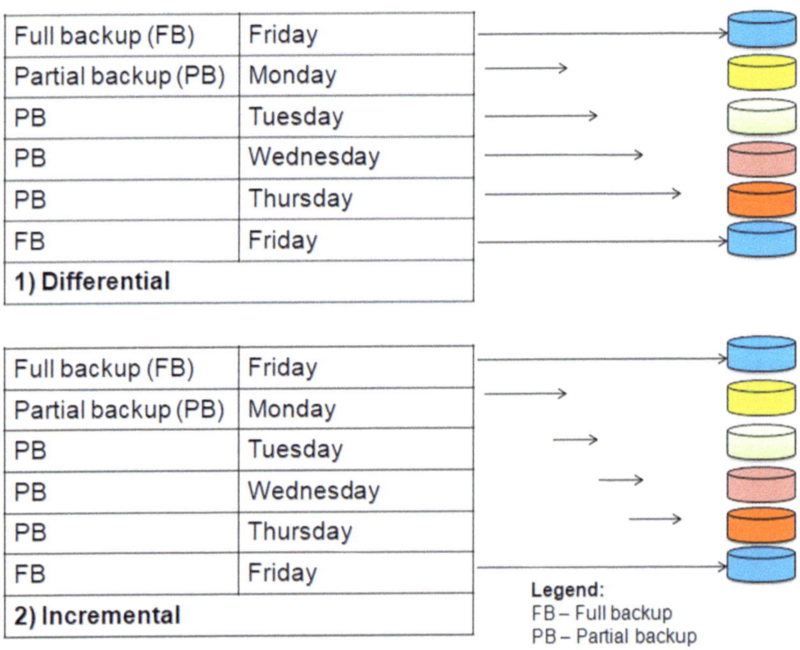

Fig. 12.28 Assignment: differential or incremental method is used?

The main disadvantages of cloud backup solutions are as follows:

- Constant fast network connection required
- Data security on the part of the provider must be guaranteed (legally, e.g. DGSVO or *General Data Protection Regulation* = GDPR, and technically)
- Existence/creditworthiness of the provider (trust in the provider to a large extent).

12.2.7 Task 15. Blockchain and Highly-Distributed Systems

Blockchain is a continuously expandable list of data records (hashes and transactions) in individual chained blocks. It serves in many applications as a decentralized ledger and is a secured alternative to the convenient PKI method for multilateral communication models.

1. What is Blockchain? What are the potential benefits of this technology? Which applications can BC support and which use of blockchain technology do you know? Give 2–3 significant examples. Specify negative examples of BC use too!Which several protection goals does Blockchain cover? Which security mechanisms are used?
2. Assign the specified protection goals to the security mechanisms known for HDS according to the application examples given:

Protection goal	Security mechanism	Example
?	Encryption	AES
Compulsoriness	?	RSA
?	Authentication	Login
Integrity	?	CRC, Hash

3. Specify pro and cons for BC!
4. Specify pro and cons for in comparison to networked DB, or so-called Warehouses.
5. Which consensus algorithms are deployed for blockchain generation? Discuss their properties! What is the main difference between a miner and a validator?
6. How many transactions N can be processed daily based on Blockchain if the following facts are known:
   - A new Bitcoin block is generated every 10 min.
   - The average number of AT of BTC transactions per block is 1718 (an average over the past 2 years).
7. Discuss the pros and cons of using and mining crypto-currencies (refer Fig. 12.29)!

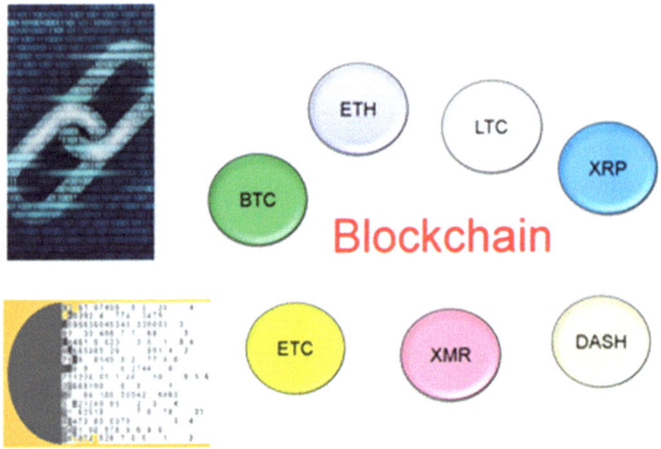

Fig. 12.29 Crypto-currencies and BC

- Specify eight factors that must be considered when calculating the profitability of mining!
- What types of devices, or so-called "rigs" (refer Fig. 12.30), can be deployed to generate hash values for crypto-currency mining?

8. Use the "AntMiner" as a mining rig (refer Fig. 12.30), which can calculate 28 TH/s with a power consumption of 1596 W. On average, this performance results in a daily return of 0.00115956 Bitcoin. The fee to the joined mining pool amounts 1.5%. Let the current value of a bitcoin equals € 3336.25. The average price per kWh let is 29.13 cent. Note: BTC-EUR exchange rate is variable!
   Please calculate the following values:
   (a) the costs for the used electricity (in EUR)
   (b) the gained Bitcoins (given in EUR)
   (c) Efficiency (given in J/TH)
   (d) Profit or loss (in EUR)
   for a reference week of mining. The purchase price is negligible here.
9. The "AntMiner" from previous subtask (9) has an energy saving (low power) operation mode too. The device can calculate 18 TH/s with a maximum power consumption only of 900 W. This results in a daily return of 0.00074543 Bitcoins. All other parameters are unchanged. Please calculate for a reference week of mining the following values:
   (a) Profit or loss (in EUR)
   (b) Efficiency (given in J/TH).
10. How high does the price for one kWh have to be so that mining is worthwhile for private individuals in the given situation? Note: please calculate for both operation modes of the "AntMiner" (high performance/low power)!

Fig. 12.30 Mining "Rig" (aggregate)

11. Differentiate HDS from DS ("classical" distributed systems). Which main components belong to the architecture (refer Fig. 12.31)? Discuss with the help of the figure below!

Discuss the benefits and cons of BC use for HDS (Highly-Distributed Systems)? How can accountability and compulsoriness be guaranteed when executing individual steps or components of today's highly distributed systems?

Solution to (1):
BC is a continuously expandable list of data records (hashes and transactions) in individual chained blocks.

BC can be decentralized used for multilateral compulsoriness instead of convenient PKI, in opposite to PKI, which operates pairwise, centralized via a CA. Modern HDS are mostly very complex, and consist of multiple communicating parts. The potential benefits of this technology:

- Software solutions decentralisation
- Compulsoriness of the workflow steps for critical applications
- Multilateral communication models like e.g. M2M.

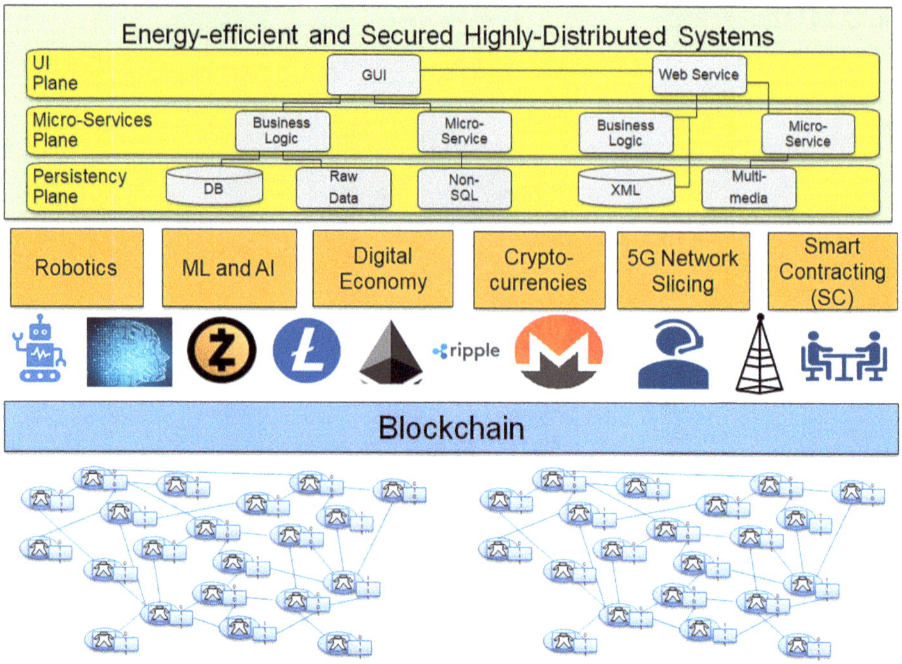

Fig. 12.31 HDS and BC

PKI and BC are based both on RSA digital signatures with hash value generation and encryption under use of private key. Decryption with public key provide the signature (block) verification.

Nowadays, multiple BC applications can be used:

- Crypto-currencies like Bitcoin, Monero, Ripple
- Smart Contracting
- Supply Chains
- 5G and Beyond applications
- Applications for HDS
- Ransomware (as a negative example of BC use): the dangerous crypto-Trojan can encrypt the data on the HDD and then require so-called "ransom", expressed in a crypto-currency (normally in BTC) for release!

Solution to (2):

The assignment of the above-mentioned protection goals to the security mechanisms known for HDS is given below:

Protection goal	Security mechanism	Example
Confidence	Encryption	AES
Compulsoriness	PKI or BC	RSA
Authorized access	Authentication	Login
Integrity	Secured data, message authentication code	CRC, Hash

Solution to (3):

BC Pro and Cons:

The use of a blockchain along a supply chain offers the possibility of seamlessly tracking the condition of the goods along the entire value chain. In this way, the above-mentioned protection goals can be achieved:

(a) Integrity: Sensor values recorded by IoT devices are stored in the BC in real time, where they are permanently and forgery proof.
(b) Availability: Every company in the value chain operates its own node in the BC P2P network and thus has access to the current and complete BC at all times. This means that problems along the supply chain can be identified immediately.
(c) Confidentiality: Cross-company blockchains are usually closed systems to which only invited parties have access. There is also the option of using encryption for the transactions, for example the public key method. Programs to regulate visibility can also be stored in the blockchain so that every company within the supply chain only sees the information that is necessary.
(d) Authenticity and accountability: Another important aspect is that it can be clearly determined at any time which participant is responsible for the written data. Here, too, the blockchain offers a solution with the use of the public key procedure.

Therefore, the BC represents an actual alternative to established technologies in the given scenarios. The cross-company collaboration also fulfils all requirements for the use of a BC:

- There are multiple participants (N>2)
- The states must be every time saved
- A third party is not always available
- All participants are known, but they are not trusted.

Solution to (4):

Specify pro and cons for in comparison to networked DB, or so-called Warehouses. In summary, there are a number of advantages, but also disadvantages, compared to conventional databases, shown in the following Table 12.6.

Table 12.6 BC and their pro and cons for in comparison to networked DB, or so-called Warehouses

Advantages	Disadvantages
The redundant storage of the blockchain content and unrestricted access by the individual participants results their maximal transparency	The usual ACID semantics of conventional networked databases (Atomicity, Consistency, Isolation, Durability paradigm) cannot be unfortunately applied. Atomicity is rather impossible due to the limited rollback functionality. Due to the simultaneous creation of blocks, consistency is also not always provided
By chaining the blocks, the use of hash values and distributing them in the network, the blockchain data is particularly forgery-proof, confident and secured	The creation of blocks, especially in the open blockchains, is associated with a very high amount of computing resource and energy consumption. The known CRUD functionality is rather affected (Create-Read-Update-Delete paradigm)
Through direct transactions and the given transparency, the intermediaries (inter-agents) previously required are totally superfluous our redundant. Decentralized use of the data without the leading instance	Due to the permanent storage of data in the blockchain, the later (post-factum) deletion is not possible too. This could lead to certain problems related to the GDPR (General Data Protection Regulation of EU)
Transactions in the blockchain can have any content, as well as programs and contracts can be stored within the P2P network	The sensible use of a blockchain possesses certain important restrictions. These affect the available number of participants, the type of exchanged data, the participants trust among each other as well as available third parties

Solution to (5):

There are several so-called consensus algorithms for creating blocks, the most important of which are listed in Table 12.7 below.

What is the main difference between a Miner and a Validator? (Table 12.8)

Solution to (6):

How many transactions N can be processed daily based on blockchain?,

- Every 10 min a new block appears, 6 × per 1 h, 24 * 6 × daily
- Transactions per block: AT = 1718 (average value fo last 2 years)
- Transaction quantity N = AT * 6 * 24 = 1718 * 6 * 24 = 247,392 BTC transactions daily!

For comparison: Number of transactions in "classic" cashless payment transactions per day: 1,764,657,534 (corresponds to 644.1 billion transactions per year worldwide according to Statista 2019).

Table 12.7 The most important blockchain consensus algorithms for creating blocks

Type of consensus algorithms	Decription	Cons	Examples
Proof-of-Work	Participants (miners) provide their computing power for a fee in order to calculate the hash values of new blocks. The difficulty of the calculation is adjusted so that blocks are created at regular time intervals	Very computationally and energy consuming	Bitcoin
Proof-of-Authority	Only certain nodes within the network can create blocks. Row-based selection	Intolerance against a manipulation	HyperledgerFabric, Ripple
Proof-of-Stake	Participants own a network's cryptocurrency. The higher the percentage, the more likely it is that you will choose to create a block for a fee	Capital accumulation can deter other participants	Stratis, Reddcoin

Table 12.8 Miner versus validator

Node type	Miner	Validator
Consensus algorithms	Proof of work	Proof of stake
Resource consumption	Higher energy expenditures	Lower (energy) expenditures
Deployment	Deployed by Bitcoin	Used by other crypto-currencies and some private bslockchains

Solution to (7):

Below we discuss the pros and cons of using and mining crypto-currencies under use of Fig. 12.29. Mining of Crypto-currency means an immense Resource Consumption in the first row. At least eight factors are necessary and must be considered when calculating the profitability of the mining:

1. *Investment* (basis device costs): hardware investment for a mining rig or cluster.
2. *Electricity Costs*: in real life currency per kWh (e.g. €/kWh).
3. *Energy Consumption*: electrical power of Mining Rig in KW.
4. *Hash Rate*: hash values computed each second.

Table 12.9 Features of a mining device (*Source:* amazon.de)

Features of AntMiner S15	High-performance mode	Low-energy mode
Hash rate	28 TH/s	18 TH/s
El. power	1596 W	900 W
Power efficiency	57 J/TH	50 J/TH
NW connection	Ethernet	Ethernet
Weight and dimensions	7 kg, 240 mm × 178 mm × 296 mm	

5. *Mining Difficulty*: factor has always a new actual value.
6. *Pool Fees*: how much in % belongs to the joined pool.
7. *Block Reward*: monetary or ideological compensation for each new computed block (or for other units).
8. *Unit Price*: exchange course for the BTC (or other units like XMR, ETH, ZEC, or LTC).

What types of "rigs" (refer Fig. 12.30) are there to generate hash values for mining? There are different multi-purpose and specialized devices for mining available, so-called Mining Rigs. Standard PCs or smartphones can also be used, but considering the energy consumption problems, they provide rather more disadvantages than justifiable use. The following types of devices can be deployed to generate the hash values in the mining process:

- CPU mining: powerful processor is required,
- GPU mining: powerful graphics card is required, or
- ASIC mining: using *A*pplication-*S*pecific *I*ntegrated *C*ircuits (cf. Table 12.9).

Solution to (8):
Using AntMiner as mining rig, which can calculate 28 TH/s with a power consumption of 1596 W: The average daily gain from this performance is 0.00115956 Bitcoin. The contribution to the joined mining pool is 1.5%. The current value of a Bitcoin is 6.640 €.[1] In Germany, the average electricity price is 0.3137 €/kWh.[2] When looking for the expectable investment to conduct a week's mining one shall consider:

- the expenditures E for the used electricity,
- gain/proceeds from the Bitcoins GB,

[1] As of 18 April 2020.

[2] National average calculated by the German Federal Grid Agency. Given price represents average as of 17 April 2020.

- efficiency Eff (in J/TH), and
- gain or losses (in €) GL per week of such mining.

The purchase costs are negligible. As a result, one can calculate:

$$
\begin{aligned}
\Rightarrow \mathrm{E} &= 1.596\ \mathrm{kW} \times 168\ \mathrm{h/week} \times 0.3137\ €/\mathrm{kWh} \\
&= 268.128\ \mathrm{kWh/week} \times 0.3137\ €/\mathrm{kWh} \\
&= 84.1117\ €/\mathrm{week} \\
\Rightarrow \mathrm{GB} &= 0.00115956\ \mathrm{Bitcoin/d} \times 7\ \mathrm{d/week} \\
&\qquad \times 6640\ €/\mathrm{Bitcoin} \\
&= 53.8963\ €/\mathrm{week} \\
\Rightarrow \mathrm{GL} &= 53.8963\ €/\mathrm{week} - 84.1117\ €/\mathrm{week} \\
&= -30.2154\ €/\mathrm{week} \\
&\text{(losses due to negative value)} \\
\Rightarrow \mathrm{Eff} &= 1596\ \mathrm{W}/28\ \mathrm{TH/s} = 57\ \mathrm{J/TH}
\end{aligned}
$$

A simple cross-multiplication of Eff shows that AntMiner rig would need to at least calculate 43.7 TH/s in order to cover the cost and turn a marginal profit.

Solution to (9):
The AntMiner rig has an extra energy saving mode. In that mode, the rig calculates only 18 TH/s with a reduced power consumption of 900 W. The process results in gain of 0.00074543 Bitcoin per day. All other parameters remain unchanged. Then the new calculation derives:

$$
\begin{aligned}
\Rightarrow \mathrm{E} &= 0.9\ \mathrm{kW} \times 168\ \mathrm{h/week} \times 0.3137\ €/\mathrm{kWh} \\
&= 47.4314\ €/\mathrm{week} \\
\Rightarrow \mathrm{GB} &= 0.00074543\ \mathrm{Bitcoin/d} \times 7\ \mathrm{d/week} \\
&\qquad \times 6640\ €/\mathrm{Bitcoin} = 34.6476\ €/\mathrm{week} \\
\Rightarrow \mathrm{GL} &= 34.6476\ €/\mathrm{week} - 47.4314\ €/\mathrm{week} \\
&= -12.7838\ €/\mathrm{week} \\
\Rightarrow \mathrm{Eff} &= 900\ \mathrm{W}/18\ \mathrm{TH/s} = 50\ \mathrm{J/TH}
\end{aligned}
$$

Once again, the rig produces a net-loss.

Solution to (10):
The interesting question for a private individual is as follows: How high does the price PKWH for 1kWh have to be, so that mining is rentable for a given situation? Note: please calculate for both operation modes of the "AntMiner" (high performance/low power)!

- High-Performance mode:

$$\begin{aligned} \Rightarrow \text{GB} &= 53.8963\ \text{€/week} \\ &= 268.128\ \text{kWh/week} \times \text{PKWH} \\ \text{PKWH} &= 0.2010\ \text{€/kWh} \end{aligned}$$

- Low-Power mode:

$$\begin{aligned} \Rightarrow \text{GB} &= 34.6476\ \text{€/week} \\ &= 151.2\ \text{kWh/week} \times \text{PKWH} \\ \text{PKWH} &= 0.2291\ \text{€/kWh} \end{aligned}$$

This practical example shows that in a lot of cases mining of the cryptocurrencies like BTC, ETH, ZEC, XMR, etc. leads to "no reward"-scenarios due to the large energy consumptions in combination with the unsustainable electricity price.

Additionally, the initial purchase costs for the rigs cannot be considered negligible.

In contrast, positive results can be reached if the powerful CPU+GPU or ASIC chip clusters are used with access to low cost renewable energy,[3] or can gain energy via waste recycling (both with minimal PKWH). A cluster with an optimal ERE value can use such recycled energy (like e.g. C&H Dresden). The gained PR can then be targeted for crypto currency mining.

Solution to (11):
To differentiate HDS from DS ("classical" distributed systems) we use Table 12.10 and refer Fig. 12.26 with HDS architecture components.

12.2.8 Task 16. IoT Efficiency: Electricity and Data Unit Costs for a WSN

A Wireless Sensor Network with so-called "one hope GW" is given (refer Fig. 12.32). As an alternative Wi-Fi 6 net can be used for the same target. The system must be energy-efficient and has to provide low data unit costs for the collected data.

1. Provide an annual costs calculation for such kind of WSN in comparison to Wi-Fi 6 (IEEE 802.11ax) for the typical operation period T=3 years by the typical dimensions of 2500 and 10,000 nodes correspondently.

[3] Such as hydro-electric power in Norway, or geo-thermal power in Iceland. Both locations are favored by many crypto currency mining operations.

Table 12.10 Comparison DS versus HDS

Features	DS (distributed systems)	HDS (highly-distributed systems)
• n-tier structure	+	
• Multi plane structure		+
• Implementation basis	Macro-services, "quasi-monolithic"	Loosely coupling, micro-services
• Use of SWT process model by implementation	Conventional and agile process models, e.g. DevOps	Mostly agile, scrum
• Interfaced in higher degree on security	+	++
• Compulsoriness of certain workflow steps	Not always guaranteed	Available via BC
• Energy efficiency		In higher degree
• New deployment fields		Smart contracting, 5G and beyond, network slicing, robotics, crypto-currencies, digitalization
• Artificial intelligence	Not compulsory	Available, ML

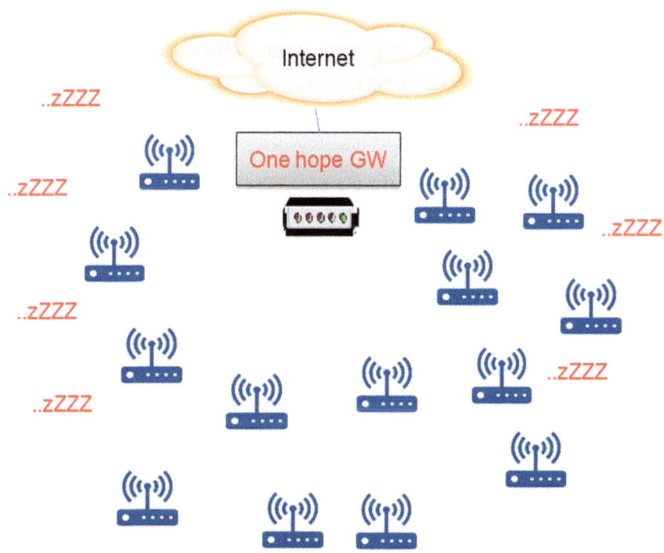

Fig. 12.32 Example of a WSN with a one hope Gateway

The key data for the calculation is as follows:

(a) Period T = 3 years
(b) N = 10,000 sensors (or of 2500 sensors)
(c) Telegram length FL = 100 Bit
(d) Frequency bands: F = 900/2400 MHz (ZigBee)
(e) Enabled power PTx = 1 mW (0dBm)
(f) Average Electricity Price: AEP = 29 ct/kWh
(g) DR = 2000 Bits/h (event frequency depended)

Reference parameters: Overall Electricity Costs per anno C and Annual Data Unit Costs (ADUC):

$$C = 25 \text{ EUR } 40 \text{ ct}$$

$$\text{ADUC} = 38 \text{ ct/ GB}$$

2. Prove the mentioned low data unit costs for the collected data for the WSN case!
3. Provide qualitative WSN comparison with other RF technologies!

Solution to (1):
The solution we represent compactly within Table 12.11.

Conclusion:
It means the energy efficiency is 100 × or, even, 200 × times higher for the WSN in comparison to **Wi-Fi 6 (IEEE 802.11 ax)**.

Table 12.11 Comparison ZigBee and Wi-Fi-6

Parameter	ZigBee (IEEE 802.15.4)	Wi-Fi 6 (IEEE 802.11 ax)	
Frequency	900 MHz, 2400 MHz	2.4 GHz	5 GHz
Power consumption	1 mW (0 dBm)	100 mW (20 dBm)	200 mW (23 dBm)
Calculation of electricity costs = 29 ct/kWh =	1m W * 24 h * 365 d = 8.76 Wh = 8.76 * 10 ** − 3 kWh Amount [ct] = 8.76 * 10 ** − 3 kWh * 29 ct/kWh = 0.254 ct/year	100 × performance, i.e Amount × 100 Amount [ct] = 25.4 ct/year	200 × power, i.e Amount × 200 Amount [ct] = 50.8 ct/year
Annual costs for 2500 sensors	6.35 €	635 €	1270 €
Annual costs for 10,000 sensors	25.40 €	2540 €	5080 €
Note:	Average electricity price in Germany: 29 ct/kWh		

Solution to (2):

Let us a ZigBee WSN consists of 10,000 sensors (refer Table 12.8). Depending on the event frequency within this WSN, the event driven, periodically, permanent, and the behavior of the sensors (push, pull), a sensor survey and data accumulation are performed. Thus, if each sensor can transfer a short telegram up to 100 Bits only, then:

- The survey for each sensor is conducted 20 times per hour: 2000 Bits/h
- x 24 h = 48000 Bits/daily = 6000 Bytes/daily for each sensor
- In general, an average sensor wireless network accumulates experimental data for 3 years × 365 days:
- × 365 × 3 = 6.57 MByte for each of the sensors
- The overall-data for the mentioned network:
- 6.57 MBytes × 10,000 sensors ~ 66 GB of raw data!

Within the 3 years the electricity costs can only reach 76 € 20 Cent (refer Table 12.1).

The unit costs per GByte for the acquired data in the given case are tiny and amounts to:

76.2 €/66 GByte = 1.15 €/GByte only (i.e. 38 ct/GByte yearly)!

Solution to (3):

A qualitative WSN comparison with other RF technologies is given. The comparison was made considering three following criteria such as: Data Rate, Distance and Energy Consumption (s. the diagram in Fig. 12.33).

12.3 Tasks on Internet of Things

12.3.1 Task 17. Introduction to IoT

1. Specify several wired and wireless smart devices and characterize their tasks!

Solution to (1):

- Wired smart devices [1–5]:
- IoT for controlling machines, IoT production process, IoT in vehicles
- Wireless smart devices [1–5]:
- IoT accompanying a product (in production, in handling and retail), movable sensors for monitoring of human health, sensors in places without a power supply (Energy Harvesting).

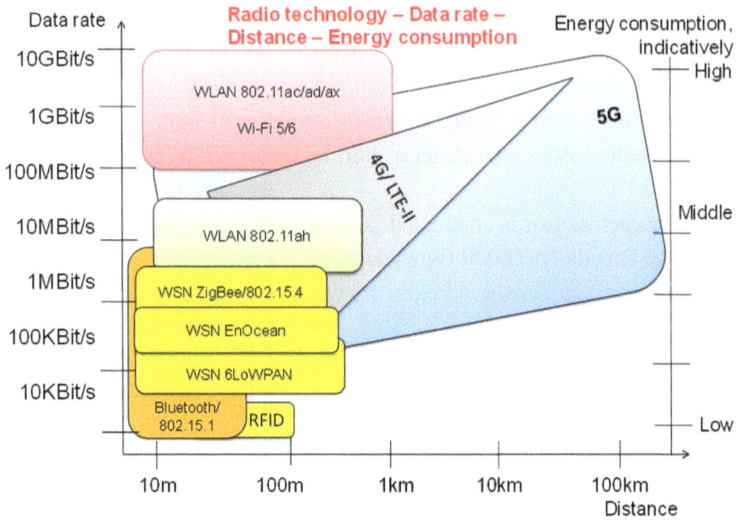

Fig. 12.33 Qualitative WSN comparison with other RF technologies

2. How do you develop Smart Services?

Solution to (2):

- Complex modelling of a service (minimization of the number of sensors/actuators and the necessary information exchange)
- selection of a suitable development system/platform
- detailed development under behaviour simulation
- testing in real operation environment

3. What is the relationship between Smart Devices and Smart Objects?

Solution to (3):

- Smart Devices are communication capable devices with a certain autonomous intelligence and ability to operate and provide some required functionality
- Smart Objects are particularly the tiny smart devices that are characterized by their small size and low power consumption.

4. Discuss some methods of (Micro-) Energy Harvesting!

Solution to (4):

- Use of mechanical energy (human movement, vibrations…)
- Use of electromagnetic energy (HF radiation, surroundings of coils,)

- Thermal energy (difference between body and environment temperature, room temperature differences…).

5. Specify the deployment areas for IoT!

Solution to (5):
- Construction and civic craft
- Smart Home
- Traffic on streets and highways
- Military affairs
- Healthcare
- Weather service.

12.3.2 Task 18. Data Acquisition for IoT

1. Which problems do appear during Data Acquisition?

Solution to (1):
- Sensor output values differ from the physical values to be measured and must be prepared before processing (calibration, value and time discretization).

2. Discuss Data Processing and Data Mining using an industrial example!

Solution to (2):
- Data processing includes specifications for data selection.
- This includes the number, type and location of sensors as well as query frequency and measurement accuracy.
- Data Mining includes mathematical processes for pattern identification with the aim of reducing the required data volumes.

3. What is the aim of determining the correlation of measurement data?

Solution to (3):
- Recognition of dependencies between different model parameters with the aim of reducing the number of parameters to be considered.
- Correlations can be used to predict the behaviour of the whole systems.

4. Discuss an example of the use of AI for technical processes.

Solution to (4):
- Image recognition
- Recognition of laws in complex systems that are difficult for modelling.

12.3.3 Task 19. Traffic Telemetric

1. Discuss the possible deployment scenarios for IoT in vehicle automation.

Solution to (1):
- Running gear support
- Vehicle maintenance support
- Increase in driving comfort
- Support for autonomous driving.

2. What features do you propose for an autonomous driving system?

Solution to (2):
- Monitoring the operability of the vehicle
- Recognition of the lane delimitation, the quality of the lane, the distance between the vehicle and other traffic participants
- Recognition of special situations (obstacles, barriers, ...)
- Predictions of changes in the traffic situation while driving (recognizing and analysing traffic signs and traffic light positions with foresight, tight curves, possibly also the intentions of other traffic participants).

12.3.4 Task 20. Smart Energy

Discuss the benefits of the following "smart" technologies: Smart Metering, Smart Grid, Smart Home, and Smart City!

Solution:
1. Smart Metering
    - Measurement of the energy consumption by the objects as precisely as possible
    - Optimal control of energy generation and distribution becomes real.
    - Economic benefits for producers and consumers
2. Smart Grid
    - Up-to-date and demand-oriented energy production
    - Inclusion of energy conversion systems and energy storage systems
3. Smart Home
    - Better building economy
    - Greater living comfort
4. Smart City
    - Better traffic concepts for cities
    - Improvement of the environmental quality
    - Support of disaster control.

12.3.5 Task 21. Sensor Networks

1. Discuss examples of sensor networks.

Solution to (1):
- multiple temperature sensors aimed to optimal climate control of a building
- individual product companions in production
- individual product companions in handling and retail
- extensive information on traffic control for big surfaces

2. Discuss sensor power supplying problems.

Solution to (2):
- Sensors should often operate autonomously over a long period. However, small, light and inexpensive batteries are only available for low capacities/time spans.
- Energy Harvesting technologies offer a way out. Available energy form in the environment is converted into electrical energy for the working sensor.

3. Discuss swarm intelligence!

Solution to (3):
- Individual sensors and actuators are unintelligent. However, since the information is distributed over multiple objects, many small pieces of information add up to form complex swarm information.
- These can be processed inside or outside the swarm.

4. What are the Blockchains in sensor networks used for?

Solution to (4):
- Multilateral communication n:m instead of 1:n
- Secure and compulsory decentralized retrieving and processing of transactions
- Storage of processing steps in production with compulsoriness
- Storage of process and cost changes due to changes in ownership.

5. Discuss location problems in sensor networks!

Solution to (5):
- Location via reception power is possible
- Location accuracy is relatively low, with many sensors working in parallel the accuracy can be improved by statistical methods.
- There are a lot of possible applications, e.g. always knowing the processing location in production technology.

References

1. Luntovskyy, Andriy, Gütter, Dietbert. Moderne Rechnernetze: Protokolle, Standards und Apps in kombinierten drahtgebundenen, mobilen und drahtlosen Netzwerken, 481 Seiten, 263 Abb., Springer Nature, Juli 2020, ISBN: 9783658256166 (https://www.springer.com/gp/book/9783658256166).
2. Luntovskyy, Andriy, Spillner, Josef. Architectural Transformations in Network Services and Distributed Systems: Current technologies, standards and research results in advanced (mobile) networks, Springer Vieweg Wiesbaden, 2017, 344p., ISBN: 9783658148409 (https://www.springer.com/gp/book/9783658148409#otherversion=9783658148423).
3. Luntovskyy, Andriy, Gütter, Dietbert. Moderne Rechnernetze–Übungsbuch: Aufgaben und Musterlösungen zu Protokollen, Standards und Apps in kombinierten Netzwerken, 145 Seiten, 44 Abb., Springer Nature, Juli 2020, ISBN: 9783658256180 (https://www.springer.com/gp/book/9783658256180).
4. Schill, Alexander, Springer, Thomas. Verteilte Systeme: Grundlagen und Basistechnologien: Kompakte Darstellung der Grundlagen und Techniken Verteilter Systeme, Springer-Verlag Berlin Heidelberg, 2012, 2.Ausgabe, ISBN: 9783642257957 (https://www.springer.com/gp/book/9783642257957#otherversion=9783642257964).
5. Luntovskyy, Andriy, Gütter, Dietbert, Melnik, Igor. Planung und Optimierung von Rechnernetzen. Methoden, Modelle, Tools für Entwurf, Diagnose und Management im Lebenszyklus von drahtgebundenen und drahtlosen Rechnernetzen: Planung von Rechnernetzwerke theoretisch anspruchsvoll und praxisnah, Springer/Vieweg+Teubner Wiesbaden, 2012, 415 Seiten, 245 Abb., ISBN: 9783834814586 (https://www.springer.com/gp/book/9783834814586#otherversion=9783834882424).

Appendix A. Key Words

Digital Age, Digitalization, Distributed Systems and Highly-Distributed Systems,
Agile Process Models, Web-Services and Micro-Services, M2M,
5G and Beyond, Inter-operability Radio Networks, 6G, Starlink,
Blockchain, Crypto-Currencies, DAO, Smart Contracting, Transactions and Workflow Compulsoriness, CIDN,
Big Data, 6 V Principle, Analytics and Data Mining, AI and ML,
Green IT, Grid, Clustering, Cloud Computing, Speedup Models, PUE and ERE, Cooling Process, Workload Migration,
Energy Autarky and Harvesting, Energy Efficiency, Robotics,
IoT and Fog Computing, Industry 4.0, Sensors and Actuators, RFID, NFC,
IoT and Robotics Platforms, SLMA, Traffic Telematics,
Smart Data, Smart Energy, Smart Grid, Smart Home.

A. Luntovskyy and D. Gütter, *Highly-Distributed Systems*,
https://doi.org/10.1007/978-3-030-92829-2

Appendix B. Related Works

Appendix B1. Relevant Papers

1. Andriy Luntovskyy, Dietbert.Guetter, Gert Pfeifer, Alexander Schill. Network Design Methodology and Workflow within the CANDY Framework, 14th International Multi-Conference ACS-AISBIS 2007, Szczecin, Poland, 17–19 Oct. 2007 (published in Polish Journal of Environmental Studies, Vol.16, No.5B (2007), pp. 186–192).
2. Luntovskyy, V. Vasyutynskyy, C. Oezluek, D. Guetter. Requirement Analysis and Optimization of combined Networks, 5th IEEE International Conference on Industrial Informatics INDIN 2007, 23–27 July 2007, Vienna, Austria, pp. 69–74, ISBN: 1-4244-0864-4 (IEEE Xplore).
3. Larisa Globa, Andriy Luntovskyy, Dietbert Guetter; Tatyana Kot. CASE Tools for IT-System Integration, 14th International Multi-Conference ACS-AISBIS 2007, Szczecin, Poland, 17–19 Oct. 2007 (published in Polish Journal of Environmental Studies, Vol.16, No. 5B (2007), pp. 152–157.
4. Andriy Luntovskyy, Volodymyr Vasyutynskyy, Klaus Kabitzsch. Dynamic Evaluation of Costs in Combined Wired and Wireless LAN, 7th IEEE International Workshop on Factory Communication Systems WFCS 2008, Dresden, 20–23 May 2008, pp. 349–356, ISBN: 978-1-4244-2349-1 (IEEE Xplore).
5. Andriy Luntovskyy, Stefan Uhlig, and Alexander Schill. CANDY Site Finder, a Planning Tool for Wireless Networks. IEEE CriMiCo Conference, Sevastopol, 8–12 Sept. 2008, paper ID = 014, pp. 356–359 (ISBN: 978-966-335-166-7) (IEEE Xplore).
6. Andriy Luntovskyy, Stefan Uhlig, Dietbert Guetter and Alexander Schill. Integrated Design Tools for Combined LAN. IEEE CriMiCo Conference, Sevastopol, 8–12 Sept. 2008, paper ID = 027, pp. 360–361 (ISBN: 978-966-335-166-7) (IEEE Xplore).

A. Luntovskyy and D. Gütter, *Highly-Distributed Systems*,
https://doi.org/10.1007/978-3-030-92829-2

7. Andriy Luntovskyy, Marius Feldmann, Volodymyr Vasyutynskyy. A Semantic Web Application for Coordination of Combined Networks Design; IEEE CriMiCo Conference, Sevastopol, 8–12 Sept. 2008, paper ID = 122, pp. 364–367 (ISBN: 978-966-335-166-7) (IEEE Xplore).
8. Volodymyr Vasyutynskyy, Andriy Luntovskyy, Klaus Kabitzsch. Limit Cycles in PI Control Loops with Absolute Deadband Sampling; IEEE CriMiCo Conference, Sevastopol, 8–12 Sept. 2008, paper ID = 123, pp. 362–363 (ISBN: 978-966-335-166-7) (IEEE Xplore).
9. Andriy Luntovskyy. Data Security Guaranteeing Web Systems, Magazine "Treffpunkt BA", Dresden, 2009, Vol.1, pp. 10–12.
10. Andriy Luntovskyy, Stefan Uhlig, Dietbert Guetter. Cross-Layered Wireless Network Planning and Modeling Methods and Tools, IWCMC 2009—5th International Wireless Communications and Mobile Comput ing Conference (sponsored by IEEE and ACM), Leipzig, 21–24 June 2009, pp. 1417–1421.
11. Andriy Luntovskyy, Alexander Schill. Functionality of Wireless Network Design Tools, IEEE CriMiCo Conference, Sevastopol, 14–18 Sept. 2009, ID = 075, pp. 343–345 (ISBN: 978-1-4244-4796-1) (IEEE Xplore).
12. Andriy Luntovskyy, Stefan Uhlig, Dietbert Guetter. Deployment of Virtual Servers in an Enterprise IT-Landscape, ACS-AISBIS'09 International Conference on Advanced Computer Systems, Szczecin, Poland, 14–16 Oct. 2009, published in Polish Journal of Environmental Studies, Vol.18, No.4A (2009), pp. 31–34.
13. Andriy Luntovskyy, Volodymyr Vasyutynskyy. On Computer-Aided Design of Energy Efficient Wireless Sensor Networks, IEEE and ACM supported IWCMC 2010, Caen, France, June 28-July 2, 2010, ID: 1569263815, pp. 311–315.
14. Andriy Luntovskyy, Volodymyr Vasyutynskyy and Klaus Kabitzsch. Propagation Modeling and Placement Algorithms for Wireless Sensor Networks, IEEE ISIE 2010, Bari, Italy, July 4–7, 2010, ID: BD-011827, pp. 3493–3497, ISBN: 978-1-4244-6390-9 (IEEE Xplore).
15. Andriy Luntovskyy, Michailo Klymash. The Service-Oriented Internet. IEEE 11th TCSET 2012 Conf. for Modern Problems of Radio Engineering, Telecommunications and Computer Science, Lviv-Slavske, 21–24 Febr. 2012, Lviv Polytechnic National University. p. 256 (ISBN: 978-617-607-208-9). (IEEE Xplore).
16. Andriy Luntovskyy, Volodymyr Vasyutynskyy. Performance, Energy and Cost Efficiency in Combined Sensor Networks. 11th International Conference on Wireless and Mobile Communications ICWMC 2012—Performance Evaluation, Simulation and Modeling of wireless networks and systems, June 24–29, Venice, Italy (IARIA), ISBN: 978-1-61208-203-5.
17. Dietbert Guetter, Andriy Luntovskyy. WWW Based Solution of Computationally Intensive Mathematical and Physical Problems with Concrete Examples in the Range of Computer Network Design. MPH 2012—Third International Conference on Mathematical Physics and Its Applications, Samara/ Volga, August 27–September 1, 2012, pp. 324–325, ISBN: 978-5-7964-1555-9.

18. Andriy Luntovskyy, Dietbert Guetter. A Concept for a Modern Virtual Telecommunication Engineering Office, International Research Journal Telecommunication Sciences, Kiev (Registration Certificate KB No 15064-3636P, ISSN 2219-9454). 2012. Issue 2, No.1, pp. 15–21.
19. Andriy Luntovskyy, Volodymyr Vasyutynskyy, Josef Spillner. RAICs as Advanced Cloud Backup Technology in Telecommunication Networks, International Research Journal Telecommunication Sciences, Kiev (Registration Certificate KB No. 15064-3636P, ISSN 2219-9454). 2012. Issue 3, No.2, pp. 30–38.
20. Schwab, R. Loeffler, A. Luntovskyy. Konzeption einer optimierten Netzwerklandschaft auf Grundlage von Leitungs- und Durchsatzmessungen, Magazine "BA Treff", Dresden, 2/2013, pp. 18–19 (in German).
21. Andriy Luntovskyy, Josef Spillner. RAIC Integration for Network Storages on Mobile Devices, 7th IEEE Int. Conf. on Next Generation Mobile Apps, Services and Technologies NGMAST-2013, Prague, 25–27 September 2013, pp. 142–147 (IEEE Xplore).
22. Andriy Luntovskyy. Energy Efficiency of the Networks. 12th Int. Conf. TCSET-2014 for Modern Problems of Radio Engineering, Telecommunications and Computer Science, Lviv Technical University (Lvivska Polytechnika), Slavsko, 25 Febr.–1 March 2014, 3 p. (IEEE Xplore).
23. G.Suessmilch, L.Jonuscheit and A.Luntovskyy. Untersuchung der bidirektionalen Kommunikation zwischen mobilen Anwendungen und der Head Unit eines Automobils unter Verwendung von aktuellen Technologien, in "Treffunkt BA", Heft 2/2014, pp. 30–31 (in German).
24. Andriy Luntovskyy, Josef Spillner, Volodymyr Vasyutynskyy. Energy-Efficient Network Services as Smart Grid Issue/ Proceedings of 19th Int. Conf. ACS-2014 for Advanced Computer Systems, West Pomeranian University of Technology of Szczecin, Miedzyzdroje, 22–24 Oct. 2014: In Soft Computing in Computer and Information Science/ Advances in Intelligent Systems and Computing; editors: A. Wilinski, J. Pejas, I. El-Fray, Springer International Publising Switzerland, Volume 342, 2015, pp. 293–308 (ISBN 978-3-319-15146-5).
25. Andriy Luntovskyy et al. Performance and Energy Efficiency in Parallel Computing, 1-st Int. Conf. AICT-2015 (For Advanced Information and Communication Technologies), Lviv, Ukraine, 29 Oct.–1 Nov. 2015, 4p.
26. Luntovskyy, L. Globa, O. Stepurin. Performance-Energy Tradeoff Models for Distributed Computing, 13th Int. IEEE Conf. TCSET'2016 (On Modern Problems of Radio Engineering, Telecommunications and Computer Science) 23–26 February 2016, Slavske-Lviv, Ukraine, pp. 613–617, DOI: https://doi.org/10.1109/TCSET.2016.7452131 (IEEE Xplore).
27. Andriy Luntovskyy, Igor Melnyk, Larisa Globa. Waste Heat Transport Models for Green Clouds, Int. IEEE Conf. BlackSeaCom-2016, 6–9 June 2016, Varna, Bulgaria, pp. 1–5 (IEEE Xplore), DOI: https://doi.org/10.1109/BlackSeaCom.2016.7901554.

28. Andriy Luntovskyy. Performance and Energy Efficiency in Distributed Computing, 20th Int. Multi-Conf. ACS-2016 (On Advanced Computer Systems), 12p., in "Hard and Soft Computing for Artificial Intelligence, Multimedia and Security", Springer Int. Publishing, edited by Shin-ya Kobayashi, Andrzej Piegat, Jerzy Pejas, Imed El Fray, Janusz Kacprzyk.
29. L. Globa, I. Svetsynska and A. Luntovskyy. Case Studies on Big Data, Journal on Theoretical and Applied Computer Science, JTACS, Polish Academy of Science, Szczecin—Gdansk, Vol. 10, No. 2, 2016, ISSN 2299-2634, pp. 41–52 (Online: http://www.jtacs.org).
30. Luntovskyy, D. Guetter, M. Klymash. Up-to-date Paradigms for Distributed Computing, Int. IEEE Conf. AICT-2017 on Advanced Information and Comm. Technologies, July 4–7, 2017, Lvyv, pp. 113–119 (IEEE Xplore), ISBN: 978-1-5386-0638-4, DOI: https://doi.org/10.1109/AIACT.2017.8020078.
31. Luntovskyy et al. Intelligent Networking and Bio-Inspired Engineering, in The Second Int. IEEE Conference UkrMiCo'2017 on Information and Telecommunication Technologies and Radioelectronics, on 11 – 15 Sept. 2017, Odessa, Ukraine, 4 p. (IEEE Xplore: http://ieeexplore.ieee.org/document/8095421/).
32. Andriy Luntovskyy, M. Klymash. A Perspective Resource Allocation Method for Future WLAN, International Research Journal Telecommunication Sciences, Kiev (Registration Certificate KB No. 15064-3636P, ISSN 2219-9454). 2017. Issue 1, No.1, 8 p.
33. Andriy Luntovskyy. Advanced Software-Technological Approaches for Mobile Apps Development, 14th Int. IEEE TCSET-2018 Conf. on Advanced Trends in Radioelectronics, Telecommunications and Computer Engineering, Lviv—Slavske, 20–24 Febr. 2018, 6 p. (IEEE Xplore: https://ieeexplore.ieee.org/document/8336168/), DOI: https://doi.org/10.1109/TCSET.2018.8336168.
34. Bohdan Buhyl, Pavlo Huskov, Orest Lavriv, Roman Bak and Andriy Luntovskyy. Maximization of Service Flows Rates as a Solution of Network Capacity Allocation Problem, in Magazine for Internet of Things (IoT) and Engineering Applications, Clausius Scientific Press, Canada (2018), Vol. 3, pp. 1–10.
35. Taras Maksymyuk, Andriy Luntovskyy, Juraj Gazda and Mykhailo Klymash. Artificial Intelligence based 5G Coverage Design and Optimization using Deep Generative Adversarial Neural Networks, IEEE UkrMiCo-2018, Odessa, Sept. 2018, 8 p., IEEE Xplore: https://ieeexplore.ieee.org/document/9047611, DOI: https://doi.org/10.1109/UkrMiCo43733.2018.9047611.
36. Luntovskyy. SLMA and Novel Software Technologies for Industry 4.0, on 21-st Int. Conf. ACS—2018 on Advanced Comp. Systems, Szczecin—Miedzyzdroje, 24–26 Sept. 2018, In: Pejas J., El Fray I., Hyla T., Kacprzyk J. (eds) Advances in Soft and Hard Computing. ACS 2018, "Advances in Intelligent Systems and Computing", vol 889. Springer, Cham, Springer Int. Publishing, 12p. (Part of the Advances in Intelligent Systems and Computing book series; AISC, vol. 889, DOI: https://doi.org/10.1007/978-3-030-03314-9_16, ISBN: 978-3-030-03313-2).

37. Andriy Luntovskyy, Dietbert Guetter. Allgegenwärtige Vernetzung: Industrie 4.0, Internet der Dinge, Fog Computing u. v. m! in Magazine of BA Saxony "Wissen im Markt", Heft 2/2018, Glauchau, 8p.
38. Luntovskyy, D. Guetter. Cryptographic Technology Blockchain and its Applications, in "Advances in Information and Communication Technologies", Springer (ISBN: 978-3-030-16770-7), LNCS "Processing and Control in Information and Communication Systems (Int. Conf. UkrMiCo- 2019)" (eds.:M.Ilchenko, L.Globa et al.), 2019, pp. 14–33 (URL: https://link.springer.com/book/10.1007/978-3-030-16770-7).
39. Luntovskyy, L. Globa. Performance, Reliability and Scalability for IoT, Int. IEEE Conf. IDT-2019, Zilina, Slovakia, 25–27 June 2019, 6 p. (IEEE Xplore: https://ieeexplore.ieee.org/document/8813679, DOI: https://doi.org/10.1109/DT.2019.8813679).
40. Luntovskyy, M. Klymash. Robotic Apps and Platforms: Mobil ity, Localization, Management and Security Aspects, Int. IEEE Conf. AICT-2019, Lviv, Ukraine, 2–6 July 2019, 6 p. (IEEE Xplore: https://ieeexplore.ieee.org/document/8847741, DOI: https://doi.org/10.1109/AI-ACT.2019.8847741).
41. Luntovskyy, L. Globa. Industry 4.0 and Modern Approaches for Big Data, International Research Journal Telecommunication Sciences, Kiev Registration Certificate KB No. 15064-3636P, ISSN 2219-9454). – No. 2 (2019), pp. 47–53 (DOI: https://doi.org/10.20535/2411-2976.22019.47-53).
42. Taras Maksymyuk, Longzhe Han, Stanislav Larionov, Bohdan Shubyn, Andriy Luntovskyy, Mykhailo Klymash. Intelligent Spectrum Management in 5G Mobile Networks based on Recurrent Neural Network, 15th IEEE International Conf. on the Experience of Designing and Application of CAD Systems (CADSM), in Polyana, Ukraine, 26 Feb.-2 March 2019, 6p. (IEEE Xplore: https://ieeexplore.ieee.org/document/8779301, DOI: https://doi.org/10.1109/CADSM.2019.8779301).
43. Luntovskyyy, B. Shubyn, I. Scherm. Blockchaining for modern HDS, BA Magazine "Wissen im Markt", Year 3, Issue Nov. 2019, pp. 38–44. (ISSN 2512-4366).
44. Luntovskyy, L. Globa, B. Shubyn. From Big Data to Smart Data: The Most Effective Approaches for Data Analytics, in Advances in Information and Communication Technology and Systems, Springer Nature, Part of the "Lecture Notes in Networks and Systems Series", MCT 2019, Vol. LNNS 152, Chapter 2, pp. 23–40, Sept. 2020 (ISBN: 978-3-030-58359-0, DOI: https://doi.org/10.1007/978-3-030-58359-0_2).
45. Luntovskyy, B. Shubyn. Highly-Distributed Systems Based on MicroServices and their Construction Paradigms, IEEE Conf. TCSET 2020, Lviv-Slavske, Ukraine, 25–29 Febr. 2020, 8p. (IEEE Xplore: https://ieeexplore.ieee.org/document/9088603l DOI: https://doi.org/10.1109/TCSET49122.2020.235378).
46. T. Hara, A. Luntovskyy, I. Braun, T. Kubica. Designing IT Curriculum to Foster Self-Regulated Learning through Peer Instruction and Audience Response Systems, IEEE Conf. TCSET 2020, Lviv-Slavske, Ukraine, 25–29 Febr. 2020, 8p. (IEEE Xplore: https://ieeexplore.ieee.org/document/9088585, DOI: https://doi.org/10.1109/TCSET49122.2020.240939).

47. Luntovskyy, B. Shubyn. Advanced Architectures for IoT Scenarios, 5th Int. IEEE Conf. on Smart and Sustainable Technologies (SpliTech), Split and Bol, Croatia, 23–26 Sept. 2021, 6 p. (online), (IEEE Xplore: https://ieeexplore.ieee.org/document/9243784, DOI: https://doi.org/10.23919/SpliTech49282.2020.9243784).
48. Luntovskyy, B. Shubyn, T. Maksymyuk, M. Klymash. Highly-Distributed Systems: What Is Inside? 2020 Int. IEEE Scientific-Practical Conf. on Problems of Infocommunications. Science and Technology PICS&T-2020, Kharkiv, Ukraine, 6–9 October 2020, 6 p. (online).
49. Luntovskyy, B. Shubyn. Energy Efficiency for IoT, IEEE Workshop RECI-2020 on “Reliability Engineering and Computational Intelligence”, Zilina, Slovakia, 27–29 Oct. 2020 (online), in Springer LNCS series “Reliability Engineering and Computational Intelligence”, Studies Comp. Intelligence”, Vol. 976, by Coen van Gulijk and Elena Zaitseva (Eds.), ISBN: 978-3-030-74555-4, Chapter 12, 16 p., July 2021.
50. Luntovskyy, B. Shubyn, T. Maksymyuk, M. Klymash. 5G Slicing and Handover Scenarios: Compulsoriness and Machine Learning, in “Current Trends in Communication and Information Technologies”, in Springer Lecture Notes in Networks and Systems series, Volume 212), by Vorobiyenko, Petro, Ilchenko, Mykhailo, Strelkovska, Iryna (Eds.), based on Conf. IPF-2020 “Infocommunications – Present and Future”, Odessa, 16–19 Nov. 2020 (online), ISBN: 978-3-030-76342-8, 32 p., July 2021.
51. Tim Zobjack, Andriy Luntovskyy. Blockchained IoT: Verbindlichkeit in der dezentralisierten Welt smarter Dinge, BA Magazine “Wissen im Markt”, Year 4, Issue 4, Nov. 2020 (in German), pp. 75–81, ISSN 2512-4366, Berufsakademie Sachsen.
52. Andriy Luntovskyy, Tenshi Hara. Green IT: Energy Efficient Constructions and Applications for Data Centers and Clusters, ACS-2021 in Bydgoszcz, 28–30 June 2021, Springer, 13 p.
53. Andriy Luntovskyy, Mykola Beshley. Designing HDS under Considering of QoS Robustness and Security for Heterogeneous IBN, IEEE AICT Conf. 2021 Lviv, 21–25 Sept. 2021, 8 p., LNEE, Springer Nature.
54. Igor Melnyk, Andriy Luntovskyy. Estimation of Energy Efficiency and Quality of Service in Cloud Realizations of Parallel Algorithms for IBN, IEEE AICT Conf. 2021 Lviv, 21–25 Sept. 2021, 8 p., LNEE, Springer Nature.
55. Andriy Luntovskyy, Tim Zobjack. Secured and Blockchained IoT, IDT-2021 Zilina, IEEE Xplore, 22–24 June 2021, in IEEE Issue “Information and Digital Technologies”, ISBN 978-1-6654-3692-2, ISSN 2575-677X, 2021, pp. 1–10 (online).
56. Luntovskyy, T. Zobjack, B. Shubyn. Energy Efficiency and Security for IoT Scenarios via WSN, RFID and NFC, IEEE UkrMico Conf. 2021 Kyiv, November 29 – December 3, 2021, 8 p., LNEE, Springer Nature.
57. Igor Melnyk, Andriy Luntovskyy. Estimation of Level of Algorithms Parallelization and Efficiency of Using Different Error-Correction Codes in Cloud Realizations of Parallel Computing Algorithms, IEEE AICT Conf. 2021 Lviv, 21–25 Sept. 2021, 4p.

Appendix B2. Specialized Books

1. Luntovskyy, D. Gütter. Moderne Rechnernetze-Uebungsbuch: Aufgaben und Musterlösungen zu Protokollen, Standards und Apps in kombinierten Netzwerken, Springer Fachmedien Wiesbaden GmbH, 2020, ISBN: 978-3-658-25619-7.
2. Luntovskyy, D. Gütter. Moderne Rechnernetze-Theoriebuch: Protokolle, Standards und Apps in kombinierten drahtgebundenen, mobilen und drahtlosen Netzwerken, Springer Fachmedien Wiesbaden GmbH, 2020, ISBN: 978-3-658-25617-3 (Vorwort A. Schill).
3. Andriy Luntovskyy, Josef Spillner. Architectural Transformations in Network Services and Distributed Systems: Service Vision. Case Studies, XXIV, 344p. 238 pict., 2017, ISBN: 978-3-658-14840-9, Springer Heidelberg Springer Nature.
4. Andriy Luntovskyy, Dietbert Guetter, Igor Melnyk. Planung und Optimierung von Rechnernetzen: Methoden, Modelle, Tools für Entwurf, Diagnose und Management im Lebenszyklus von drahtgebundenen und drahtlosen Rechnernetzen // Springer Vieweg Wiesbaden, 2012, 411 p. (ISBN: 978-3-8348-1458-6).
5. Dietbert Guetter, A. Luntovskyy. Programmiersprachen—1: Studienhefte "C-Programmierung (Hefte I/ II/ III) mit Online-Aufgaben und Musterloesungen", Fernstudiengang E-Technik der HS OWL, Springer Heidelberg. 2017, Vol.1,2,3 (in German).
   Heft 01, 117 Seiten, ISBN: 978-3-662-54281-1;Heft 02, 141 Seiten, ISBN: 978-3-662-54283-5;Heft 03, 113 Seiten, ISBN: 978-3-662-54285-9.
6. Andriy Luntovskyy. Integration Concepts for Computer-Aided Design Tools for Wired and Wireless Local-Area Networks, Shaker Verlag Aachen, 2008, 196 p., (ISBN 978-3-8322-6968-5).

Appendix B3. Springer Book Chapters

1. Andriy Luntovskyy, Bohdan Shubyn. Energy Efficiency for IoT, based on IEEE Workshop RECI-2020 on "Reliability Engineering and Computational Intelligence", Zilina, Slovakia, 27–29 Oct. 2020, in Springer LNCS series "Reliability Engineering and Computational Intelligence—Studies Comp. Intelligence", Vol. 976, by Coen van Gulijk and Elena Zaitseva (Eds.), ISBN: 978-3-030-74555-4, pp. 199–215, on July, 2021, 507420_1_En, (Chap. 12).
2. Andriy Luntovskyy, Tenshi Hara. Green IT: Energy Efficient Constructions and Applications for Data Centers and Clusters, ACS-2021 in Bydgoszcz, 28–30 June 2021 (online), in Springer Lecture Notes in Networks and Systems series, "Progress in Image Processing, Pattern Recognition and Communication Systems", Proceedings of CORES, IP&C, ACS-2021 (by M.Choraś, R.S.Choraś, M.Kurzyński, P.Trajdos, J.Pejas, T.Hyla, Eds.), due: 6 November 2021, ISBN 978-3-030-81522-6, 13p.

3. Luntovskyy, Andriy (et al.) 5G Slicing and Handover Scenarios: Compulsoriness and Machine Learning. In Current Trends in Communication and Information Technologies. DOI https://doi.org/10.1007/978-3-030-76343-5_12, © 2021.
4. Andriy Luntovskyy, Dietbert Guetter. From Big Data to Smart Data: The Most Effective Approaches for Data Analytics In Advances in Information and Communication Technology and Systems.
5. Luntovskyy, Andriy (et al.) Cryptographic Technology Blockchain and Its ApplicationsIn Advances in Information and Communication Technologies. DOI https://doi.org/10.1007/978-3-030-16770-7_2, © 2019.
6. Luntovskyy, Andriy SLMA and Novel Software Technologies for Industry 4.0. In Advances in Soft and Hard Computing. DOI https://doi.org/10.1007/978-3-030-03314-9_16, © 2019.
7. Luntovskyy, Andriy Performance and Energy Efficiency in Distributed Computing. In Hard and Soft Computing for Artificial Intelligence, Multimedia and Security DOI https://doi.org/10.1007/978-3-319-48,429-7_26, © 2017.
8. Luntovskyy, Andriy (et al.) Energy-Efficient Network Services as Smart Grid Issue.In Soft Computing in Computer and Information Science DOI https://doi.org/10.1007/978-3-319-15147-2_25, © 2015.
9. Luntovskyy, Andriy, Dyadenko, Alexander (et al.) Multimedia Services Differentiation in 4G Mobile Networks under Use of Situational Priorities. In Energy-Aware Communications. DOI https://doi.org/10.1007/978-3-642-23541-2_21, © 2011.

Further Literature

1. Schill, Alexander, Springer, Thomas. Verteilte Systeme: Grundlagen und Basistechnologien: Kompakte Darstellung der Grundlagen und Techniken Verteilter Systeme, Springer-Verlag Berlin Heidelberg, 2012, 2.Ausgabe, ISBN: 9783642257957 (https://www.springer.com/gp/book/9783642 257957#otherversion=9783642257964).
2. Schill, Alexander, Springer, Thomas. Verteilte Systeme: Grundlagen und Basistechnologien: Kompakte Darstellung der Grundlagen und Techniken Verteilter Systeme, Springer-Verlag Berlin Heidelberg, 2007, 1.Ausgabe, ISBN: 9783540684718 (https://www.springer.com/gp/book/9783540 684718).
3. Luntovskyy, Andriy, Spillner, Josef. Architectural Transformations in Network Services and Distributed Systems: Current technologies, standards and research results in advanced (mobile) networks, Springer Vieweg Wiesbaden, 2017, 344p., ISBN: 9783658148409 (https://www.springer.com/gp/book/9783658148409#otherversion=9783658148423).
4. Luntovskyy, Andriy, Gütter, Dietbert, Melnik, Igor. Planung und Optimierung von Rechnernetzen. Methoden, Modelle, Tools für Entwurf, Diagnose und Management im Lebenszyklus von drahtgebundenen und drahtlosen Rechnernetzen: Planung von Rechnernetzwerke theoretisch anspruchsvoll und praxisnah, Springer/Vieweg+Teubner Wiesbaden, 2012, 415 Seiten, 245 Abb., ISBN: 9783834814586 (https://www.springer.com/gp/book/9783834814586#otherversion=9783834882424).
5. Luntovskyy, Andriy, Gütter, Dietbert. Moderne Rechnernetze: Protokolle, Standards und Apps in kombinierten drahtgebundenen, mobilen und drahtlosen Netzwerken, 500 Seiten, 263 Abb., Springer Nature, Juli 2020, ISBN: 9783658256166 (https://www.springer.com/gp/book/9783658256166).
6. Luntovskyy, Andriy, Gütter, Dietbert. Moderne Rechnernetze - Übungsbuch: Aufgaben und Musterlösungen zu Protokollen, Standards und Apps in kombinierten Netzwerken, 150 Seiten, 44 Abb., Springer Nature, Juli 2020, ISBN: 9783658256180 (https://www.springer.com/gp/book/978 3658256180)
7. Sam Newman. Building Micro Services, Publishing by O'Reilly Media, USA, 2015, ISBN: 978-1-491-95035-7, 473p.
8. Mahmood Zaigham (Ed.): Fog Computing: Concepts, Frameworks and Technologies, Springer 2017, London, ISBN 978-3-319-94890-4.
9. Jamil Y. Khan, Mehmet R. Yuce (Eds.). Internet of Things (IoT): Systems and Applications, 2019, New York, Jenny Stanford Publishing, ISBN 9780429399084, 366 p.

A. Luntovskyy and D. Gütter, *Highly-Distributed Systems*,
https://doi.org/10.1007/978-3-030-92829-2

MIX
Papier aus verantwortungsvollen Quellen
Paper from responsible sources
FSC® C105338

If you have any concerns about our products,
you can contact us on
ProductSafety@springernature.com

In case Publisher is established outside the EU,
the EU authorized representative is:
Springer Nature Customer Service Center GmbH
Europaplatz 3, 69115 Heidelberg, Germany

Printed by Libri Plureos GmbH
in Hamburg, Germany